W9-BNI-716

THE FOLLY OF FOOLS

THE FOLLY OF FOOLS

The Logic of Deceit and Self-Deception in Human Life

Robert Trivers

BASIC BOOKS
A MEMBER OF THE PERSEUS BOOKS GROUP
New York

Published by Basic Books,
A Member of the Perseus Books Group

Books published by Basic Books are available at special discounts for bulk purchases in the United States by corporations, institutions, and other organizations. For more information, please contact the Special Markets Department at the Perseus Books Group, 2300 Chestnut Street, Suite 200, Philadelphia, PA 19103, or call (800) 810-4145, ext. 5000, or e-mail special.markets@perseusbooks.com.

Designed by Timm Bryson

Library of Congress Cataloging-in-Publication Data
Trivers, Robert.
The folly of fools : the logic of deceit and self-deception in human life / Robert Trivers.
p. cm.
ISBN 978-0-465-02755-2 (alk. paper)—ISBN 978-0-465-02805-4 (e-book) 1. Self-deception. 2. Deception—Psychological aspects. 3. Deception—Social aspects. I. Title.
BF697.5.S426T76 2011
153.4—dc23
2011028453

10 9 8 7 6 5 4 3 2 1

In memory of Dr. Huey P. Newton,
Black Panther and dear friend

CONTENTS

Preface, xv

1. THE EVOLUTIONARY LOGIC OF SELF-DECEPTION 1

The Evolution of Self-Deception, 3

Deception Is Everywhere, 6

What Is Self-Deception?, 8

Detecting Deception in Humans via Cognitive Load, 9

Self-Deception Is Older Than Language, 13

Nine Categories of Self-Deception, 15

The Hallmarks of Self-Deception, 27

2. DECEPTION IN NATURE 29

The Coevolutionary Struggle Between Deceiver and Deceived, 30

Frequency-Dependent Selection in Butterflies, 30

An Epic Coevolutionary Struggle, 31

Intelligence and Deception, 36

Female Mimics, 38

False Alarm Calls, 39

Camouflage, 40

Death and Near-Death Acts, 41

Randomness as a Strategy, 42

Deception May Induce Anger, 43
Animals May Be Conscious of Deception, 45
Deception as an Evolutionary Game, 48
A Deeper Theory of Deception, 50

3. NEUROPHYSIOLOGY AND LEVELS OF IMPOSED SELF-DECEPTION 53
The Neurophysiology of Conscious Knowledge, 54
The Neurophysiology of Thought Suppression, 56
The Irony of Trying to Suppress One's Thoughts, 57
Improving Deception Through Neural Inhibition, 58
Unconscious Self-Recognition Shows Self-Deception, 59
Can One Half of the Brain Hide from the Other?, 61
Imposed Self-Deception, 63
Implicit Versus Explicit Self-Esteem, 64
False Confessions, Torture, and Flattery, 66
False Memories of Child Abuse, 67
Is Self-Deception the Psyche's Immune System?, 68
The Placebo Effect, 70

4. SELF-DECEPTION IN THE FAMILY— AND THE SPLIT SELF 77
Parent/Offspring Conflict, 80
Cases of Extreme Abuse, 81
Genomic Imprinting, 82
Internal Conflict from Oppositely Imprinted Genes, 84
Parental Manipulation and Imprinting, 85
The Effect of Marital Conflict on Genetic Conflict, 86
Imprinting and Self-Deception, 87
Deception in Children, 88
Parental Effects on Children's Deception, 92

5. DECEIT, SELF-DECEPTION, AND SEX 95

Why Sex?, 96

Two Sexes—Two Coevolving Species, 97

Deception and Self-Deception at Courtship, 99

Whose Baby Is It?, 100

Male Response to Female Infidelity, 101

Deceit and a Woman's Monthly Cycle, 103

Men's Self-Deceit About Female Interest, 105

Male Denial of Homosexual Tendencies, 106

Is Self-Deception Good or Bad for Marriage?, 107

The Appeal and Danger of Fantasy, 109

The Pain of Betrayal, 110

6. THE IMMUNOLOGY OF SELF-DECEPTION 115

The Immune System Is Expensive, 117

The Importance of Sleep, 120

Trade-Offs with Immunity, 121

Writing About Trauma Improves Immune Function, 125

Homosexuality and the Effects of Denial, 128

Positive Affect and Immune Function, 130

The Effects of Music, 132

Positivity in Old Age, 133

An Immunological Theory of Happiness, 135

7. THE PSYCHOLOGY OF SELF-DECEPTION 139

Avoiding Some Information and Seeking Out Other, 140

Biased Encoding and Interpretation of Information, 142

Biased Memory, 143

Rationalization and Biased Reporting, 145

Predicting Future Feelings, 146

Are All Biases Due to Self-Deception?, 147
Denial and Projection, 148
Denial Is Self-Reinforcing, 150
Your Aggression, My Self-Defense, 151
Cognitive Dissonance and Self-Justification, 151
Social Effects of Cognitive Dissonance Reduction, 154
Cognitive Dissonance in Monkeys and Young Children, 155

8. SELF-DECEPTION IN EVERYDAY LIFE 157
Sex Differences in Overconfidence, 157
Metaphors in the Stock Market, 160
Manipulative Metaphors in Life, 161
The Name-Letter Effect, 164
Deceiving Down and Dummying Up, 167
Face-ism, 168
Spam Against Anti-Spam, 170
Humor, Laughter, and Self-Deception, 172
Drugs and Self-Deception, 173
Vulnerability to Manipulation by Others, 175
Professional Con Artists, 176
Lie-Detector Tests, 180

9. SELF-DECEPTION IN AVIATION AND SPACE DISASTERS 183
Air Florida Flight 90—Doomed by Self-Deception?, 184
Disaster 37,000 Feet Above the Amazon, 189
Eldar Takes Command—Aeroflot Flight 593, 191
Simple Pilot Error—or Pilot Fatigue?, 192
Ice Overpowers the Pilots; Airlines Overpower the FAA, 194
The US Approach to Safety Helps Cause 9/11, 198

The *Challenger* Disaster, 201
The *Columbia* Disaster, 205
Egypt and EgyptAir Deny All, 209
Saved by *Lack* of Self-Deception?, 212

10. FALSE HISTORICAL NARRATIVES 215
The US False Historical Narrative, 218
Control Through Small Wars and Installed Proxies, 222
US History Textbooks, 224
Larger View of US History, 225
The Rewriting of Japanese History, 227
Turkey's Holocaust Denial, 230
A Land Without People for a People Without Land, 233
The Founding of the State of Israel, 235
Voluntary Flight or Ethnic Cleansing?, 237
Arab Deceit and Self-Deception, 239
Christian Zionism, 241
First Line of Defense: Cry "Anti-Semite", 243
alse Historical Narratives?, 245

DECEPTION AND WAR 247
anzee Raiding → Human Warfare, 249
eception Encourages Warfare, 252
ation of Others → Fatal Overconfidence, 255
03 US War on Iraq, 257
Creating Knowledge and Then Walling It Off, 263
Can Wars Be Won Through Bombing?, 265
Bombing to Eradicate History and to Reinforce It, 268
Carnage in Gaza, 270
Self-Deception and the History of War, 274

12. RELIGION AND SELF-DECEPTION 277

Cooperation Within the Group, 280

Religion: A Recipe for Self-Deception, 282

Religion and Health, 285

Parasites and Religious Diversity, 288

Why the Bias Against Women?, 291

Power Corrupts, 292

Religions Impose Mating Systems, 294

Religion Preaches Against Self-Deception, 295

Intercessory Prayer—Does It Work?, 299

Religion and Support for Suicide Attacks, 300

Religion → Self-Righteousness → Warfare, 301

13. SELF-DECEPTION AND THE STRUCTURE OF THE SOCIAL SCIENCES 303

Precedence of Justice Over Truth?, 304

Success of Science Is Based on Anti-Self-Deception Devices, 305

The More Social the Discipline, the More Retarded Its Development, 307

Self-Deception in Biology, 308

Is Economics a Science?, 310

Cultural Anthropology, 313

Psychology, 315

Psychoanalysis: Self-Deception in the Study of Self-Deception, 317

Self-Deception Deforms Disciplines, 319

14. FIGHTING SELF-DECEPTION IN OUR OWN LIVES 321

To Fight One's Own Self-Deception or Not?, 323

A Series of Minor Victories Followed by a Major Disaster, 324

Signals of Underlying Mental Screw-Ups, 324

Correcting for Our Own Biases, 326
Why Are We So Compulsive?, 327
The Value of Being Conscious, 329
The Danger of Fantasy in Propagating Deception, 330
The Benefits of Prayer and Meditation, 331
Value of Friends and Counselors, 332
An Invitation to Self-Deception and Personal Disaster, 333
A Never-Ending Extravaganza, 334

Acknowledgments, 339
Notes, 341
Bibliography, 355
Index, 385

PREFACE

The time is ripe for a general theory of deceit and self-deception based on evolutionary logic, a theory that in principle applies to all species but with special force to our own. We are thoroughgoing liars, even to ourselves. Our most prized possession—language—not only strengthens our ability to lie but greatly extends its range. We can lie about events distant in space and time, the details and meaning of the behavior of others, our innermost thoughts and desires, and so on. But why, why *self*-deception? Why do we possess marvelous sense organs to detect information only to distort it after arrival?

Evolutionary biology provides the foundation for a functional view of the subject—in this case, we lie to ourselves the better to lie to others—but many other aspects are involved. Self-deception sits squarely within psychology, but if you restrict yourself to that subject, you may well go blind (and crazy as well) long before you discern the underlying principles. In many situations, an understanding of daily life is more valuable than findings from the lab, but our understanding of daily life is easily colored by ignorance and our own deceit and self-deception. This may be especially true where politics and international relations are concerned, but to leave out these topics would be a foolish omission, as if because of potential bias we best remain silent. Since the analysis of self-deception begins at home, I have included some personal stories. Naturally, I have tried to strike a balance between what can be scientifically shown with some certainty and what is provocative but far from certain, and I have tried to make clear which is which.

My hope is to engage you in applying these concepts to your own life and developing them further. I have tried not to linger unduly over points of uncertainty but to draw attention to them where they occur and move on. Some real fraction of what I write must inevitably be wrong, but I hope that the logic being advanced and facts asserted will easily invite improvements toward a deeper, integrated science of self-deception.

The topic is a negative one. This book is about untruth, about falsehoods, about lies, inward and outward. At times, it is a depressing subject but, surely more than most, deceit and self-deception deserve to see the light of day, to enjoy the benefits of explicit scientific analysis and study. It is a dark and opaque side to ourselves, one that we leave untreated at our own peril, but it is also a source of endless humor and amazement, so we can also enjoy the subject as we suffer it.

I have written the book in a certain order—evolutionary logic and deception in nature first, neurophysiology, imposed self-deception, the family, two sexes, immunology, and social psychology next, then self-deception in daily life, including airplane crashes, false historical narratives, war, religion, and the social sciences, before offering final thoughts on how we may fight self-deception in ourselves.

But in fact, after the first chapter, the book can be read in almost any order. I have made an effort to refer back and forward to related material so that if you skip over material that is later necessary you will often know at once where to find it. For any fact or theory asserted in the text, it is easy to locate the appropriate source by going to the endnotes, which begin after the last chapter and are pegged to page number and content. I have occasionally added references to related material. Full references can then be found in the bibliography.

Everyone can participate in building a science of self-deception. We all have something to add. The logic is very simple and most of the evidence, easy to grasp. The topic is universal and its many subareas carry us into every corner of human life.

CHAPTER 1

The Evolutionary Logic of Self-Deception

In the early 1970s, I busied myself trying to construct social theory based on natural selection. I wanted to understand the evolution of our basic social relationships—parent/offspring, male/female, relative/friend, in-group member/out-group one, whatever. Natural selection, in turn, was the key to understanding evolution, and the only theory that answered the question, what is a trait designed to achieve? Natural selection refers to the fact that in every species, some individuals leave more surviving offspring than do others, so that the genetic traits of the reproductively successful tend to become more frequent over time. Since this process knits together genes associated with high reproductive success (RS = number of surviving offspring), all living creatures are expected to be organized accordingly, that is, to attempt to maximize personal RS. Because the replicating units are actually genes, this also means that our genes are expected to promote their own propagation.

When applied to social behavior, natural selection predicts a mixture of conflicting emotions and behavior. Contrary to widespread beliefs of the time (and even sometimes now), parent/offspring relations are not expected to be free of conflict, not even in the womb. At the same time, reciprocal relations are easily exploited by cheaters, that is, nonreciprocators, so that a sense of fairness may naturally evolve to regulate such relations in a protective manner. Finally, a coherent and unbiased theory for the evolution of sex differences can be built on the concept of relative parental investment—how much time and effort each parent puts into creating the offspring—as well as an understanding of selection acting on their relative numbers (the sex ratio). This work gives us a deeper view of the meaning of being a male or a female.

The general system of logic worked perfectly well for most subjects I encountered, but one problem stood out. At the heart of our mental lives, there seemed to be a striking contradiction—we seek out information and then act to destroy it. On the one hand, our sense organs have evolved to give us a marvelously detailed and accurate view of the outside world—we see the world in color and 3-D, in motion, texture, nonrandomness, embedded patterns, and a great variety of other features. Likewise for hearing and smell. Together our sensory systems are organized to give us a detailed and accurate view of reality, exactly as we would expect if truth about the outside world helps us to navigate it more effectively. But once this information arrives in our brains, it is often distorted and biased to our conscious minds. We deny the truth to ourselves. We project onto others traits that are in fact true of ourselves—and then attack them! We repress painful memories, create completely false ones, rationalize immoral behavior, act repeatedly to boost positive self-opinion, and show a suite of ego-defense mechanisms. Why?

Surely these biases are expected to have negative effects on our biological welfare. Why degrade and destroy the truth? Why alter information after arrival so as to reach a conscious falsehood? Why should natural selection have favored our marvelous organs of perception, on the one hand, only to have us systematically distort the information gathered, on the other? In short, why practice self-deception?

During a brainstorm on parent-offspring conflict in 1972, it occurred to me that deception of others might provide exactly the force to drive deception of self. The key moment occurred when I realized that parent-offspring conflict extended beyond how much parental investment is delivered to the behavior of the offspring itself. Once I saw conflict over the offspring's personality, it was easy to imagine parental deceit and self-deception molding offspring identity for parental benefit. Likewise, one could imagine parents not just practicing self-deception but also *imposing* it—that is, inducing it in the offspring—to the offspring's detriment but to parental advantage. After all, the parent is in the position of advantage—larger, stronger, in control of the resources at issue, and more practiced in the arts of self-deception.

Applied more broadly, the general argument is that we deceive ourselves the better to deceive others. To fool others, we may be tempted to reorganize information internally in all sorts of improbable ways and to do so largely unconsciously. From the simple premise that the primary function of self-deception is offensive—measured as the ability to fool others—we can build up a theory and science of self-deception.

In our own species, deceit and self-deception are two sides of the same coin. If by deception we mean only consciously propagated deception—outright lies—then we miss the much larger category of unconscious deception, including active self-deception. On the other hand, if we look at self-deception and fail to see its roots in deceiving others, we miss its major function. We may be tempted to rationalize self-deception as being defensive in purpose when actually it is usually offensive. Here we will treat deceit and self-deception as a unitary subject, each feeding into the other.

THE EVOLUTION OF SELF-DECEPTION

In this book we take an evolutionary approach to the topic. What is the biological advantage to the practitioner of self-deception, where advantage is measured as positive effects on survival and reproduction? How does self-deception help us survive and reproduce—or, slightly more accurately, how does it help our genes survive and reproduce? Put differently, how does natural selection favor mechanisms of self-deception? We shall see that we have a large set of such mechanisms and that they may have important costs. Where is the benefit? How do such mechanisms increase individual reproductive and genetic success?

Although the biological approach defines "advantage" in terms of survival and reproduction, the psychological approach often defines "advantage" as feeling better, or being happier. Self-deception occurs because we all want to feel good, and self-deception can help us do so. There is some truth to this, as we shall see, but not much. The main biological objection is this: Even if being happier is associated with higher survival and reproduction, as expected, why should we use such a dubious—and potentially costly—mechanism as self-deception to regulate our happiness? Lying to ourselves has costs. We are basing conscious activity on falsehoods, and in many situations this can turn around and bite us, as we shall see many, many times in this book. Whether during airplane crashes, the planning of stupid offensive wars, personal romantic disasters, family disputes, whatever, we shall see time and again that self-deception brings with it the expected costs of being alienated from reality, although, alas, there is a tendency for other people to suffer disproportionately the costs of our self-deception, while the benefits, such as they are, go to ourselves. So how does self-deception pay for itself biologically? How does it actually improve survival and reproduction?

The central claim of this book is that self-deception evolves in the service of deception—the better to fool others. Sometimes it also benefits deception by saving on cognitive load during the act, and at times it also provides an easy defense against accusations of deception (namely, I was unconscious of my actions). In the first case, the self-deceived fails to give off the cues that go with consciously mediated deception, thus escaping detection. In the second, the actual process of deception is rendered cognitively less expensive by keeping part of the truth in the unconscious. That is, the brain can act more efficiently when it is unaware of the ongoing contradiction. And in the third case, the deception, when detected, is more easily defended against—that is, rationalized—to others as being unconsciously propagated. In some cases, self-deception may give a direct personal advantage by at least temporarily elevating the organism into a more productive state, but most of the time such elevation occurs without self-deception.

In short, this book will attempt to describe a science of self-deception that is actually built on preexisting science—in this case, biology. The book will showcase what seem to be some of the most important features of the subject. The field is in its infancy, and surely many mistakes will be made here, but if the underlying logic is sound, and is linked by evidence and logic to the rest of biology, then corrections should come very quickly and we may rapidly grow a mature science that this book seeks only to outline.

The dynamics of deception and its detection have been studied in a broad range of other species (see Chapter 2), with the advantage that we can see things in others that we can't easily see in ourselves. This enterprise also greatly extends our range of evidence and leads to a few general principles of some considerable value. Deceiver and deceived are trapped in a coevolutionary struggle that continually improves adaptations on both sides. One such adaptation is intelligence itself. The evidence is clear and overwhelming that both the detection of deception and often its propagation have been major forces favoring the evolution of intelligence. It is perhaps ironic that dishonesty has often been the file against which intellectual tools for truth have been sharpened.

Regarding underlying mechanisms, some interesting work in neurophysiology shows that the conscious mind is more of an observer after the fact, while behavior itself is usually unconsciously initiated (see Chapter 3). Knocking out activity in deception-related areas of the brain improves the quality of deception, while suppression of memory can be achieved consciously by inhibiting brain activity in relevant areas. The classic experiment demonstrating human self-deception shows that we often unconsciously recognize our own voices while consciously failing to do so, and this tendency can be manipulated. An important concept is that of imposed self-deception, in which we act out self-deceptions others have imposed on us. The possibility that self-deception evolves as a purely defensive device to make us feel better is addressed and rejected, with some latitude for self-deception that directly benefits self (without fooling others). The placebo effect provides an interesting example.

Our logic also applies with special force to family and sexual interactions (see Chapters 4 and 5), each involving both conflict and cooperation over reproduction, life's key aim. Family interactions can select for a divided self, in which our maternal half is in conflict with the paternal half, leading to a kind of "selves deception" between the two halves. Sexual relations are likewise fraught with conflict—and deceit and self-deception—from courtship to long-term life partnerships.

And there is an intimate association between our immune system and our psyches, such that self-deception is often associated with major immune effects, all of which must be calculated if we are to understand the full biological effects of our mental lives (see Chapter 6). There is a whole world of social psychology that shows how our minds bias information, from initial avoidance, to false encoding, memory, and logic, to incorrect statements to others—from one end to the other (see Chapter 7). Key mechanisms include denial, projection, and perpetual efforts to reduce cognitive dissonance.

The analysis of self-deception illuminates daily life, whether the evidence is embedded in personal experience or unconscious and uncovered only through careful study (see Chapter 8). One example from everyday life that has an entire chapter devoted to it is airplane and spacecraft crashes—they permit the cost of self-deception to be studied intensively under almost controlled conditions (see Chapter 9).

Self-deception is intimately tied to false historical narratives, lies we tell ourselves about our past, usually in the service of self-forgiveness and aggrandizement (see Chapter 10). Self-deception plays a large role in the launching of misguided wars (see Chapter 11) and has important interactions with religion, which acts as both an antidote to self-deception and an accelerant (see Chapter 12). We are hardly surprised to note that nonreligious systems of thought—from biology to economics to psychology—are affected by self-deception according to the rule that the more social a discipline, the more its development is retarded by self-deception (see Chapter 13). Finally, as individuals, we can choose whether to fight our own self-deceptions or to indulge them. I choose to oppose my own—with very limited success so far (see Chapter 14).

DECEPTION IS EVERYWHERE

Deception is a very deep feature of life. It occurs at all levels—from gene to cell to individual to group—and it seems, by any and all means, necessary. Deception tends to hide from view and is difficult to study, with self-deception being even worse, hiding itself more deeply in our own unconscious minds. Sometimes the subject must be ferreted out before it can be inspected, and we often lack key pieces of evidence, given the complexity of the subterfuges and our ignorance of the internal physiological mechanisms of self-deception.

When I say that deception occurs at all levels of life, I mean that viruses practice it, as do bacteria, plants, insects, and a wide range of other animals. It is everywhere. Even *within* our genomes, deception flourishes as selfish genetic elements use deceptive molecular techniques to over-reproduce at the expense of other genes. Deception infects all the fundamental relationships in life: parasite and host, predator and prey, plant and animal, male and female, neighbor and neighbor, parent and offspring, and even the relationship of an organism to itself.

Viruses and bacteria often actively deceive to gain entry into their hosts, for example, by mimicking body parts so as not to be recognized as foreign. Or, as in HIV, by changing coat proteins so often as to make mounting an enduring defense almost impossible. Predators gain from being invisible to their prey or resembling items attractive to them—for example, a fish that dangles a part of itself like a worm to attract other fish, which it eats—while prey gain from being invisible to their predators or mimicking items noxious to the predator, for example, poisonous species or a species that preys on its own predator.

Deception within species is expected in almost all relationships, and deception possesses special powers. It always takes the lead in life, while detection of deception plays catch-up. As has been said regarding rumors, the lie is halfway around the world before the truth puts its boots on. When a new deception shows up in nature, it starts rare in a world that often lacks a proper defense. As it increases in frequency, it selects for such defenses in the victim, so that eventually its spread will be halted by the appearance and spread of countermoves, but new defenses can always be bypassed and new tricks invented.

Truth—or, at least, truth detection—has been pushed back steadily over time by the propagation of deception. It always amazes me to hear some economists say that the costs of deceptive excesses in our economy (including white-collar crime) will naturally be checked by market forces.

Why should the human species be immune to the general rule that where natural selection for deception is strong, deception can be selected that extracts a substantial net cost (in survival and reproduction) every generation? Certainly there is no collective force against this deception, only the relatively slow generation and evolution of counterstrategies. These lines were written in 2006, two years before the financial collapse that resulted from such practices and beliefs. I know nothing about economics and—from evolutionary logic—could not have predicted a thing about the collapse of 2008, but I have disagreed for thirty years with an alleged science called economics that has resolutely failed to ground itself in underlying knowledge, at a cost to all of us (see Chapter 13).

As for the notion that deception is naturally constrained to be of modest general cost, consider the case of stick insects (or Phasmatodea), a group that has given itself over to imitating either sticks (three thousand species) or leaves (thirty species). These forms have existed for at least fifty million years and achieve a remarkably precise resemblance to their models. In those forms resembling sticks, there is apparently strong evolutionary pressure to produce a long, thin (sticklike) body, even if doing so forces the individual to forgo the benefits of bilateral symmetry. Thus, to fit the internal organs into a diminishing space, one of two organs has often been sacrificed, leaving only one kidney, one ovary, one testis, and so on. This shows that selection for successful deception has been powerful enough not only to remold the creature's external shape but to remold its internal organs as well—even when this is otherwise disadvantageous to the larger creature, as loss of symmetry must often be. Likewise, as we shall see in the next chapter, selection can evolve a male fish that lives its entire adult life pretending to be a female and hooks up with territory-holding males in order to steal paternity of eggs laid in their territories by real females.

WHAT IS SELF-DECEPTION?

What exactly is self-deception? Some philosophers have imagined that self-deception is a contradiction in terms, impossible at the outset. How can the self deceive the self? Does that not require that the self knows what it does not know (p/~p)? This contradiction is easily sidestepped by defining the self as the conscious mind, so that self-deception occurs when the conscious mind is kept in the dark. True and false information may be simultaneously stored, only with the truth stored in the unconscious mind and falsehood in the conscious. Sometimes this involves activities of the conscious mind itself, such as active memory suppression, but usually the processes themselves are unconscious yet act to bias what we are conscious of. Most animals also have a conscious mind (not usually self-conscious), in the sense of a light being turned on (when awake) that allows integrated ongoing concentration on the outside world via their sense organs.

So the key to defining self-deception is that true information is preferentially excluded from consciousness and, if held at all, is held in varying degrees of unconsciousness. If the mind acts quickly enough, no version of the truth need be stored. The counterintuitive fact that needs to be explained is that the false information is put into the conscious mind. What is the point of this? One would think that if we had to store true and false versions of the same event simultaneously, we would store the true version in the conscious mind, the better to enjoy the benefits of consciousness (whatever they may be), while the false information would be kept safely out of sight somewhere in the basement. The hypothesis of this book is that this entire counterintuitive arrangement exists for the benefit of manipulating others. We hide reality from our conscious minds the better to hide it from onlookers. We may or may not store a copy of that information in self, but we certainly act to exclude it from others.

DETECTING DECEPTION IN HUMANS VIA COGNITIVE LOAD

If the main function of self-deception is to make deception more difficult to detect, we are naturally led to how humans detect consciously propagated deception. What cues do we use when we do it? When interactions are anonymous or infrequent, behavioral cues cannot be read against a background of known behavior, so more general attributes of lying must be used. Three have been emphasized:

Nervousness: Because of the negative consequences of being detected, including being aggressed against and also possibly guilt, people are expected to be more nervous when lying.

Control: In response to concern over appearing nervous (or concentrating too hard) people may exert control, trying to suppress behavior, with possible detectable side effects such as overacting, overcontrol, a planned and rehearsed impression, or displacement activities. More to the point, tensing ourselves up almost inevitably increases the pitch of our voices. When asked to create a painful reaction or suppress it, for example in response to cold, children and adults are more successful suppressing than inventing—they tend to overact.

Cognitive load: Lying can be cognitively demanding. You must suppress the truth and construct a falsehood that is plausible on its face and does not contradict anything known by the listener, nor likely to be known. You must tell it in a convincing way and you must remember the story. This usually takes time and concentration, both of which may give off secondary cues and reduce performance on simultaneous tasks.

Cognitive load often appears to be the critical variable among the three, with a minor role for control and very little for nervousness. At least, this seems to be true in real criminal investigations as well as experimental situations designed to mimic them. Absent well-rehearsed lies, people who are lying have to think too hard, and this causes several effects, some of which are opposite to those of nervousness.

Consider, for example, blinking. When nervous, we blink our eyes more often, but we blink less under increasing cognitive load (for example, while solving arithmetic problems). Recent studies of deception suggest that we blink *less* when deceiving—that is, cognitive load rules. Nervousness makes us fidget more, but cognitive load has the opposite effect. Again, contra usual expectation, people often fidget *less* in deceptive situations. And consistent with cognitive load effects, men use fewer hand gestures while deceiving and both sexes often employ longer pauses when speaking deceptively. An absurd example of the latter occurred the other day on my property in Jamaica when I questioned a young man just arriving on a motorcycle, intent (in my opinion) on either extorting money or robbing me. What was his name, I wanted to know. "Steve," he said. "And what is your last name?" Pause. "It is not supposed to take a long time to remember your own last name." Quick as you can say "Jones," he said, "Jones." So it was "Steve Jones"—not an entirely unlikely pair of names in Jamaica—but less believable on its face than his actual name, which turned out to be Omar Clarke. The point is that cognitive load gave him away at once. The most recent work shows that there is by no means always a delay prior to lying. It depends on the kind of lie. Denial is apt to be quicker than the truth, and so are well-rehearsed lies.

Efforts at controlling oneself can also reveal deception. A nice example is pitch of voice. Deceivers tend to have higher-pitched voices. This is a very general finding and is a natural consequence of stress or of any effort to suppress behavior by becoming more rigid. Tensing up the body inevitably tends to raise the pitch of voice, and this tensing will naturally increase the closer the liar comes to the key word. For example, someone denying a sexual relationship with "Sherri" may see her voice shoot up upon mention of the key person's name: "You think I am there with SHERri." Well, I had been leaning toward that theory, but now I had a fresh piece of evidence.

Another effect of suppression is the production of displacement activities. As classically described in other animals, these are irrelevant activities often seen when two opposing motivations are simultaneously aroused. Since neither impulse can express itself, the blocked energy easily activates irrelevant behavior, such as a twitch. For this reason, displacement activities in primates reliably indicate stress. For example, I once tried to slip a minor lie by a female friend at a bar and saw my left arm twitch involuntarily. Since we had by then been dating for some time, her eyes shot at once to the twitching arm. A few months later, the situation happened again, only with the roles reversed. If this had been a tennis match, the referee would have said on each occasion, "Advantage, your opponent."

Nervousness is almost universally cited as a factor associated with deception, both by those trying to detect it as well as by those trying to avoid it, yet surprisingly enough, it is one of the weaker factors in predicting deception in scientific work. This is partly because, with no ill effects of having their deception detected, many experiments do not make people nervous. But also in real-life situations (for example, criminal investigations), being suspected of lying can make you nervous regardless of whether you lie and, perhaps more important, because we are conscious of our nervousness as a factor, suppression mechanisms may be almost as well developed as the nervousness itself, especially in those experienced in lying. And as we saw earlier, the effects of cognitive load involved in lying are often opposite to those of nervousness.

The point about cognitive load (and pitch of voice) is that there is no escape. If suppressing your nervousness increases pitch of voice, then trying to suppress that effect may only increase pitch further. If it is cognitively expensive to lie, there is no obvious way to reduce the expense, other than to increase unconscious control. Mechanisms of denial and repression may serve to reduce immediate expense, but with ramifying costs later on.

Separately, it is worth pointing out that cognitive load has important effects across a broad range of psychological processes, according to the rule that the greater the cognitive load, the more likely the unconscious processes will be revealed. For example, under cognitive load, people will more often blurt out something they are trying to suppress and will more often express biased opinions they are otherwise hiding. In short, cognitive load does more than slow down your responses—in a whole host of ways, it tends to reveal unconscious processes. These predominate when conscious degree of control is minimized because of cognitive load.

Verbal details of lies can also be revealing. Excellent work, aided by computer analysis, has demonstrated several common verbal features of lies. We cut down on the use of "I" and "me" and increase other pronouns, as if disowning our lie. We cut down on qualifiers, such as "although." This streamlines the lie, lowering both our immediate cognitive load and later need to remember. A truth teller might say, "Although it was raining, I still walked to the office"; a liar would say, "I walked to the office." Negative terms are more common, perhaps because of guilt or because lies more frequently involve denial and negation.

It is difficult to measure the frequency with which lies are detected in everyday life. Interviews of people in the United States show that they believe their lies are detected 20 percent of the time and that another 20 percent may be detected. Of course, the 60 percent of lies they feel are successful may contain some detections where the detector hides his or her knowledge of the deception.

SELF-DECEPTION IS OLDER THAN LANGUAGE

How biologically deep is the subject we are discussing? Many people imagine that self-deception is, almost by definition, a human phenomenon, the "self" suggesting the presence of language. But there is no reason to suppose that self-deception is not far deeper in evolutionary history, as it does not require words. Consider self-confidence, a personal variable that others can measure. It can be inflated to deceive them, with self-deception making the act more plausible. This feature probably extends far back in our animal past.

In nature, two animals square off in a physical conflict. Each is assessing its opponent's self-confidence along with its own—variables expected to predict the outcome some of the time. Biased information flow within the individual can facilitate false self-confidence. Those who believe their self-enhancement are probably more likely to get their opponent to back down than those who know they are only posing. Thus, nonverbal self-deception can be selected in aggressive and competitive situations, the better to fool antagonists. Much the same could be said for male/female courtship. A male's false self-confidence may give him a boost some of the time. A biased mental representation can be produced, by assumption, without language. Note, of course, that self-deception tends to work only with plausible limits to self-inflation.

The above is meant to demonstrate that in at least two widespread contexts—aggressive conflict and courtship—selection for deception may easily favor self-deception even when no language is involved. There are undoubtedly many other such contexts, for example, parent/offspring. On top of that, as we shall see, very clever recent work demonstrates in monkeys forms of self-deception that are well-known in humans: a consistency bias, for example, as well as implicit in-group favoritism, both being shown by the same kinds of experiments that reveal them in humans. As we shall see, men are more prone to overconfidence than are women, just as expected, and in rational situations such as stock trading, where fooling others is rarely involved, men do correspondingly worse.

Self-confidence is an internal variable and thus especially prone to deception. I can inflate my apparent size by muscling up, but this is fairly obvious to observers, and increasing my apparent symmetry, another important variable, is very difficult to achieve. But pretending to be more confident than I am is more easily achieved and more strongly selects for self-deception, especially when self-confidence may be as important as apparent size in predicting aggressive outcomes. Thus, I believe that overconfidence is one of the oldest and most dangerous forms of self-deception—both in our personal lives and in global decisions, such as going to war.

On the other hand, language certainly greatly expanded the opportunities for deceit and self-deception in our own lineage. If one great virtue of language is its ability to make true statements about events distant in space and time, then surely one of its social drawbacks is its ability to make false statements about events distant in space and time. These are so much less easily contradicted than statements about the immediate world. Once you have language, you have an explicit theory of self and of social relationships ready to communicate to others. Numbers of new true assertions are matched by an even greater number of false ones.

A very disturbing feature of overconfidence is that it often appears to be poorly associated with knowledge—that is, the more ignorant the individual, the more confident he or she may be. This is true of the public when asked questions of general knowledge. Sometimes this phenomenon varies with age and status, so that senior physicians, for example, are both more likely to be wrong and more confident they are right, a potentially lethal combination, especially among surgeons. Another case with tragic consequences concerns eyewitness testimony—witnesses who are more mistaken in eyewitness identification and more confident that they are right, and this in turn has a positive effect on jurors. It may be that a rational approach to the world is nuanced and gray, capable of accommodating contradictions, all of which leads to hesitancy and a lack of certainty, as is indeed true. An easy shortcut is to combine ignorance with straight-out endorsement of ignorance—no signs of rational inquiry but, more important, no signs of self-doubt or contradiction.

NINE CATEGORIES OF SELF-DECEPTION

We begin with simple cases of self-inflation and derogation of others. We consider the effects of in-group feelings, a sense of power, and the illusion of control. Then we imagine false social theories, false internal narratives, and unconscious modules as additional sources of self-deception.

Self-Inflation Is the Rule in Life

Animal self-inflation routinely occurs in aggressive situations (size, confidence, color) as well as in courtship (same variables). Self-inflation is also the dominant style in human psychological life, adaptive self-diminution appearing in both animals and humans as an occasional strategy (see Chapter 8). Much of this self-inflation is performed in the service of what one psychologist aptly called "beneffectance"—appearing to be both beneficial and effective to others. Subtle linguistic features may easily be involved. When describing a positive group effect, we adopt an active voice, but when the effect is negative, we unconsciously shift to a passive voice: this happened and then that happened and then costs rained down on all of us. Perhaps a classic in the genre was the man in San Francisco in 1977 who ran his car into a pole and claimed afterward, as recorded by the police: "The telephone pole was approaching. I was attempting to swerve out of the way, when it struck my front end." Perfectly legitimate, but it shifts the blame to the telephone pole. And self-bias extends in every direction. If you question BMW owners on why they own that brand of car, they will tell you it had nothing to do with trying to influence others but will see others as owning one for exactly that reason.

Self-inflation results in people routinely putting themselves in the top half of positive distributions and the lower half of negative ones. Of US high school students, 80 percent place themselves in the top half of students in leadership ability. This is not possible. But for self-deception, you can hardly beat academics. In one survey, 94 percent placed themselves in the top half of their profession. I plead guilty. I could be tied down to a bed in a back ward of some hospital and still believe I am outperforming half my colleagues—and this is not just a comment on my colleagues.

When we say we are in the top 70 percent of people for good looks, this may be only our mouths talking. What about our deeper view? A recent methodology gives a striking result. With the help of a computer, individual photos were morphed either 20 percent toward attractive faces (the average of fifteen faces regarded as attractive out of a sample of sixty) or 20 percent toward unattractive ones (people with cranial-facial syndrome, which produces a twisted face). Among other effects, when a person tries to quickly locate his or her real face, the 20 percent positive face, or the 20 percent negative one, each embedded in a background of eleven faces of other people, he or she is quickest to spot the positive face (1.86 seconds), 5 percent slower for the real face (2.08 seconds), and another 5 percent slower for the ugly one (2.16 seconds). The beauty is that there has not been the usual verbal filter—what do you think of yourself?—only a measure of speed of perception. When people are shown a full array of photos of themselves, from 50 percent more attractive to 50 percent less attractive, they choose the 20 percent better-looking photo as the one they like the most and think they most resemble. This is an important, general result: self-deception is bounded—30 percent better looking is implausible, while 10 percent better fails to gain the full advantage.

I hardly need the above result to convince myself, because if I am in a big city, I experience the effect almost every week. I am walking down the street with a younger, attractive woman, trying to amuse her enough that she will permit me to remain nearby. Then I see an old man on the other side of her, white hair, ugly, face falling apart, walking poorly, indeed shambling, yet keeping perfect pace with us—he is, in fact, my reflection in the store windows we are passing. Real me is seen as ugly me by self-deceived me.

Is the tendency toward self-inflation really universal in humans? Some cultures, such as in Japan and China, often value modesty, so that if anything, people might be expected to compete to show lack of self-inflation. Certainly in some domains modesty rules, but in general it seems that one can still detect tendencies toward self-inflation, including self over other in terms of good and bad. Likewise, as in other cultures, inflation often applies to friends, who are seen as better than average (though less strongly than self in some cultures and more so in others).

By the way, recent work has located an area of the brain where this kind of self-inflation may occur. Prior work has shown that a region called the medial prefrontal cortex (MPFC) seems often to be involved in processing self-related information. Even false sensations of self are recorded there, and the region is broadly involved in deceiving others. One can suppress neural activity in this region (by applying a magnetic force to the skull where the brain activity takes place), deleting an individual's tendencies toward self-enhancement (while suppression in other regions has no effect).

An extreme form of self-adulation is found among so-called narcissists. Though people in general overrate themselves on positive dimensions, narcissists think of themselves as special and unique, entitled to more positive outcomes in life than others. Their self-image is good in dominance and power (but not caring or morality). Thus, they seem especially oriented toward high status and will seek out people of perceived status apparently for this reason. Though people in general are overconfident regarding the truth of their assertions, narcissists are especially so. Because they are overconfident, narcissists in the laboratory are more likely to accept bets based on false knowledge and hence lose more money than are less narcissistic people. They are persistent in their delusions as well. They predict high performance in advance, guess they have done well after the fact when they have not, and continue to predict high future performance despite learning about past failure—a virtuoso performance indeed. Calling someone a narcissist is not a compliment—it suggests someone whose system of self-enhancement is out of control, to the individual's disadvantage.

Derogation of Others Is Closely Linked

In one sense, derogation of others is the mirror image of self-inflation; either way, you look relatively better. But there is an important difference. For self-inflation, you need merely change the image of yourself to achieve the desired effect, but for derogation of others, you may need to derogate an entire group. Exactly when would we expect this to be advantageous to you? Perhaps especially when your own image has been lowered—suddenly it becomes valuable to deflect attention onto some disliked group—so that by comparison, you do not look as bad as they do.

This is precisely what social psychology appears to show—derogation of others appears more often as a defensive strategy that people adopt when threatened. Contrast two sets of college students who have been told (at random) that they scored high or low on an IQ test. Only those scoring low later choose to denigrate a Jewish woman (but not a non-Jewish) woman on a variety of traits. Apparently association with intellectual achievement is sufficient reason to denigrate the woman if one's own intellectual powers are in doubt. Likewise, the same "low scorers" (as they are told they are) are more likely to complete "duh" and "dan" as "dumb" and "dangerous" when subliminally primed with a black face. So let us say that there is some evidence that I am stupid (in fact, fictitious). I apparently lash out by denigrating members of allegedly intelligent groups (against which there may be other biases) while calling attention to negative stereotypes of allegedly less gifted ones. Incidentally, the derogation does make me feel better afterward, as measured by an interview, so the act may fool me as well.

As we shall see later (Chapter 11), derogation of others—including racial, ethnic, and class prejudices—can be especially dangerous when contemplating hostile activity, such as warfare.

In-Group/Out-Group Associations Among Most Prominent
Few distinctions bring quicker and more immediate psychological responses in our species than in-group and out-group—almost as much as, if not sometimes more than, for self and other. Just as you are on average better than others, so is your group—just as others are worse, so are out-groups. Such groups, in and out, are pathetically easy to form. You need not stoke Sunni or Catholic fundamentalism to get people to feel the right way; just make some wear blue shirts and others red and within a half-hour you will induce in-group and out-group feelings based on shirt color.

Once we define an individual as belonging to an out-group, a series of mental operations are induced that, often quite unconsciously, serve to degrade our image of the person, compared with an in-group member. The words "us" and "them" have strong unconscious effects on our thinking. Even nonsense syllables (such as "yaf," "laj," and "wuhz"), when paired with "us," "we," and "ours," are preferred over similar syllables paired with "they," "them," and "theirs." And these mechanisms can be primed to apply to artificial groups, experimentally created—those with different-colored shirts, for example. We easily generalize bad traits in an out-group member while reserving generalization for good traits performed by an in-group member. For example, if an out-group member steps on my toes, I am more likely to say, "He is an inconsiderate person," though with an in-group member I will describe the behavior exactly: "He stepped on my toes." In contrast, an out-group member acting nicely is described specifically—"she gave me directions to the train station"—while an in-group member is described as being "a helpful person." Similar mental operations serve to derogate others compared to self. Even minor positive social traits, such as a smile, are imputed unconsciously more often to in-group members than to out-group ones.

This bias begins early in life, among infants and young children. They divide others into groups based on ethnicity, attractiveness, native language, and sex. By age three, they prefer to play with in-group members and also begin to display explicit negative verbal attitudes toward out-group members. They also share with adults a strong tendency to prefer groups to which they have been randomly assigned, to believe that their own group is superior to others, and to begin to treat out-group members in a harmful fashion.

Recent work shows a similar mental architecture in monkeys regarding in-groups and out-groups. When a test is performed on a monkey in which it responds visually to matched facial pictures of in-group and out-group members, corrected for degree of experience with them, there is a clear tendency to view the out-group member longer—a measure of concern and hostility. Likewise, a monkey will attach an out-group orientation to an object an out-group group member is looking at and vice versa for an in-group member. Finally, male monkeys (but not female) more readily associate out-group members with pictures of spiders but in-group members with pictures of fruits. The beauty of this work is that the monkeys were migrating in and out of different groups at various times, so one could control exactly for degree of familiarity. In-group members, for example, tend to be more familiar, but independent of familiarity, they are preferred over out-group members. That males more readily associate out-group members with negative stimuli, and in-group with positive, is consistent with work on humans in which men typically are relatively more prejudiced against out-group than in-group members.

The Biases of Power

It has been said that power tends to corrupt and absolute power, absolutely. This usually refers to the fact that power permits the execution of ever more selfish strategies toward which one is then "corrupted." But psychologists have shown that power corrupts our mental processes almost at once. When a feeling of power is induced in people, they are less likely to take others' viewpoint and more likely to center their thinking on themselves. The result is a reduced ability to comprehend how others see, think, and feel. Power, among other things, induces blindness toward others.

The basic methodology is to induce a temporary state of mind via a so-called prime, which can be conscious or unconscious and as short as a word or considerably more detailed, as in this case. The power prime consists of asking some people to write for five minutes about a situation in which they felt powerful, supplemented by having the subjects apportion candy among a group, while the low-power prime group writes about the opposite situation and is allowed only to say the amount of candy they hope to receive.

This modest prime produced the following striking results. When the subjects were asked to snap their right-hand fingers five times in succession and quickly write the letter *E* on their foreheads, an unconscious bias was revealed. Those who had been primed to feel powerless were three times as likely to write the *E* so that *others* could read it, compared to those primed to feel powerful. This effect was equally strong for the two sexes. The basic shift in focus from other to self with power was confirmed in additional work. When compared with those with a neutral prime, those with the power prime were less able to discriminate among common human facial expressions associated with fear, anger, sadness, and happiness. Again, the sexes responded similarly to the power prime, but in general women are better at making the emotional discriminations, and men are more likely to be overconfident. In short, powerful men suffer multiple deficits in their ability to apprehend the world of others correctly, due to their power and their sex. And since, at the national level, it is powerful men who usually decide for war, they have an in-built bias in the wrong direction, less oriented toward others, less inclined to value their viewpoint, with, alas, often tragic effects all the way around (see Chapter 11).

There must be a thousand examples of power inducing male blindness, but why not look at Winston Churchill? He experienced highs and lows in his life that were often nearly absolute. One moment he was the prime minister of the UK during World War II—one of the most powerful prime ministers ever—and the next moment he is an ex–prime minister with almost no political power at all. Similar reverses were associated with World War I. At the heights of his power, he was described as dictatorial, arrogant, and intolerant, the stuff of which tyrants are made; at low power, he was seen as introspective and humble.

Moral Superiority

Few variables are as important in our lives as our perceived moral status. Even more than attractiveness and competence, degree of morality is a variable of considerable importance in determining our value to others—thus it is easily subject to deceit and self-deception. Moral hypocrisy is a deep part of our nature: the tendency to judge others more harshly for the same moral infraction than we judge ourselves—or to do so for members of other groups compared to members of our own group. For example, I am very forgiving where my own actions are concerned. I will forgive myself in a heartbeat—and toss in some compassionate humor in the bargain—for a crime that I would roast anybody else for.

Social psychologists have shown these effects with an interesting twist. When a person is placed under cognitive load (by having to memorize a string of numbers while making a moral evaluation), the individual does not express the usual bias toward self. But when the same evaluation is made absent cognitive load, a strong bias emerges in favor of seeing oneself acting more fairly than another individual doing the identical action. This suggests that built deeply in us is a mechanism that tries to make universally just evaluations, but that after the fact, "higher" faculties paint the matter in our favor. Why might it be advantageous for our psyches to be organized this way? The possession of an unbiased internal observer ought to give benefits in policing our own behavior, since only if we recognize our behavior correctly can we decide who is at fault in conflict with others.

The Illusion of Control

Humans (and many other animals) need predictability and control. Experiments show that occasionally administering electrical shocks at random creates much more anxiety (profuse sweating, high heart rate) than regular and predictable punishment. Certainty of risk is easier to bear than uncertainty. Controlling events gives greater certainty. If you can control, to some degree, your frequency of being shocked, you feel better than if you have less control over less frequent shocks. Similar effects are well known for other animals, such as rats and pigeons.

But there is also something called an illusion of control, in which we believe we have greater ability to affect outcomes than we actually do. For the stock market, we have no ability to affect its outcome by any of our actions, so any notion that we do must be an illusion. This was measured directly on actual stockbrokers. Scientists set up a computer screen with a line moving across it more or less like the stock market average, up and down—jagged—initially with a general bias downward but then recovering to go into positive territory, all while a subject sits in front of the screen, able to press a computer mouse, and told that pressing it "may" affect the progress of the line, up or down. In fact, the mouse is not connected to anything. Afterward, people are asked how much they thought they controlled the line's movement, a measure of their "illusion of control."

A very interesting finding emerged when those taking the tests were stockbrokers (105 men and 2 women) whose firms provided data both on internal evaluation and on salaries paid. In both cases, those with a higher illusion of control did worse. They were evaluated by their superiors as being less productive and, more important, they earned less money. Cause and effect is not certain, of course. But if the direction of effect were such that poor performers responded to their own failure by asserting greater control over external events, then they would be blaming themselves more for failure than success, contrary to the well-documented human bias to rationalize away one's failures. The alternative scenario then seems much more likely—that imagining one has greater control over events than one actually has leads to poorer performance: being a worse stockbroker. Note the absence of a social dimension here. One has no control over the movement of markets and scarcely much knowledge. There seems little possibility to fool your superiors along these lines when they can measure your success easily and directly. It is not at all clear that such an illusion in other situations may not give some social benefits—or even individual ones, as in prompting greater effort toward achieving actual control.

It is interesting to note that lacking control increases something called illusory pattern recognition. That is, when individuals are induced to feel a lack of control, they tend to see meaningful patterns in random data, as if responding to their unfortunate lack of control by generating (false) coherence in data that would then give them greater control.

The Construction of Biased Social Theory

We all have social theories, that is, theories regarding our immediate social reality. We have a theory of our marriages. Husband and wife may agree, for example, that one party is a long-suffering altruist while the other is hopelessly selfish, but disagree over which is which. We each have a theory regarding our employment. Are we an exploited worker, underpaid and underappreciated for value given—and therefore fully justified in minimizing output while stealing everything that is not nailed down? We usually have a theory regarding our larger society as well. Are the wealthy unfairly increasing their share of resources at the expense of the rest of us (as has surely been happening) or are the wealthy living under an onerous system of taxation and regulation? Does democracy permit us to reassert our power at regular intervals or is it largely a sham exercise controlled by wealthy interests? Is the judicial system regularly biased against our kinds of people (African Americans, the poor, individuals versus corporations)? And so on. The capacity for these kinds of theories presumably evolved not only to help understand the world and to detect cheating and unfairness but also to persuade self and others of false reality, the better to benefit ourselves.

The unconscious importance of biased social theory is revealed most vividly perhaps when an argument breaks out. Human arguments feel so effortless because by the time the arguing starts, the work has already been done. The argument may appear to burst forth spontaneously, with little or no preview, yet as it rolls along, two whole landscapes of information lie already organized, waiting only for the lightning of anger to reveal them. These landscapes have been organized with the help of unconscious forces designed to create biased social theory and, when needed, biased evidence to support them.

Social theory inevitably embraces a complex set of facts, which may be only partially remembered and poorly organized, the better to construct a consistent, self-serving body of social theory. Contradictions may be far afield and difficult to detect. When Republicans in the US House of Representatives bemoaned what the Founding Fathers would have thought had they known a future president (Clinton) would have sex with an intern, the black American comedian Chris Rock replied that they were having sex not with their interns but with their *slaves*. This of course is an important function of humor—to expose and deflate hidden deceit and self-deception (see Chapter 8).

False Personal Narratives

We continually create false personal narratives. By enhancing ourselves and derogating others, we automatically create biased histories. We were more moral, more attractive, more "beneffective" to others than in fact we were. Recent evidence suggests that forty- to sixty-year-olds naturally push memories of negative moral actions roughly ten years deeper into their past than memories of positive ones. Likewise, there is a similar but not so pronounced bias regarding nonmoral actions that are positive or negative. An older self acted badly; a recent self acted better. I am conscious of this in my own life. When saying something personal, whether negative or positive, I displace it farther in the past, as if I am not revealing anything personal about my current self, but this is especially prominent for negative information—it was a former self acting that way.

When people are asked to supply autobiographical accounts of being angered (victim) or angering someone else (perpetrator), a series of sharp differences emerges. The perpetrator usually describes angering someone else as meaningful and comprehensible, while victims tend to depict such an event as arbitrary, unnecessary, or incomprehensible. Victims often provide a long-term narrative, especially one emphasizing continuing harm and grievance, while perpetrators describe an arbitrary, isolated event with no lasting implications. One effect of this asymmetry between victim and perpetrator is that when the victim suppresses anger at a provocation, only to respond after an accumulation of slights, the perpetrator sees only the final, precipitating event and easily views the victim's angry response as an unwarranted overreaction.

There is also something called false internal narratives. An individual's perception of his or her own ongoing motivation may be biased to conceal from others the true motivation. Consciously, a series of reasons may unfold to accompany actions so that when they are challenged, a convinced alternative explanation is at once available, complete with an internal scenario—"but I wasn't thinking that at all; I was thinking . . ."

Unconscious Modules Devoted to Deception

Over the years, I have discovered that I am an unconscious petty thief. I steal small objects from you while in your presence. I steal pens and pencils, lighters and matches, and other useful objects that are easy to pocket. I am completely unconscious of this while it is going on (as are you, most of the time) even though I have been doing it for more than forty years now. Perhaps because the trait is so unconscious, it appears to have a life of its own and often seems to act directly against my own narrow interests. I steal chalk from myself while lecturing and am left with no chalk with which to lecture (nor do I have a blackboard at home). I steal pens and pencils from my office, only to offload them at home—leaving me none the next day at the office—and so on. Recently I stole a Jamaican principal's entire set of school keys from the desk between us. No use to me, high cost to him.

In summary, there appears to be a little unconscious module in me devoted to petty thievery, sufficiently isolated to avoid interfering with ongoing activity (such as talking). I think of a little organism in me looking out for the matches, the ideal moment to seize them, the rhythm of the actual robbery, and so on. Of course, this organism will study the behavior of my victim but it will also devote time to my own behavior, in order best to integrate the thievery while not giving off any clues. Noteworthy features of this little module in my own life are that the behavior has changed little over my lifetime, and that increasing consciousness of the behavior after the fact has done little or nothing to increase consciousness prior to, during, or immediately after the behavior. The module also appears to misfire more often the older I get. Incidentally, the only time I can remember getting caught is by my brother, born a year after me—we were raised as twins. We each had an ability to read deception in the other that others in the family could not match. Once when we were both in our late forties, I began to pocket his pen, but he grabbed my hand halfway to my pocket and the pen was his again.

I think I never pilfer from someone's office when it is empty. I will see a choice pen and my hand moving toward it but will say, "Robert, that would be stealing," and stop. Perhaps if I steal from you in front of your face, I believe you have given implicit approval. When I stole the principal's keys, I was simultaneously handing him some minor repayment for a service performed and thinking I might be paying too much. Perhaps I said to myself, "Well this is for you, so *this* must be for me," and he went along with the show.

How many of these unconscious modules operate in our lives? The only way I know about this one is that my pockets fill up with contraband, and I get occasional questions from friends. Stealing ideas will not leave much evidence and is very common in academia. I once wrote a paper that borrowed heavily from a well-known book, a fact I had forgotten by the time I finished the paper. Only when I reread my copy of the book did I see where the ideas had come from—these sections were heavily underlined, with many marginal notations.

It also seems certain that unconscious ploys to manipulate others in specific ways must be common. Specialized parts of ourselves look out for special opportunities in others. The value of this is precisely that two or more activities can go on simultaneously, with little or no interference. If an independent unconscious module studies for opportunities to steal or lie, it need not interfere (except slightly) with other, ongoing mental activities. We really have no idea how common this kind of activity may be.

THE HALLMARKS OF SELF-DECEPTION

In summary, the hallmark of self-deception in the service of deceit is the denial of deception, the unconscious running of selfish and deceitful ploys, the creation of a public persona as an altruist and a person "beneffective" in the lives of others, the creation of self-serving social theories and biased internal narratives of ongoing behavior, as well as false historical narratives of past behavior that hide true intention and causality. The symptom is a biased system of information flow, with the conscious mind devoted (in part) to constructing a false image and at the same time unaware of contravening behavior and evidence.

Of course, it must usually be advantageous for the truth to be registered somewhere, so that mechanisms of self-deception are expected often to reside side-by-side with mechanisms for the correct apprehension of reality. The mind must be constructed in a very complex manner, repeatedly split into public and private portions, with complicated interactions between them.

The general cost of self-deception is the misapprehension of reality, especially social, and an inefficient, fragmented mental system. As we shall learn, there are also important immune costs to self-deception, and there is something called imposed self-deception, in which an organism works unconsciously to further the interests of the organism inducing the self-deception costs on all sides, the worst of all possible worlds. At the same time, as we shall also see in Chapter 3, there is sufficient slack in the system for people to sometimes deceive themselves for direct advantage (even immunological). Before we turn to that, we will review the subject of deception in nature. There is an enormous literature on this subject and a few principles of genuine importance.

CHAPTER 2

Deception in Nature

Before we take a deeper look at self-deception, let us examine deception in other species. It is often easier to see patterns of importance if we cast our net of evidence widely—in this case, to include all species, not just our own. What can we learn about deception by viewing it in an evolutionary context? The evolutionary approach to deception is to study deception in all its forms while looking for general principles. So far, the forms of deception turn out to be very numerous and the principles very few. Deception hides from view, so its secrets often have to be pried out by meticulous study and analysis, of which, fortunately, there has been a lot, and several important principles have emerged that apply across species. First, there is a tremendous premium on novelty that in turn generates an enormous variety of deceptive ploys. Since novel tricks—almost by definition—lack defenses against the tricks, they usually spread quickly. This is the beginning of a so-called coevolutionary struggle between deceiver and deceived, acted out over evolutionary time. This struggle leads to complexity on both sides—to the evolution of bizarre, intricate, and beautiful examples of deception, as well as the ability to spot them. In general, but especially in birds and mammals, this evolutionary struggle also favors intelligence on both sides. Consider the simple matter of picking out an object against a background. If the object has not been selected to match the background, it should be easy to detect, differing in numerous random details. But if there has been selection to match, detection is an entirely different matter. Selection will have obliterated many of the random mismatches, leaving a much more complex cognitive problem for the observer to solve.

THE COEVOLUTIONARY STRUGGLE BETWEEN DECEIVER AND DECEIVED

The most important general principle is that deceiver and deceived are locked into a coevolutionary struggle. Since the interests of the two are almost always contrary—what one gains by perpetrating a falsehood, the other loses by believing it—a struggle (over evolutionary time) takes place in which genetic improvements on one side favor improvements on the other. One key is that these effects are "frequency dependent"—deception fares well when rare and poorly when frequent. And detection of deception fares well when deception is frequent but not when it is rare. This means that deceiver and deceived are locked into a cyclic relationship, in the sense that neither can drive the other extinct. Over time the relative frequencies of deceiver and deceived oscillate, but they do so within bounds that prevent either from disappearing. Likewise, in a verbal species like our own, we will be warned about new tricks more often by others the more frequent the tricks become. Note that no role is exclusive to some and not others—all of us are both deceiver and deceived, depending on context.

FREQUENCY-DEPENDENT SELECTION IN BUTTERFLIES

You don't have to look far to find evidence of frequency-dependent selection in systems of deception between prey and their predators. For example, in model/mimic systems, such as are found in butterflies (and snakes), a distasteful or poisonous species (model) evolves bright coloration to warn predators that it is distasteful. This selects for mimics, species that are perfectly tasty and harmless but gain protection by resembling the model. In West Africa, there is a genus of butterfly that is distasteful and as many as five species of the genus, all differing in coloration, may be found in the same forest. It turns out that there is a single species capable of mimicking all five model species. That is, females of the mimetic species can lay five kinds of eggs, each of which grows up to resemble one of the poisonous species.

This unusual system of mimicry provides striking evidence of frequency-dependent selection. Here, one species is delicious but mimics any one of five related poisonous species. These differ in color and pattern, and so do their respective mimics. When several poisonous species are found in the same forest with their mimics, the frequency of each mimic within this species matches the frequency of the model among its group of related, distasteful species. This could have been brought about only by frequency-dependent selection, where each mimetic form loses value when it becomes too common relative to its own model. If all the tasty butterflies looked the same, the predatory birds would rapidly specialize on that one form, decimating it.

One implication of frequency dependency is a perpetual premium on novelty. Indeed, in the above example, novel forms are more common in the mimetic species the more it outnumbers its model. That is, the more frequent the deceivers are, the more they begin to diversify, the better to avoid detection. Every new deception, by definition, starts rare and thereby gains an initial advantage. Only with success will one's disguise become part of the backdrop against which another novelty can begin rare and flourish. We can also see how easily break-off forms in the mimic might happen to resemble a second poisonous species, leading to two forms mimicking two species.

AN EPIC COEVOLUTIONARY STRUGGLE

A very rich illustration of coevolutionary principles is found in the relationships between brood parasites and their disadvantaged hosts, especially in birds but also in ants. A surprising percentage of all bird species, about 1 percent (usually cuckoos and cowbirds but also including a species of duck), is entirely dependent on other species to raise their young. Naturally this arrangement is rarely to the advantage of the "host" birds, who may end up raising unrelated young in addition to their own—or worse still, as is often the case, unrelated young *instead* of their own. This particular host/parasite relationship has been studied in unusual detail. Indeed, it is mentioned almost as early as human writing permits, some four thousand years ago in India, later described by Aristotle, and recently studied intensively by very clever field experiments designed to tease apart how the relationship works.

The first move is for the deceiver to lay one of its eggs in the victim's nest. This selects in the victim for the ability to recognize a strange-looking egg and eject it. This, in turn, selects for egg mimicry in the brood parasite—the tendency to produce eggs that have the same spotting and coloration as the eggs of the species whose parental care is being borrowed. Some parasitic species lay in the nests of multiple species, with individual species specialized to lay eggs that match in coloration the eggs of the species in whose nest they are laying. It is now advantageous for the host to be able to count total number of eggs, and reject nests with one too many. This is especially valuable if the parasite's young hatches before those of the host, ejecting all of its eggs so as to monopolize parental investment, leaving the host no offspring of its own to rear. Better for the host to start over. This selects for parasites that remove one egg for each one laid, leaving total number the same, and the egg is eaten or moved some distance from the nest, perhaps to hide the crime.

Once the egg has safely hatched, selection may favor brood-parasite mouth colors that resemble those of the host species, since parents feed more strongly mouth colors that resemble those of their own species. Within their own brood, evidence from other birds suggests that mouth color may be brighter for healthier chicks, so it is interesting that brood parasites make their mouth colors especially bright. By pushing out its foster siblings, the host young can monopolize parental investment, but since parents adjust their feeding to the total begging calls they hear, a single cuckoo chick may evolve to mimic the calls of an entire brood of the host. In an even more bizarre twist, a species of hawk cuckoo that parasitizes a hole-nesting species in Japan has evolved inner-wing patches that resemble the throat coloration of its host, so that when begging for food, a chick can flap its wings and simulate the begging of three offspring instead of one. The wing patches are even occasionally fed, a case of deception being too convincing for its own good.

A very important selective factor is errors in recognizing a host's own offspring—so-called false positives—that are an inevitable feature of any system of discrimination (see spam versus anti-spam in Chapter 8). For weak systems of discrimination, a host rarely rejects itself, but it is fooled too often into accepting cowbird chicks. Stronger systems of discrimination cut down on the host's loss due to the cowbirds but also impose a cost on the host, as it inevitably accidentally rejects its own offspring more of the time. In reed warblers, parents learn the appearance of their own eggs and then reject those differing by a certain amount. If their nests are parasitized about 30 percent of the time, it makes evolutionary sense for them to reject strange eggs, but if they are parasitized less often, the cost in destruction of their own eggs is too great. Sure enough, reed warblers are parasitized only 6 percent of the time in the UK and do not reject new eggs—unless a cuckoo is seen near the nest at about the right time (perhaps pushing probability above 30 percent). In one population, a drop in parasitism rate from 20 percent to 4 percent was matched by a one-third reduction in rejection rate, an effect too rapid to be genetic, so reed warblers probably often adjust their degree of discrimination to evidence of ongoing brood parasitism.

Note the important frequency-dependent effect. When almost all eggs are their own, discrimination will result in the warblers destroying some percent—say, 10 percent of their clutches—with only rare gain. But at 30 percent parasite frequency, they risk harming themselves only 7 percent of the time, while with perfect discrimination, they save themselves a substantial cost (in nurturing other species almost 30 percent of the time). At low frequency, deceivers are hardly worth detecting—only at high frequency are important defenses expected to kick in.

There is one striking peculiarity in the entire system. Birds repeatedly fail to evolve the ability to see that the cuckoo or cowbird chick bears no resemblance to their own chicks beyond mouth color and begging call. In size, a cuckoo chick is often six times or more larger than its host, so that a foster parent may perch on the shoulder of the chick it is about to feed. Since it would seem beneficial to note this absurd size discrepancy and act accordingly, why are birds, in species after species, unable to do so? The answer to the mystery is by no means certain, but there are some interesting possibilities. Failure to make the appropriate discrimination happens preferentially in species in which the brood parasite ejects its foster siblings before they hatch. Thus, if the parent learns the appearance of its own chicks by imprinting on the first ones produced, this will work fine if the first brood is its own, but it will prove fatal if at their first attempt they are parasitized. The host will imprint on the brood parasite and kill its own young whenever it sees them. This will wipe out the host's entire lifetime reproductive success, since it will now see all of its own chicks as foreign.

More generally, some of the brood parasite's characteristics are superoptimal from the foster parents' standpoint. We expect parents often to favor the larger of their chicks as being healthier, stronger, and more likely to provide a good return on investment. This may make foster parents vulnerable to implausibly large chicks that nevertheless release the bias that bigger is better. More to the point, many brood parasites have evolved begging calls that are louder than the host's and hence presumably harder to resist. Likewise, parasite mouth colors are especially brightly colored. These signals are less costly to magnify than is body size.

There is yet another explanation for hosts' not discriminating against obvious mimics—fear of the consequences. "Mafia-like" behavior has been described in a couple of bird species, in which a cuckoo or cowbird punishes those hosts who eject their eggs by destroying their entire nest. It becomes a matter of accepting a degree of parasitism or being really badly treated—like a demand payoff (tax) instead of an outright killing. Good evidence from one system shows that accepting the mafia tax leads to greater reproductive success than fighting it—and ending up with a destroyed nest.

Recently it has been shown that there is something resembling cultural transmission of knowledge regarding brood parasites. At least reed warblers can learn from the mobbing behavior of their neighbors toward models of cuckoos (whereas they do not bother to learn from induced mobbing toward innocuous species, such as parrots). Warblers are attracted to the sound of nearby mobbing and approach to observe. If it is a cuckoo being mobbed, they are more likely to approach quickly a model of a cuckoo in their own territory and to mob it. This social learning permits a much more rapid spread of defenses against brood parasites than can occur through genetic change alone. A brood parasite has also evolved to resemble a local hawk, and this resemblance reduces the degree to which it is be mobbed by potential hosts.

Birds are not the only group subject to brood parasites. Ants spend an enormous amount of energy raising large broods that are highly attractive as nurseries to other species. There are as many species of social parasites on ants as there are ant species (about ten thousand of each). Even though the nest may be fiercely defended, parasites have ways of gaining entry, usually by mimicking some part of the ant's communication system. Caterpillars of one butterfly species manage to get into an ant's nest by curling up in a ball and emitting the smell of ant larvae. They are then carried into the ant nest, where they imitate the sounds of a queen ant, the very sounds that lead actual queens to be preferentially fed and protected. When food is short, workers will feed young larvae to the pseudoqueens and, when the nest is disturbed, will rescue them over ant larvae. The caterpillars are even sometimes treated as rivals by the real ant queen. This is another example of a deception being too effective for its own good. These kinds of relations have been described for dozens of butterfly species that parasitize ant nests.

In sum, each move is met by a new countermove, resulting in principle in an evolutionary struggle that may last millions and millions of years. This is especially true of relationships between different species, where issues of relatedness no longer apply, but it may be true of many similar relationships within species as well. The two sexes, for example, are partly cooperative and partly in conflict, with move being matched by countermove, locking them into a tight frequency-dependent relationship that usually stabilizes at equal number of the two sexes (see Chapter 5).

Deception can be beautiful, complex, and very amusing. It can also be very, very painful. To be victimized by systematic deception in your own life can cause deep pain. Even watching another species victimized by deception can sear your heart. Every spring in Jamaica, I watch a few dove couples trying to reproduce by raising their young in my trees. These are birds I love to watch, and I wish them every success. Emerge from stage left the anis—large, black, ominous-looking birds that prey on the nestlings of other birds, eagerly gobbling up the chicks. Arriving in groups of about six to twelve, the anis are noisy and fast-moving, saturating their terrain and relying on one heartless trick. One ani gives a loud call that mimics the generic chick begging call of other species—a plaintive kind of squawk that the chick is most likely to give when it is hungry and the parent is nearby. Now the victim chick hears the ani's begging call and promptly begs itself, the better to outcompete its imaginary sibling. The ani (or one of its group members) makes a beeline to the chick and gobbles it up, along with any other nestlings. Your heart goes out to the victim, fooled by its naive tendency to beg when it hears a begging call. Or worse, you suffer for its silent and completely innocent siblings, only fated to have a fool in their nest. I spent one long evening flinging stones at anis who were about to devour a nest they had detected through this deception. They stayed nearby overnight and consumed the nest contents first thing in the morning.

INTELLIGENCE AND DECEPTION

Deception spawns the mental ability to detect it. In the above case, this includes the ability to discriminate very similar objects, the ability to count, the ability to adjust discriminatory powers to contextual factors, and the ability to act as if making multiple inferences: eggshells on ground, egg destroyed, nest parasitized, investment best curtailed, and so on.

These improved intellectual abilities select for more subtle means of deception, which, in turn, select for greater abilities to detect the deception. In short, deception continually selects for mental ability in the deceived. Since the target of apprehension is a moving target—that is, evolves away from your ability to detect it—ever-new discriminations proliferate. The ability to see through a deception requires special talents unnecessary for discriminating a target that has no ability or interest in hiding. Thus, deception has probably been a major factor favoring intelligence, certainly in highly social species.

Intelligence also helps deceivers. In behavioral deception, intelligence presumably increases the range and quality of the deception displayed. In humans, at one extreme, the behaviorally retarded will largely be limited to nonverbal forms of deception—a lunge in one direction when the opposite is intended—but rarely sophisticated patterns of verbal deception. By contrast, the very bright can lie in multiple dimensions. Thus, deception selects for intelligence on both sides, though more reliably on the perceptual side. For example, a moth's back comes to more and more exactly represent tree bark. This requires no new mental abilities on the part of the moth but implies growing powers of discrimination in its visual predators, such as birds and lizards. Not so for behavioral deception.

The best evidence for a robust role of intelligence in deception comes from a study of monkey and ape brains. The size of the neocortex (so-called social brain)—or better still, its relative portion of total brain—is positively associated with the use in nature of tactical deception, which includes any kind of deception that can be seen to give an advantage. The relative size of the neocortex is, in turn, a good measure of relative intelligence, especially social intelligence. Scientists used published studies of monkey and ape behavior in nature to assemble a large set of examples of deception, then solicited a still larger sample of unpublished studies. They next made sure the evidence was not biased by group size, or degree to which a species had been studied, or applied only to some monkeys and apes but not others. The strong conclusion was that among monkeys and apes, the smarter the species, the more often deception occurs. So perhaps does self-deception. We shall see later that the brighter children are, for a given age, the more often they lie. The importance of this can't be overemphasized. We often think that greater intelligence will be associated with less self-deception—or at least intellectuals imagine this to be true. What if the reverse is true, as I believe it is—smarter people on average lie and self-deceive more often than do the less gifted?

FEMALE MIMICS

You would think that telling the sexes apart would evolve easily and reliably, but in an extraordinary number of cases, one sex imitates the other (or the same sex of another species). In each case, females are being mimicked, as in the following three examples. In many groups of fireflies, particular species have evolved to prey on others by sexual mimicry. A predatory female of one species responds to the courtship flash of a male of another species by giving not her own flash of interest but that of a female of *his* species. He turns toward her, expecting to enjoy sex, and is seized and eaten instead. Sex is a very powerful force and especially in males often selects for "indiscriminate eagerness," which provides fertile ground for deception to parasitize.

In another example, that of orchids, fully one-third of all species are pollinated through deception—that is, the plant offers no actual reward to its pollinators, only the illusion of one. Most species mimic the smell of their pollinators' food without supplying any. A smaller number (about four hundred species) mimics an adult female of the pollinator species in both appearance and smell, so as to induce pseudo-copulation by the aroused male. The plant takes care not to give the male a full copulation with ejaculation, presumably to keep him in a perpetually aroused state, driven to seek out new "female" after new "female," pollinating the flowers all the way. Males who find pseudo-females do not linger and test nearby flowers as do males in plant species that have just given a nectar reward. Instead they fly immediately to a new patch of flowers, presumably in search of actual rewards. Thus, sexual mimics tend to be more outbred than closely related species that offer a real reward—a side effect of being deceived that may actually benefit the species itself.

Selection has also repeatedly favored males who mimic females within their species to fool territorial males into thinking they are females so they can get close enough to steal paternity of some or all of the eggs about to be laid by real females. These eggs will be cared for by the territorial male as his own. Sometimes selection for deception has been strong enough to mold morphs that are permanently committed to deception, that is, morphological forms whose strategy depends entirely on a life spent deceiving others. A classic example occurs in the bluegill sunfish, where a specialized male form has evolved that mimics a female in appearance and behavior, being one-sixth the size of a territorial male and roughly the size of an actual female. This female-mimic seeks out a territorial male, permits himself to be courted, and responds enough to keep the other male interested, so that when a true female spawns, the pseudo-female is ready nearby to help fertilize the eggs. It is as if the territorial male imagines he is in bed with two females when in fact he is in bed with one female and one male. The female almost certainly knows the truth.

The two kinds of males appear to be distinct forms that never turn into each other. To have persisted for so long, their long-term reproductive success must be identical—that is, over evolutionary time, the deceiver is doing exactly as well as the deceived—and this equality must, in turn, be enforced by frequency-dependent selection. When the female-mimic is relatively rare, he will do relatively well; when common, less so. Whether the female expresses any kind of preference for either male is unknown, but in general, females prefer rare males, that is, the less frequent of two choices. Perhaps one of the most spectacular cases of sexual mimicry is performed by a tiny blister beetle, itself a parasite on a solitary bee. To achieve dispersal, one hundred to two thousand individuals aggregate in groups that mimic in size, color, and perching location a single female of the host bee species, even moving as a unit up and down a tree. So here a kaleidoscopic falsehood is produced, its individual parts one-hundredth or less the size of the picture they are creating. In turn, a male bee copulating with the picture will serve to disperse the beetles to future bee nests since the beetles attach to him.

FALSE ALARM CALLS

Alarm calls occur in a variety of species, especially birds, and serve to warn other individuals (often relatives) that a predator is nearby. An alarm call is obviously a key moment—with little room for error on the receiving end. Thus, it is not surprising that true alarm calls have served as a template for the repeated evolution of false alarm calls. In mixed-species flocks of birds found in the tropics, an individual will give a false warning call when another bird has caught and is about to eat a large, tasty insect. Half the time, this causes the bird to drop the insect and dive for cover. In the other half of the cases, the bird is not fooled—while it always responds to a true alarm call with immediate flight. Thus, the birds have evolved to tell false from real alarm calls half the time.

In skuas, false warning calls by parents are used to frighten warring offspring into separating and fleeing for cover, at which point the parents intervene to prevent further strife. In swallows, males apparently use false alarm calls to guard their paternity. They will give an alarm call when they spot their mate near another male, often causing both birds to dive for cover. Males breeding in colonies almost always give such calls when returning to an empty nest during egg laying (when female copulations outside the pair are frequent and threaten his paternity of the offspring) but not at other times (even swallows do not wish to cry "wolf"). Antelopes have been discovered playing the same trick. After a male has spent a day or two in sexual consort with an adult female, he will give a warning bark if the female seeks to move on, as if signaling that a predator lurks nearby and she should remain with him.

CAMOUFLAGE

Camouflage is so common in nature as almost to escape notice. Most creatures are selected at the very least to blend in to their backgrounds, with stick and leaf insects merely extreme examples. But at the behavioral level, octopuses and squid are so advanced as to be worth special note.

Octopuses and squid are fat, tasty creatures without a protective shell, so they are naturally sought after by a wide range of predators, mostly fish but also mammals and diving birds. Their only defense (beyond ink clouds and biting) is camouflage, and here they have evolved a remarkable system in which each skin-color cell is innervated by a single neuron, thus cutting out all synaptic delays and permitting a near-perfect adjustment to the background in about two seconds. While feeding, the animal can move very slowly across a great range of backgrounds, continuously remaining nearly invisible to others by adjusting its color to each new surface—sand, mud flats, coral reefs, rocks, sea-grass beds, and so on. Octopuses look as if they are slowly rolling while continuously adjusting to what is below. When they want to swim fast, they mimic flounders, in shape, color, swimming movements, and speed, darting swiftly along the sea bottom.

At intermediate speeds (when foraging), they adopt a most unusual strategy of randomly displaying variant phenotypes at about the rate of three per minute, for hours at a time, as if they are shuffling through a deck of cards featuring different camouflaged versions of themselves. This helps prevent the predator from forming a specific search image for any particular version. Just as the predator recognizes potential prey, the prey has morphed into a novel camouflaged form. One species of squid has also evolved a female mimic, one so good that he sometimes fools even fellow female mimics, who approach in search of copulation. This is yet another case of deception being too convincing for its own good.

DEATH AND NEAR-DEATH ACTS

It has long been known in predator/prey relations that deception can work anywhere from first detection until final consumption. Consider two examples near the time of death itself. The feigning of death typically occurs after the prey is caught, and is thought to inhibit the final death-dealing strike. The bird acts dead, lifeless, but remains conscious and alert so that often the only sign of life is its open eyes. Chickens run at the first opportunity, typically when the predator lets go, but a duck threatened by a fox often remains immobile for some time after release, especially if other foxes appear to be present. The fox's counteradaptations are to kill some prey immediately upon capture and to disable the remaining ones by severing a wing on each.

In the broken-wing display, a bird near its nest tries to distract a potential predator by acting like an injured bird, with one broken and extended wing. The bird moves awkwardly near the predator with wing extended but flies away quickly when attacked. This display is much more dramatic the closer the predator is to the nest. Birds have a variety of other acts they conduct when their nest is threatened. Crakes, ground-nesting birds, will mimic rats scurrying away from their nest, their backs slightly hunched, with both wings partly open and drooping to mimic a fat rat scurrying away in the wide open—an easy prey that looks tasty to various mammals and birds, but one that can suddenly take to the air when attacked. At other times, among reeds, the crake will drop like a stone into the water, creating a big splash, and then move loudly through the reeds, much like a frog staying at the surface. What is noteworthy is that the crake calls attention to itself while acting as if it is not. It must not be such a good rat or frog that it remains undetected, yet it must act like a target trying to avoid detection. Thus, movements are outwardly furtive but louder than usual.

RANDOMNESS AS A STRATEGY

We use patterns to detect deception, and randomness is the absence of pattern. It is often not appreciated how valuable randomness is as part of a deceptive strategy designed to avoid detection. Consider a couple of examples. Fake butterfly eggs are actually plant structures evolved to prevent butterflies from laying their eggs—since butterflies avoid laying eggs where they see some have already been laid. The fake eggs appear at random on the surface of the plant's leaves. Yet in closely related species, where the plant structures serve their original function, they are symmetrically located on each side of the leaf. Thus, natural selection created the randomness, presumably since butterflies had evolved to treat symmetrical patterns of eggs as if they were not really eggs (as indeed they are not). An ongoing struggle for randomness occurs in a pronghorn antelope. The pronghorn mother leaving her offspring hidden between nursings while she eats initially orients herself away from her offspring, then for much of the time she faces in random directions. Finally, only just before returning to nurse does the mother reveal the offspring's position by facing it.

Now consider a human example. In the old days, when customs officers routinely searched most bags in the owner's presence, a tried-and-true method to detect smuggling was to poke around randomly while watching the owner out of the corner of their eye. Whenever the owner became agitated or showed undue attention, the customs officer eliminated the rest of the bag and concentrated on the suspicious section. Again, by poking around (and paying close attention), the officer allowed the owner to guide him or her to the problem, presumably something illegal. Note that lack of preparation for this eventuality—being caught—only heightens one's anxiety and inadvertent information leakage.

For years, I have been well aware of the importance of information limitation. I have not used it with customs officials, but if a police officer is searching the trunk of my car, I simply turn my back. The officer may think I have something to hide, but he or she will learn nothing from me about where it is, if indeed there is something. Of course, when being watched for other purposes, we may also busy ourselves with semi-random behavior to hide the truth.

Once, when trying to get readmitted to Harvard after a medical leave, I had to take the famous "What do you see in this inkblot?" (Rorschach) test. I had learned that results were graded based on whether you saw a picture or told a story, whether it was in color, whether the story was coherent, and so on, but I had forgotten what the "appropriate" answers were supposed to look like to signify "normal," so I simply randomized my responses, figuring absence of a pattern was my best hope. Sometimes they got a story, sometimes a snapshot, sometimes in color, and so on. At least I did not appear to be rigid or compulsive. I was readmitted.

It may, indeed, be that a certain degree of randomness is built into the very core of our behavior. Not only will others not detect a pattern, but neither shall we—thus preventing us from inadvertently revealing ourselves.

DECEPTION MAY INDUCE ANGER

How do animals react when they detect deception directed at them? Studies from a range of species—wasps, birds, and monkeys—suggest they often get angry and seek immediate retribution. At least this seems to be true of several species in which individuals have what appear to be arbitrary symbols that confer status—so-called badges—such as greater melanin (darker color) on the chest feathers of sparrows or the mouthparts (clypeus) of wasps. In each case, the signals are on the part of the body most visible in a face-to-face encounter, and each is positively associated with body size and dominance. How is the association maintained between the arbitrary badge of status and the status itself? In wasps, for example, less than 1 percent of the body's melanin is found on the clypeus. Why do cheaters not invade the system and produce higher-status badges than their size warrants? Precisely because they are immediately attacked and are usually unable to defend themselves. Those whose clypeuses are painted to look more dominant do not become more dominant but are attacked six times as often by truly dominant individuals, while wasps painted to look less dominant are attacked twice as often as nonpainted controls. And it is interesting that subordinate wasps attack those painted to look dominant more often than they attack those who look dominant to begin with. A key perceptual factor is incongruity between appearance and behavior—when individuals are painted darker and made more aggressive via hormone treatment, they gain in dominance, but when made more aggressive without the change in appearance, the wasps fail to establish stable dominance relations, presumably because others are continually tempted to challenge them.

When a sparrow's chest is painted blacker to enhance the apparent badge size and, thus, status of the sparrow, the effect on status is usually the opposite. The altered bird is attacked more frequently than before, especially by those with the same apparent badge size or larger. The result is a drop in status—or ostracism from the group—for the individual with the deceptive badge. By contrast, those who, in effect, deceive downward—that is, who are bleached to appear less dark than they really are—often become hyperaggressive, whirling around and attacking their near neighbors who now act disrespectfully by standing too close to them based on their new (diminished) badge.

That deception might induce anger and attack was suggested to me very forcefully in my own life some thirty years ago. I was taking a walk, carrying my one-year-old son in my arms, when I spotted a squirrel in a tree. The problem was that my son did not see the squirrel, so I whistled as melodically as I could to draw the squirrel closer to us and, sure enough, the squirrel crept forward, but my son still could not see it. So I decided to reverse my relationship with the squirrel and mimic an attack. I suddenly lunged at it. I expected it to scamper away from me. I would have ruined a budding friendship but allowed my son to see the squirrel as it rushed away from us. Instead, the squirrel ran straight at us, chittering in apparent rage, teeth fully exposed, jumping to the branch closest to me and my son. *Now* my son saw the squirrel, and I had the fright of my life, quickly running several steps away.

For my folly, the squirrel could have killed my son with a leap to my shoulders and two expert bites to his neck. Had I begun the relationship hostile, I believe the squirrel never would have become so angry. It was the betrayal implied by beginning friendly, only then to attack (deception), that triggered the enormous anger. There is nothing quite like the humility you feel as you sneak your son back into your home, not telling your wife, of course, that in a little pseudo-scientific work on the side, you had managed to enrage a squirrel to the point of putting her child at risk. I had no plans to try that stunt again anytime soon.

The importance of aggression following knowledge of deception is that it may greatly increase the costs of deceptive behavior and the benefits of remaining undetected. Fear of aggression can itself become a secondary signal suggesting deception, and its suppression an advantage for self-deception. Of course, aggression is not the only social cost of detected deception. A woman may terminate a relationship upon learning of a lie, usually a crueler punishment than her giving you a good beating, assuming she is capable. Detected deception may lead to social shame—bad reputation, loss of credibility and status, so that there will always be pressure on the deceiver to hide the deception, not only to make it successful but to avoid the larger consequences of detection.

ANIMALS MAY BE CONSCIOUS OF DECEPTION

Naturally one must be careful in imputing particular kinds of consciousness to other species, but some situations strongly suggest that animals are conscious of ongoing deception in some detail. Ravens, for example, have evolved a set of elaborate behaviors surrounding their tendency to cache (that is, bury and hide) food for future consumption, which can be enjoyed by another bird who happens to view the caching. Accordingly, ravens who are about to hide food seem very sensitive to just this possibility. They distance themselves from others and often cache behind a structure that obstructs others' view. They regularly interrupt caching to look around. At any evidence they are being observed, they will usually retrieve the cached food and wait to rebury somewhere else, preferably while not under observation. If they do cache food, they will often return within a minute or two. The watchers, in turn, stay at a safe distance, often hiding behind a tree or other object. They stop looking if the other stops caching and wait a minute or more after the bird has left before going for the cache. Hand-reared ravens, in turn, can follow the human gaze by repositioning themselves to see around an obstacle. This suggests the possibility that ravens can project the sight of another individual into the distance. Likewise, when jays are caching in others' presence, they maximize their distance from others and cache in the shade and in a confusing pattern, moving caches frequently. Experimental work shows that they remember who has watched them cache in the past and when being observed by such individuals are more likely to re-cache than when they are being watched by a newcomer—another example of intelligence evolving in the context of deception.

In the presence of other squirrels, gray squirrels cache farther apart, build false caches, and build with their backs turned to the other squirrels; no such responses are shown to crows who may be watching. Turning one's back often shows up in other mammals, as well. A chimpanzee male displaying an erection to a female may turn his back when a more dominant male arrives, until his erection has subsided. Children as young as sixteen months will turn their backs to conceal the object in hand or what they are doing. I personally find it very hard in the presence of a woman with whom I am close to receive a phone call from another woman with whom I may have, or only wish to have, a relationship, without turning my back to pursue the conversation. This occurs even though there is nothing visual to hide and the act of turning gives me away. Perhaps this is a case of reducing cognitive dissonance—and cognitive load—by not having to watch one woman watch you while you pretend not to talk to another woman.

In ravens, the pilferers avoid searching for known caches when in the presence of those who cache but will go immediately to the caches in the presence of a noncaching bird (that is unlikely to defend). In addition, they actively search away from the cache in the presence of the cacher, as if hiding their intentions. In one experiment, when ravens were introduced into an area where food was hidden, a subordinate male quickly developed the ability to find food, which the most dominant quickly learned to parasitize. This in turn led the subordinate to first search in areas where no food was present, to lure the dominant away, at which point the subordinate moved quickly to the food itself.

Mantis shrimps are hard-shelled and their claws dangerous for seven weeks out of eight. On the eighth week, they are molting, and their body and claws are soft; they are unable to attack others and are vulnerable to attack by them. When encountered at this time, they greatly increase their rate of claw threats, sometimes combined with insincere lunges at the opponent. About half the time, this scares off their opponent. The other half, the soft-shelled shrimp runs for its life. The week before a mantis shrimp becomes soft-shelled, it increases its rate of claw threats but also increases the rate at which these threats are followed by actual attack, as if signaling that threats will be backed up by aggressive action just before the time when they will not.

In fiddler crabs, the male typically has a large claw used to fight and threaten other males and to court females. Should he lose this claw, he regenerates one very similar in appearance but less effective than the original. The size of the first claw does indeed correlate (independent of body size) with claw strength as well as ability to resist being pulled from one's burrow, but the size of the replacement claw does not, and males can't distinguish between the two kinds of claws in an opponent.

In primates, hiding information from others may take very active forms. For example, in both chimpanzees and gorillas, individuals will cover their faces in an apparent attempt to hide a facial expression. Gorillas in zoos have been seen to cover "play faces" (facial expressions meant to invite play) with one or both hands, and these covered faces are less likely to elicit play than uncovered play faces. Of course, a play face hidden in this fashion is hardly undetectable and may easily become a secondary signal. Chimpanzees will hide objects behind their backs that they are about to throw. They will also throw an object to one side of a tree to frighten another chimp into moving to the opposite side, where his opponent awaits him.

DECEPTION AS AN EVOLUTIONARY GAME

An important part of understanding deception is to understand it mathematically as an evolutionary game, with multiple players pursuing multiple strategies with various degrees of conscious and unconscious deception (in a fine-grained mixture). Contrast this with the problem of cooperation. Cooperation has been well modeled as a simple prisoner's dilemma. Cooperation by both parties benefits each, while defections hurt both, but each is better off if he defects while the other cooperates. Cheating is favored in single encounters, but cooperation may emerge much of the time, if players are permitted to respond to their partner's previous moves. This theoretical space is well explored.

The simplest application of game theory to deception would be to treat it as a classical prisoner's dilemma. Two individuals can tell each other the truth (both cooperate), lie (both defect), or one of each. But this cannot work. One problem is that a critical new variable becomes important—who believes whom? If you lie and I believe you, I suffer. If you lie and I disbelieve you, you are likely to suffer. By contrast, in the prisoner's dilemma, each individual knows after each reciprocal play how the other played (cooperate or defect), and a simple reciprocal rule can operate under the humblest of conditions—cooperate initially, then do what your partner did on the previous move (tit for tat). But with deception, there is no obvious reciprocal logic. If you lie to me, this does not mean my best strategy is to lie back to you—it usually means that my best strategy is to distance myself from you or punish you.

The most creative suggestion I have heard to mathematically model deception is to adapt the ultimatum game (UG) to this problem. In the UG, a person proposes a split of, say, $100 (provided by the experimenter)—$80 to self, $20 to the responder. The responder, in turn, can accept the split, in which case the money is split accordingly, or the responder can reject the offer, in which case neither party gets any money. Often the game is played as a one-shot anonymous encounter. That is, individuals play only once with people they do not know and with whom they will not interact in the future. In this situation, the game measures an individual's sense of injustice—at what level of offer are you sufficiently offended to turn it down even though you thereby lose money? In many cultures, the 80/20 split is the break-even point at which one-half of the population turns down the offer as too unfair.

Now imagine a modified UG in which there are two possible pots (say, $100 and $400) and both players know this. One pot is then randomly assigned to the proposer. Imagine the proposer offers you $40, which could represent 40 percent of a $100 pot (in which case you should accept) or 10 percent of a $400 pot (most people would reject). The proposer is permitted to lie and tell you that the pot is the smaller of the two when in fact it is the larger. You can trust the proposer or not, but the key is that you are permitted to pay to find out the truth from a (disinterested) third party. This measures the value you place in reducing your uncertainty regarding the proposer's honesty.

If you then discover that the proposer lied, you should have a moral (or, at least, moralistic) motive to reject the offer, and the other way around for the truth—all compared to uncertainty, or not paying to find out. Note that from a purely economic point of view, there is no benefit in finding out the truth, since it costs money and may lead to an (otherwise) unnecessary loss of whatever is offered. The question can then be posed: How much would a responder be prepared to pay to reduce the uncertainty and go for a possibly inconvenient truth? Note that the game can be played in real life with varying degrees of anonymity and also multiple times, as in the iterated prisoner's dilemma. As ability to discriminate develops, the other person will benefit more from your honesty (quickly seen as such) and suffer less from deception (spotted and discarded).

When we add self-deception, the game quickly becomes very complicated. One can imagine actors who are:

- Stone-cold honest (cost: information given away, naive regarding deception by others).
- Consciously dishonest to a high degree but with low self-deception (cost: higher cognitive cost and higher cost when detected).
- Dishonest with high self-deception (more superficially convincing at lower immediate cognitive cost but suffering later defects and acting more often in the service of others).

And so on.

A DEEPER THEORY OF DECEPTION

Those talented at the mathematics of simple games or studying them via computer simulation might find it rewarding to define a set of people along the lines just mentioned, and then assign variable quantitative effects to explore their combined evolutionary trajectory. Perhaps results will be trivial and trajectories will depend completely on the relative quantitative effects assigned to each strategy, but it is much more likely that deeper connections will emerge, seen only when the coevolutionary struggle is formulated explicitly. The general point is, of course, that there are multiple actors in this game, kept in some kind of frequency-dependent equilibrium that itself may change over time. We choose to play different roles in different situations, presumably according to the expected payoffs. Of course it is better to begin with very simple games and only add complexity as we learn more about the underlying dynamics.

It stands to reason that if our theory of self-deception rests on a theory of deception, advances in the latter will be especially valuable. I have known this for thirty years but have not been able to think of anything myself that is original regarding the deeper logic of deception, nor have I seen much progress elsewhere. Yes, signals in male/female courtship interactions may evolve toward costlier ones that are more difficult to fake (for example, antler size, physical strength, and bodily symmetry), but there is always room for deception, and many systems do not obey this simple rule regarding cost.

CHAPTER 3

Neurophysiology and Levels of Imposed Self-Deception

Although study of the neurophysiology of deceit and self-deception is just beginning, there are already some interesting findings. Evidence suggests a greatly diminished role for the conscious mind in guiding human behavior. Contrary to our imagination, the conscious mind seems to lag behind the unconscious in both action and perception—it is much more observer of action than initiator. The precise details of the neurobiology of active thought suppression suggest that one part of the brain has been co-opted in evolution to suppress another part, a very interesting development if true. At the same time, evidence from social psychology makes it clear that trying to suppress thoughts sometimes produces a rebound effect, in which the thought recurs more often than before. Other work shows that suppressing neural activity in an area of the brain related to lying appears to improve lying, as if the less conscious the more successful.

There is something called induced self-deception, in which the self-deceived person acts not for the benefit of self but for someone who is inducing the self-deception. This can be parent, partner, kin group, society, or whatever, and it is an extremely important factor in human life. You are still practicing self-deception but not for your own benefit.

Among other things, it means that we need to be on guard to avoid this fate—not defensive via self-deception but via greater consciousness.

Finally, we have treated self-deception as part of an offensive strategy, but is this really true? Consider the opposite—and conventional—view, that self-deception serves a purely defensive function, for example, protecting our degree of happiness in the face of reality. An extreme form is the notion that we would not get out of bed in the morning if we knew how bad things were—we levitate ourselves out via self-deception. This makes no coherent sense as a general truth, but in practicing self-deception, we may sometimes genuinely fool ourselves for personal benefit (absent any effect on others). Placebo effects and hypnosis provide unusual examples, in that they show direct health benefits from self-deception, although this typically requires a third party, either hypnotist or doctor-model. And people can almost certainly induce positive immune effects with the help of personal self-deception, as we shall see in Chapter 6.

THE NEUROPHYSIOLOGY OF CONSCIOUS KNOWLEDGE

Because we live inside our conscious minds, it is often easy to imagine that decisions arise in consciousness and are carried out by orders emanating from that system. We decide, "Hell, let's throw this ball," and we then initiate the signals to throw the ball, shortly after which the ball is thrown. But detailed study of the neurophysiology of action shows otherwise. More than twenty years ago, it was first shown that an impulse to act begins in the brain region involved in motor preparation about six-tenths of a second before consciousness of the intention, after which there is a further delay of as much as half a second before the action is taken. In other words, when we form the conscious intention to throw the ball, areas of the brain involved in throwing have already been activated more than half a second earlier.

Much more recent work, from 2008, gives a more dramatic picture of preconscious neural activity. The original work involved a neural area, the supplementary motor area involved in late motor planning. An important distinction is whether preparatory neural activity is related to a particular decision (throw the ball) or just activation in general (do something). A novel experiment settled the matter. While seeing a series of letters flash in front of him or her, each a half-second apart, an individual is asked to hit one of two buttons (with left or right index finger) whenever he or she feels like it and to remember which letter was seen when the conscious choice was made. After this, the subject had to choose which of four letters was the one he or she saw when consciously deciding to press the button. This served roughly to demarcate when conscious knowledge of the decision is made, since each letter is visible for only half a second and conscious knowledge of intention occurs about one second before the action itself.

What about prior unconscious intention? Computer software can search through fMRI images (showing blood flow associated with neural activity) taken in various parts of the brain during intervals prior to action. Most strikingly, a full seven seconds before consciousness of impending action, activity occurs in the lateral and medial prefrontal cortex, quite some distance from the supplementary motor area and the motor neurons themselves. Given the slowness of the fMRI response, it is estimated that fully ten seconds before consciousness of intent, the neural signals begin that will later give rise to the consciousness and then the behavior itself. This work also helps explain earlier findings that people develop anticipatory skin conductance responses to risky decisions well before they consciously realize that such decisions are risky.

One point is well worth emphasizing. From the time a person becomes conscious of the intent to do something (throw a ball), he or she has about a second to abort the action, and this can occur up to one hundred milliseconds before action (one-tenth of a second). These effects can themselves operate below consciousness—that is, subliminal effects operating at two hundred milliseconds before action can affect the chance of action. In that sense, the proof of a long chain of unconscious neural activity before conscious intention is formed (after which there is about a one-second delay before action) does not obviate the concept of free will, at least in the sense of being able to abort bad ideas and also being able to learn, both consciously and unconsciously, from past experience.

On the flip side, it is now clear that consciousness requires some time for perception to occur. Put another way, a neural signal travels from the toe to the brain in about twenty milliseconds but takes twenty-five times as long, a full five hundred milliseconds (half a second) to register in consciousness. Once again, consciousness lags reality and by a large amount, plenty of time for unconscious biases to affect what enters consciousness.

In short, the best evidence shows that our unconscious mind is ahead of our conscious mind in preparing for decisions, that consciousness occurs relatively late in the process (after about ten seconds), and that there is ample time for the decision to be aborted after consciousness (one second). In addition, incoming information requires about half a second to enter consciousness, so that the conscious mind seems more like a post-hoc evaluator and commentator upon—including rationalizing—our behavior, rather than the initiator of the behavior. Chris Rock, the comedian, says that when you meet him for the first time (conscious mind and all), you are not really meeting him—you are only meeting his representative.

THE NEUROPHYSIOLOGY OF THOUGHT SUPPRESSION

One particular kind of self-deception—consciously mediated efforts at suppressing true information from consciousness—has been studied by neurophysiologists in a most revealing way. The resulting data are striking in our context: different sections of the brain appear to have been co-opted in evolution to suppress the activity of other sections to create self-deceptive thinking.

Consider the active conscious suppression of memory. In real life, we actively attempt to suppress our thoughts: *I won't think about this today*; *please, God, keep this woman from my mind*, and so on. In the laboratory, individuals are instructed to forget an arbitrary set of symbols they have just learned. The effect of such efforts is highly variable, measured as the degree of memory achieved a month later when attempting to recall the symbols. This variation turns out to be associated with variation in the underlying neurophysiology. The more highly the dorsolateral prefrontal cortex (DLPFC) is activated during directed forgetting, the more it suppresses ongoing activity in the hippocampus (where memories are typically stored) and the less is remembered a month later. The DLPFC is otherwise often involved in overcoming cognitive obstacles and in planning and regulating motor activity, including suppressing unwanted responses. One is tempted to imagine that this area of the brain was co-opted for the new function of suppressing memories because it was often involved in affecting other brain areas, in particular, suppressing behavior. There is a physical component to this—I know it well. When I experience an unwanted thought and act to suppress it, I often experience an involuntary twitch in one or both of my arms, as if trying to push something down and out of sight.

THE IRONY OF TRYING TO SUPPRESS ONE'S THOUGHTS

The neurophysiological work employed meaningless strings of letters or numbers during short periods of memorization followed by short periods of attempted forgetting, results measured a month later. But another factor operates if we try to suppress something meaningful. One might easily suppose that a conscious decision to suppress a thought (don't think of a white bear) could easily be achieved, each recurrence of the thought suppressed more deeply so that soon enough the thought itself fails to recur. But this is not what happens. The mind seems to resist suppression, and under some conditions we do precisely what we are trying to suppress. For example, we may blurt out the very truth we are trying to hide from others, as if involuntarily or contra-voluntarily. The suppressed thought often comes back to consciousness, sometimes at the rate of once per minute, and often for days. As with the neurophysiology of thought suppression, some people are better at thought suppression and some try harder. But few people are completely successful.

Two processes are thought to work simultaneously. On the one hand, there is an effort to consciously suppress the undesired thought, initially and whenever it reappears. On the other hand, an unconscious process to search for the prohibited word, as if looking for errors, that is, thoughts that need additional suppression. This process is itself subject to errors, especially when we are under cognitive load. When one is distracted or overburdened mentally, the unconscious search for the thought is not combined with suppression of it, so that the suppressed thought may burst forth *more* often than expected.

IMPROVING DECEPTION THROUGH NEURAL INHIBITION

The first great advances in neurophysiology came from the ability to measure ongoing brain activity in space and time, first crudely through EEG and then more precisely through fMRI and PET scans. Now a recent method (as we saw in Chapter 1) has taken the opposite approach and selectively knocked out brain activity in particular parts of the brain to see the effects. This was achieved by applying external electrical stimulation on the scalp to inhibit brain activity directly underneath. For example, stimulation can be applied to a brain area involved in deception (at the anterior prefrontal cortex, aPFC) while a person chooses whether to lie in response to a series of questions designed to determine whether she was involved in the mock crime of stealing money from a room. Although in general we expect any artificially induced effect on life—for example, rapping a person hard on his or her knee—to be negative much more often than positive, this intervention was clearly positive where deception was concerned. At least three key components were altered in an advantageous direction. Reaction time while lying was decreased under inhibition, as was physiological arousal. So people were quicker and more relaxed. The electrical inhibition also appeared to reduce the moral conflict during lying. That is, people felt less guilt under inhibition, and the less guilt they felt, the quicker their response times. In addition, people with this area knocked out lied more frequently on relevant questions and less on irrelevant ones, thus more finely tuning their lying.

This is a very striking result. Artificially suppressing mental activity improves performance. This provides an analogy to self-deception, because the suppression of mental activity can come externally via a magnetic device applied to the skull or internally via neuronal suppression emanating from elsewhere in the brain—via self-deception in service of deceit. The only thing we do not know is whether the external inhibition also knocked out consciousness to aspects of the deception, as we might well expect.

Incidentally, two recent studies in China suggest that the brains of those regarded as pathological liars show more white matter in the areas of the brain believed to be involved in deception. "White matter" refers not to the neurons themselves but to the supporting glial cells that nourish the neurons, especially their long, thin dendritic extensions. We know from work on jugglers that the more they practice, the more white matter shows up in the "juggling center" of their brains, so this correlation with lying may result from repeated practice.

UNCONSCIOUS SELF-RECOGNITION SHOWS SELF-DECEPTION

The classic experimental work demonstrating self-deception took place some thirty years ago and involved (largely unconscious) verbal denial or projection of one's own voice. In a brilliant series of experiments, true and false information was shown to be simultaneously stored within an individual, but with a strong bias toward the true information being hidden in the unconscious mind and the false in the conscious. In turn, people's tendency to deny (or project) their voices could be affected by making them feel worse or better about themselves, respectively. Thus, one could argue that the self-deception was ultimately directed toward others.

The experiment was based on a simple fact of human biology. We are physiologically aroused by the sound of a human voice but more so to the sound of our own voice (for example, as played from a tape recorder). We are unconscious of these effects. Thus one can play a game of self-recognition, in which people are asked whether a voice is their own (conscious self-recognition) while at the same time recording (via higher arousal) whether unconscious self-recognition has been achieved.

Here is how it worked. People were asked to read the same paragraph from a book. These recordings were chopped into two-, four-, six-, twelve-, and twenty-four-second segments, and a master tape was created consisting of a mixture of these segments of their own and other voices (matched for age and sex). Meantime, each individual was hooked up to a machine measuring his or her galvanic skin response (GSR), a measure of arousal that is normally twice as great for hearing one's own voice as hearing someone else's. People were asked to press a button to indicate that they thought the recording was of themselves and another button to indicate how sure they were.

Several interesting facts were discovered. Some people denied their own voices some of the time; this was the only kind of mistake they made and they seemed to be unconscious of making it (when interviewed later, only one was aware of having made this mistake). And yet the skin had it correct—that is, it showed the large increase in GSR expected upon hearing one's own voice. By contrast, another set of people heard themselves talking when they were not—they projected their voice, and this was the only error they made. Although half were aware later that they had sometimes made this mistake, the skin once again had it correct. This is unconscious self-recognition shown to be superior to conscious recognition. There were two other categories: those who never made mistakes and those who made both kinds, sometimes fooling even their skin, but for simplicity we neglect these two categories (about which nothing more is known, in any case).

It is well known that making people feel bad about themselves leads to less self-involvement (e.g., looking in the mirror). In the above experiment, people made to feel bad by a poor score on a pseudo-exam just taken (in fact, with grades randomly assigned) started to deny their voices. Made to feel good by a good score, they started to hear themselves talking when they were not. It was as if self-presentation was expanding under success and contracting in response to failure.

Another interesting feature—never analyzed statistically—was that deniers also showed the highest levels of arousal to all stimuli. It was as if they were primed to respond quickly, to deny the reality, and get it out of sight. By contrast, inventing reality (projecting) seems a more relaxed enterprise, with more relaxed arousal levels typical of those who make no mistakes. Perhaps reality that needs to be denied is more threatening than is the absence of reality one wishes to construct. Also, denial can be dealt with quickly, with low cognitive load, but requires an aroused state for quick detection and deletion.

There is a parallel in the way in which the brain responds to familiar faces. Some people have damage to a specific part of their brain that inhibits their ability to recognize familiar faces consciously. When asked to choose familiar over unfamiliar faces or match names with faces, the individual performs at chance levels. He or she nonetheless recognizes familiar faces unconsciously, as shown through changes in brain activity and skin conductance. When asked to state which face he or she *trusts* more, choice is above chance in the expected direction. Thus, there is some access to unconscious knowledge, but not much.

Can we study this in other animals? Some birds show the human pattern exactly. In playback experiments, they show greater physiological arousal to hearing their own species' song (compared to that of others) but a stronger response still to their own voices. These birds could easily be trained to peck at a button when they recognized their own voice (this would be analogous to verbal self-recognition), while measures of physiological arousal would reveal something closer to unconscious self-recognition (GSR in humans). When birds are made to lose fights, do they start avoiding pecking to their own voice (denial) and when made to win fights, show the opposite effect?

CAN ONE HALF OF THE BRAIN HIDE FROM THE OTHER?

Our left and right brain are connected by a corpus callosum, an ancient vertebrate symmetry that has important effects on daily life. The brains partly receive information independently (left ear, right brain) and also act independently (left brain runs right hand). I have often noticed that my right brain may not actively engage in a search unless the left brain makes the goal explicit by saying it out aloud. That is, I will be searching for an object in the visual world or in my pockets, including left pocket, and I will not find it until I say the word out loud ("lighter"), then suddenly I spot it in my left visual field or feel it in my left pocket (this is a consequence of the brain being cross-wired—left-side information goes primarily to the right brain, which in turn controls movements by the left side). This happens, I believe, because the information I am searching for is not shared freely across the corpus callosum between the two sides of the brain but is apprehended by the right brain only when it hears the name of what is being searched for. Then suddenly the left visual field and left tactile side—under control of the right brain—are open to inspection.

Does this curious fact have anything to do with deceit and self-deception? I believe it does, because when I want to hide something from myself—for example, keys just lifted unconsciously from another person—they are promptly stored in my left pocket, where they will be slow to be discovered even when I am consciously searching for them. Likewise, I have noticed that "inadvertent" touching of women (that is, unconscious prior to the action) occurs exclusively with my left hand and comes as a surprise to my dominant left brain, which controls the right side of my body. In effect, the left brain, the linguistic side, is associated with consciousness; the right side (left hand) is less conscious.

This is supported by evidence that processes of denial—and subsequent rationalization—appear to reside preferentially in the left brain and are inhibited by the right brain. People with paralysis on the right side of the body (due to a stroke in the left brain) never or very rarely deny their condition. But a certain small percentage of those with left-side paralysis deny their stroke (anosognosia) and when confronted with strong counterevidence (film of their inability to move their left arm), they indulge in a remarkable series of rationalizations denying the cause of their paralysis (due to arthritis, not feeling very mobile today, overexercise). This is especially common and strong in individuals with large lesions to the right central side of the brain, and it is consistent with other evidence that the right brain is more emotionally honest and the left actively engaged in self-promotion. Normally people show a shorter response time to threatening words, but those with anosognosia show a longer time, demonstrating that they implicitly repress information regarding their own condition.

IMPOSED SELF-DECEPTION

So far we have spoken of self-deception evolving in the service of the actor, hiding deception and promoting an illusory self. Now consider effects of others on us. We are highly sensitive to others, and to their opinions, desires, and actions. More to the point, they can manipulate and dominate us. This can result in self-deception being imposed on us by others (with varying degrees of force). Extreme examples are instructive. A captive may come to identify with his or her captor, an abused wife may take on the worldview of her abuser, and molested children may blame themselves for the transgressions against them. These are cases of imposed self-deception, and if they are acting functionally from the standpoint of the victimized (by no means certain), they probably do so by reducing conflict with the dominant individual. At least this is often the theory of the participants themselves. An abused wife may be deeply frightened and may rationalize acquiescence as the path least likely to provoke additional severe assaults—this is most effective if actually believed.

The situations need not be nearly as extreme. Consider birds. In many small species, the male begins dominant—he has the territory into which the female settles. And he can displace her from preferred feeding sites. But as time goes on, his dominance drops, and when she reaches the stage of egg-laying, there is a reversal: she now displaces him from preferred sites. The presumption is that risk of extra-pair paternity and the growing importance of female parental investment shifts the dominance toward her. The very same thing may often be true in human relationships.

This finding caught my attention many years ago because it appeared to capture exactly so many of my own relationships with women, one after the other—I was initially dominant but thoroughly subordinate at the end. It was only later that I noticed that the ruling system of self-deception had changed accordingly—from mine to hers. Initially, discussions were all biased in my favor, but I hardly noticed—wasn't that the way it should be? Then came a short time when we may have spoken as equals, followed by rapid descent into *her* system of self-deception—I would apologize to her for what were, in fact, *her* failings.

Sex, for example, is an attributional nightmare—who is causing what effect on whom?—so sexual dysfunction on either or both sides can easily be seen as caused by the other person. Whether manipulated by guilt or fear of losing the relationship, you may now be practicing self-deception on behalf of someone else, not yourself—a most unenviable position.

IMPLICIT VERSUS EXPLICIT SELF-ESTEEM

Let us consider another example of imposed self-deception, one with deeper social implications. It is possible to measure something called a person's explicit preference as well as an implicit one. The explicit simply asks people to state their preferences directly—for example, for so-called black people over white (to use the degraded language of the United States), where the actor is one or the other. The implicit measure is more subtle. It asks people to push a right-hand button for "white" names (Chip, Brad, Walter) or "good" words ("joy," "peace," "wonderful," "happy") and left for "black" names (Tyrone, Malik, Jamal) or "bad" words ("agony," "nasty," "war," "death")—and then reverses everything, white or bad, black or good. We now look at latencies—how long does it take an individual to respond when he or she must punch white or good versus white or bad—and assume that shorter latencies (quicker responses) means the terms are, by implication, more strongly associated in the brain, hence the term "implicit association test" (IAT). Invented only in 1998, it has now generated an enormous literature, including (unusual for the social sciences) actual improvements in methodology. Several websites harvest enormous volumes of IAT data over the Internet (for example, at Harvard, Yale, and the University of Washington), and these studies have produced some striking findings.

For example, black and white people are similar in their explicit tendency to value self over other, blacks indeed somewhat more strongly so. But when it comes to the implicit measures, whites respond even more strongly in their own favor than they do explicitly, while blacks—on average—prefer white over black, not by a huge margin but, nevertheless, they prefer other to self. This is most unexpected from an evolutionary perspective, where self is the beginning (if not end) of self-interest. To find an organism valuing (unrelated) other people more than self on an implicit measure using generic good terms, such as "pleasure" and "friend," versus bad, such as "terrible" and "awful," is to find an organism not obviously oriented toward its own self-interest.

This has the earmarks of an imposed self-deception—valuing yourself less than you do others—and it probably comes with some negative consequences. For example, priming black students for their ethnicity strongly impairs their performance on mental tests. This was indeed one of the first demonstrations of what are now hundreds of "priming" effects. Black and white undergraduates at Stanford arrived in a lab to take a relatively difficult aptitude test. In one situation, the students were simply given the exams; in the other, each was asked to give a few personal facts, one of which was their own ethnicity. Black and white students scored equally well with no prime. With a prime, white scores were slightly (but not significantly) better, while black scores plummeted by nearly half. You can even manipulate one person's performance in opposite directions by giving opposing primes. Asian women perform better on math tests when primed with "Asian" and worse when primed with "woman." No one knows how long the effect of such primes endures, nor does anyone know how often a prime appears: how often is an African American reminded that he or she is such? Once a month? Once a day? Every half-hour?

The strong suggestion, then, is that it is possible for a historically degraded and/or despised minority group, now socially subordinate, to have an implicit self-image that is negative, to prefer other to self—indeed, oppressor to self—and to underperform as soon as they are made conscious of the subordinate identity. This suggests the power of imposed or induced self-deception—some or, indeed, many subordinate individuals adopting the dominant stereotype regarding themselves. Not all, of course, and the latter presumably are more likely to oppose their subjugation since they are conscious of it. In any case, revolutionary moments often seem to occur in history when large numbers of individuals have a change in consciousness, regarding themselves and their status. Whether there is an accompanying change in IAT is unknown.

FALSE CONFESSIONS, TORTURE, AND FLATTERY

A few more forms of induced self-deception are worth mentioning. It is surprisingly easy to convince people to make false confessions to major crimes even though this may—and often does—result in incarceration for long periods of time. All that is required is a susceptible victim and good old-fashioned police work applied 24/7: isolation of the victim from others, sleep deprivation, coercive interrogation in which denial and refutation are not permitted, false facts provided, and hypothetical stories told—"we have your blood on the murder weapon; perhaps you woke in a state of semiconsciousness and killed your parents without intending to or being aware of it"—with the implication that a confession will end the interrogation when, in fact, it will only begin the suspect's misery. People differ in how susceptible they are to these pressures and in how much self-deception is eventually induced. Some go on to create false memories to back up their false confessions—with no obvious benefit to themselves.

There is also a kind of imposed self-deception that could be considered *defensive* self-deception. Consider an individual being tortured. The pain can be so great that something called disassociation occurs—the pain is separated from other mental systems, presumably to reduce its intensity. It is as if the psyche or nervous system protects itself from severe pain by objectifying it, distancing it, and splitting it off from the rest of the system. One can think of this as being imposed by the torturer but also as a defensive reaction permitting immediate survival under most unfavorable circumstances. We know from many, many personal accounts that this is but a temporary solution and that the torture itself and utter helplessness against it endure long afterward as psychological and biological costs. Of course, there are much more modest forms of disassociation from pain than those of torture—such as a mother distracting her child by tickling him or her.

A relatively gentle form of imposed self-deception is flattery, in which the subordinate gains in status by massaging the ego or self-image of the dominant. In royal courts, the sycophant has ample time to study the king, while the latter pays little attention to the former. The king is also presumed to have limited insight into self on general grounds; being dominant, he has less time and motivation to study his own self-deception.

Imposed self-deceptions are sometimes involved in "cons," deliberate attempts to extract resources through deception (Chapter 8). For example, in one situation, the con artist's success depended on him inducing in his victim the conviction that they knew each other already. This was accomplished by wrapping his arms around the shoulders of his (male) victim, and saying, "What have you been up to, old bean?" The victim, if deferential, may quickly create a memory of when they might have met, supplying facts that the con artist can use later as evidence that they did indeed know each other.

One form of induced self-deception is widespread and *very* important. The ability of leaders to induce self-deception in their subjects has had large historical effects. As we shall see in Chapter 10, false historical narratives widely shared within a population can easily be exploited to arouse sentiments in favor of war. At the same time, political success often may turn on the ability of leaders to arouse the belief in people that something is in their self-interest when it is not.

FALSE MEMORIES OF CHILD ABUSE

In the late 1970s and the 1980s, the emerging evidence of the sexual abuse of children and women set off two epidemics of false accusations, with immense costs to innocent people who were either imprisoned or tried for nonexistent crimes, or publicly accused and shamed. All of these consequences were based on the implantation of false memories, a case of imposed self-deception with large social costs.

The two epidemics were linked. One claimed a high incidence of past childhood sexual abuse in women—discovered only through "recovered memory therapy," a variety of techniques specifically designed to elicit such memories (or create them). Women went to see a therapist for other reasons, with no past memory of abuse, and emerged convinced that they had been subjected to repeated, sustained abuse. Suggestions from the therapist, leading questions, hypnosis in an effort to retrieve the memories—these were some of the tools that managed to instill what turned out to be false memories.

The second epidemic was a natural outgrowth of the first. If so much unsuspected sexual abuse had been going on in the past, then surely it must be continuing in the present. In 1983 in California, teachers at a preschool were accused of the usual sexual abuse of children, but also of subjecting them to Satanic rituals involving the slaughter of pet rabbits, and even subjecting them to an airplane ride where similar activities took place. This was a common feature of both epidemics—you can impose false memories on other people but you cannot keep the newly freed memory from making up whatever it wishes. The increasingly unlikely "memories" eventually led to the collapse of these movements. But not before dozens of communities had gone through the wrenching trauma of learning that their children had been sexually abused, attacked by robots and lobsters, and forced to eat live frogs.

Some people were imprisoned for imaginary abuses, while some innocent parents had to endure the public shame of others believing they had practiced pedophilia on their own children. Alas, there was no lack of clinical psychologists willing to play the fool and testify in court that in their expert opinion, the women and children were telling the truth.

IS SELF-DECEPTION THE PSYCHE'S IMMUNE SYSTEM?

The major alternative view of self-deception that comes out of psychology is that self-deception is defensive, whether against our primitive unconscious urges (the Freudian system) or against attacks on our happiness (social psychology). In the latter view, happiness is treated as an outcome in its own right, a part of our mental health. Thus, it is an outcome worth protecting, and for this purpose we have a "psychological immune system" to protect our mental health just as the actual immune system protects our physical health. Healthy people are happy and optimistic, feel a greater sense of control over their lives, and so on. Since self-deception can sometimes create these effects, it is directly selected to do so. We cook the facts, we bias the logic, we overlook the alternatives—in short, we lie to ourselves. Meanwhile, we apparently have a "reasonability center" that determines just how far we will be permitted to protect our happiness via self-deception (without, for example, looking ridiculous to others or becoming dangerously delusional). Why was evolution unable to produce a more sensible way of regulating such an important emotion as happiness?

Regarding the evidence, of course successful organisms are expected to feel happier, more optimistic, and more in control. They are also more likely to show self-enhancement. Does this mean that the self-enhancement is causing the happiness, optimism, and sense of control? Hardly. Depressed people show much less self-enhancement on common traits than do happier souls—they may even show self-deprecation. This is sometimes used to argue that without self-deception, we would all be depressed. This almost certainly inverts cause and effect. A time of depression is not a good time for self-inflation, especially if this inflation is oriented toward others—depression seems instead better suited to opportunities for self-examination.

Before turning to the imaginary psychological immune system, it is well to remember that the real immune system deals with a major problem common to all of life: that of parasites, organisms that eat us from the inside (see Chapter 6). The immune system uses a variety of direct reality-based molecular mechanisms to attack, disable, engulf, and kill a veritable zoo of invading organisms—thousands of species of viruses, bacteria, fungi, protozoa, and worms—themselves using techniques honed over hundreds of millions of years of intense natural selection. The immune system also stores away an accurate and large library of previous attacks, with the appropriate counterresponse programmed in advance.

By contrast, the psychological immune system works not by fixing what makes us unhappy but by putting it in context, rationalizing it, minimizing it, and lying about it. If the physical immune system worked this way, it would do so by telling you, "Okay, you have a bad cold, but at least you don't have the flu the fellow down the street has." Thus, the real psychological immune system must be the one that causes us to go out and fix the problem. Guilt motivates us toward reparative altruism, unhappiness toward efforts to improve our lives to diminish the unhappiness, laughter to appreciate the logical absurdities in life, and so on. Self-deception traps us in the system, offering at best temporary gains while failing to address real problems.

It is true that as a highly social species, we are very sensitive to the actions and opinions of others and can be deeply affected by them—lowering our self-opinion and our happiness—but, again, why adopt something as dubious as self-deception to solve this problem? Note that a defensive view of self-deception is congenial to an inflated moral self-image—I am not lying to myself the better to deceive you, but rather I lie to myself to defend against your attacks on myself and my happiness.

There is some slack in the system. You are also part of your own social world. The eye that beholds you could be your eye studying your own behavior. What does it see? First, your conscious act, then your unconscious self? Let us initially assume so. Can fooling this inner eye help in fooling some other part of yourself, sometimes to your benefit? I believe so. We can also try to suppress painful memories about events we cannot affect. A man's daughter is murdered by an unknown killer: "When she died, I wrapped her memory in blankets and tried to forget it." Presumably the recurring painful memory serves no purpose and there is no loss in forgetting. There are also various efforts to mold our consciousness that are not, by definition, self-deceptive. They can involve us in various self-improvement projects, including meditation, prayer, optimism, a sense of purpose, meaning, and control, so-called positive illusions. As we shall see in Chapter 6, one important benefit of such projects is improved immune function. Here I wish to discuss two related examples in some depth: the placebo effect and hypnosis. Both demonstrate that belief can cure.

THE PLACEBO EFFECT

The placebo effect and the benefits of hypnosis, including self-hypnosis, are examples of self-beneficial self-deception that usually requires a third party—a person in a lab coat with a stethoscope in the first case and someone swinging a watch and talking to you in a rhythmic way in the second. The "placebo" refers to the fact that a chemically inert or innocuous substance administered as if it were a medicine often produces beneficial—even medicinal—effects. This effect is so consistent and strong that all medical research trials on a new medicine routinely have a placebo control. That is, if you are testing whether a pill helps people with arthritis, you must give an equal number of people a similar-looking pill lacking the key chemical. Only if your medicine works better than the placebo can it be said to have any effect of its own. Of course it would be nice to add a third category to the analysis—no placebo, no medicine—to measure more precisely the placebo effect itself, but doctors have been slow to realize the value of doing this.

What such work does reveal is that a sizable minority of people do *not* show a placebo effect, while others enjoy strong self-induced effects. This is consistent with what we know about hypnosis, as well as the ability to destroy memory of nonsense material. Presumably this variation is positively associated with the ability to be manipulated by others (indeed, all three examples above involve third-party effects). This suggests that an ability to self-deceive for positive effect is vulnerable to parasitism by others, allowing them to manipulate your suggestibility to their own benefit.

The following effects are very pronounced and demonstrate a clear connection between cost and perceived benefit. The placebo effect is stronger

- the larger the pill,
- the more expensive it is,
- when given in capsule form instead of a pill,
- the more invasive the procedure (injection better than pill, sham surgery is good),
- the more the patient is active (rubbing in the medicine),
- the more it has side effects, and
- the more the "doctor" looks like one (white lab coat with stethoscope).

The color of pills affects their effectiveness in different situations: white for pain (through association with aspirin?); red, orange, and yellow for stimulation; and blue and green for tranquilizers. Indeed, blue placebos can increase sleep via the blueness alone with probable immediate immune benefits (Chapter 6).

The general rules of the placebo effect are consistent with cognitive dissonance theory (Chapter 7)—the more a person commits to a position, the more he or she needs to rationalize the commitment, and greater rationalization apparently produces greater positive effects. Surgery offers repeated examples of the placebo effect. One of the great classics is the case of angina (heart pain) treated surgically in the United States in the 1960s by a minor chest operation in which two arteries near the heart were fused to (allegedly) increase blood flow to the heart, thereby reducing pain. It did the trick—pain was reduced, patients were happy, and so were the surgeons. Then some scientists did a nice study. They subjected a series of people to the same operation, opening the chest and cutting near the arteries, but they did not join any together. Everyone was sewn up the same way and nobody knew who had received which "operation" when later effects were evaluated. The beneficial effects were identical to those of the original operation. In other words, the entire effect seems to be that of a placebo. The joining of the two arteries had nothing to do with any beneficial effect.

Surgery appears to be unusually prone to placebo effects—presumably because of the great cost and the apparent massing of group support. In any case, some interventions are dubious in advance and with potential for future complications—to be corrected by further surgery—for example, think of Michael Jackson's face. So there are built-in incentives for an entire subdiscipline to develop in unhealthy ways. Remunerectomies, for example, are performed solely to remove a patient's wallet. Consider arthroscopic surgery, meant to correct defects in the knee, often due to osteoarthritis. A small study suggested that sham operations—with all the features of real ones—produced virtually the same benefits as the actual operations, suggesting that these were mainly beneficial as placebos. The actual operations were associated with greater maximum pain than the placebos, presumably because they were more invasive, but for overall level of pain and other measures, the placebo and surgery produced remarkably similar effects.

For effects on pain, the placebo has been studied in some detail, and there is no question that in some individuals, the mere belief that a pain reliever has been received is sufficient to induce the production of endorphins that, in turn, reduce the sensation of pain. That is, what the brain expects to happen in the near future affects its physiological state. It anticipates, and you can gain the benefit of that anticipation. The tendency of Alzheimer's patients not to experience placebo effects may be related to their inability to anticipate the future.

Expectancy can create strong placebo effects through a mixture of past experiences of genuine medical effects and placebos. As one author has put it:

> The medical treatment that people receive can be likened to conditioning trials. The doctor's white coat, the voice of a caring person, the smell of a hospital or a practice, the prick of a syringe or the swallowing of a pill have all acquired a specific meaning through previous experience, leading to an expectation of pain relief.

Depression seems especially sensitive to the placebo effect. Numerous studies have shown that genuine antidepressants account for about 25 percent of the improvement, while the placebo effect accounts for the remaining 75 percent. Believing you are getting something to help you is more than half the battle. After all, depression is marked by hopelessness, and placebos offer nothing if not hope. I always think about this when I am being given an antidepressant. I am told not to wait for an effect for at least three or four weeks—"it needs to build up." In other words, expect no direct test of utility anytime soon, and the usual rule of regression to the mean—or, things get better after they have gotten worse—will give you all the evidence you later need. In the meantime, get with the program! The most recent meta-analysis (2010) reveals a striking (and very welcome) fact. Placebos work as well as antidepressants for mild depression, but for severe depression, there is a sharp bifurcation: real medicine shows strong benefits and placebos almost none. This, as we have noted, is a characteristic feature of self-deception directed toward others: a modest amount works, but a great deal fails to impress.

The ability to produce autostimulatory effects is nicely illustrated by work on female sexuality. Women who appear to be sexually dysfunctional in failing to respond orgasmically can be induced to greater arousal by giving them false feedback on the blood flow to their pelvis (a correlate of arousal) to sexual stimuli. They appear to be talking themselves into greater arousal, somewhat like the sight of a man's own erection may increase his sexual desire.

There is no doubt that placebo effects operate in athletics as well. Trials have shown that cyclists respond positively to word that they have been given caffeine (without getting any) about half as well as to the caffeine itself (along with word they are getting it). Merely telling the cyclists they are getting a heavier dose of caffeine produces a stronger positive athletic response. Even that cliché of working out—no pain, no gain—has a built-in placebo effect.

One can even induce a placebo effect out of a placebo effect. That is, you can tell someone with irritable bowel syndrome that he or she will now receive a placebo—an inert chemical with no medicine in it—but then tell the person that the placebo effect is powerful, often involuntary, helped by a positive attitude, and finally, that taking the pills faithfully is critical. With this much helpful verbiage, it is not surprising that a placebo identified as such still produces benefits.

The analogy with religion is strong and tempting. Both involve strong belief. Both involve a series of conditioned associations, including common doctor or pastoral elements. And, indeed, until very recently (up to about five thousand years), medicine and religion were one and the same. You can easily imagine that regular religious attendance (especially if the music is good!) would intensify placebo and other immune benefits, just as regular visits to a caring and sensible doctor or adviser might.

A striking feature of placebo effects is that they are highly variable across a population. Typically roughly one-third show very strong effects, perhaps one-third moderate, and one-third none. This is an example of what we have emphasized repeatedly, that the deceit and self-deception system must be an evolving one, with important genetic variation for forms and degree of self-deception. We do not know how much of the variation just mentioned is genetic, but recent work shows that people with depressive disorders differ in the degree to which they show a placebo effect based on particular genes.

What else correlates with a tendency to show a placebo effect? For one thing, suggestibility, as in ease of being hypnotized, is a trait that also shows high variability, some people being highly resistant and others easily manipulated. It should hardly surprise us that ease of being hypnotized and the placebo reaction should co-vary strongly and positively. Each is a kind of self-deception requiring a third party, a hypnotist or "doctor." When people are divided into those who are easily hypnotized versus those who are not, then hypnotizing the susceptible to concentrate only on the color in which words are printed in the Stroop test (recognizing words denoting color that are written in different colors), causes them to show no interference from the words themselves. But people who are not susceptible show no improvement on the Stroop test. This, then, is a benefit from ease of being hypnotized: greater ability to concentrate or tolerate cognitive load.

We began this chapter with the illusion of conscious control. We then moved successively into deeper and subtler forms of external control—imposed self-deception in general, torture with its disassociations, false accusations of others and of self, the placebo effect, and hypnosis. It would now be valuable to tie these kinds of conflicts into our two major social relationships: the family (Chapter 4) and the two sexes (Chapter 5). When do we impose self-deceptions on family members and on sexual partners, and when and how are these imposed on us?

CHAPTER 4

Self-Deception in the Family—and the Split Self

We usually begin our lives—the first twenty years, at least—embedded in a family, typically one or both parents and one or more siblings. This is often part of a larger extended family including grandparents, uncles, cousins, and so on. The key to the biology of all this is genetic relatedness (r). That is, family members are all related to one another in the sense that there is a chance that any given gene in any one individual has an identical copy in another by direct descent from a common ancestor. A typical gene in a parent is found in its offspring half the time (hence r to offspring = ½), while a typical gene in the offspring is also found in either parent half of the time. Siblings are related by ½ but half-siblings by only ¼ and so on. This leads to "Hamilton's rule," which states that the benefit of an altruistic act toward a relative times the relevant degree of relatedness must be greater than the cost suffered by the altruist in order for selection to favor the altruism. For example, if you are helping your half-sister, then (other things being equal) the benefit to her had better be greater than four times the cost to you. Likewise, selection will oppose a selfish act that harms her four times more than it benefits you. In sum, degrees of relatedness in families are high—which tends to induce investment and restrict conflict—but degrees of relatedness are far from unitary (r = 1) so that conflict is also expected between the actors. For our purposes, the key is that relatedness adds an extra dimension and logic to the kinds of deception and self-deception that will evolve.

Parents can pretend to base their actions on shared relatedness to the child (parental investment) when, in fact, it is based on the unrelated part (parental exploitation). They may be unconscious of this bias. In turn, offspring may pretend greater need in order to induce more parental investment than is optimal for the parent, and they may be more effective when they believe it themselves. And so on. Relatedness in fact leads to a series of ramifying complexities where deceit and self-deception are concerned. These have to do with misrepresentation, manipulation, and internal bifurcation. Let us look at each in turn.

Since an individual is selected to act both altruistically and selfishly toward family members, there are chances for misrepresentations regarding motivation and orientation that are deeper than those occurring toward more distantly related people. For example, there is no presumption that a person with a low r to you is programmed to act in your self-interest, but that is precisely the presumption with related individuals—and the more so the more closely they are related to you. So your relatives can pretend an interest in you that is plausible on its face, even if their real motivation is completely manipulative. A relative can also lay a claim to you. Aren't I related to you by one-fourth, so if you are messing up in life, aren't you messing with my quarter interest in it? Get yourself together for *both* our sakes.

Or consider the following. Although one is selected to invest parental care in one's offspring, one is not selected to give as much as requested, or always to give anything at all. Hence, deeper—and, probably often, more painful—misrepresentations are possible between close relatives. Are you investing in a child or exploiting it? Do you love the child or not? Do you have in mind a separate self-interest in the child that you are willing to support or is the child entirely conceived as instrumental to your larger projects? It makes a whale of a difference to the offspring which of these is true, and there is plenty of scope for deceit and self-deception on the part of the parent, as well as of the offspring.

Second, with the added factor of a long period of parental investment soaked with language, there are many opportunities for conscious and unconscious manipulation, including induced self-deception, in which the parent can induce a pattern of self-deception in the offspring that serves the parent's interests but not those of the offspring expressing the self-deception. The child may grow to believe that its parents are acting in its true interests when in fact they are not. The offspring may not be in a position to free itself of such an imposed self-deception until it no longer requires parental investment, giving an added reason for emotional turbulence in late adolescence, along with open hostility toward parents. Adults, in turn, may differ in the degree to which they suffer costly effects from earlier parental manipulation. In addition, parents are not a unit; they are a father and a mother, with different interests in offspring manipulation because the manipulation affects them and their differing sets of relatives.

Third and unexpectedly, relatedness considerations automatically split the organism into multiple selves, with differing interests, the most important for our selves being our maternal self and our paternal one. Formerly, we used to believe that an organism had a single self-interest. It had a unitary aim—to maximize its genetic reproduction. Kinship theory says this can't be true. Different genes within us have differing rules of inheritance, and this will give them contradictory interests. For example, the Y chromosome is always passed father to son. It is not selected to have any interest in daughters. Does that mean we expect fathers to be at least slightly biased toward their sons? Not at all. The male's X chromosome is passed only to his daughters and it is more than ten times as gene-rich as his Y, so if anything, men should show a slight genetic bias toward their daughters. No one knows whether this is true, but there is some evidence that paternal grandmothers favor their granddaughters over their grandsons according to the differing chances that their X chromosome will be found in them (½ versus 0).

The Y and the X are only small parts of the whole genome. The main genetic split within us is between our maternal and paternal halves, which are equally strong. There are a few hundred genes in us that are active only if inherited from our mother, so-called maternally active genes, and about an equal number from the father, so-called paternally active ones. Maternally active genes are selected to promote maternal interests and paternally active, paternal. This generates internal genetic conflict in which two separate genetic selves compete for control of our behavior and larger phenotype. This conflict has two important effects. We expect deception between these two halves—not directed toward outsiders but toward each other. For example, maternal genes in you may overemphasize the benefits to the organism as a whole of acting on its special relatedness to others (when these are maternally biased), while paternal genes may be selected to discount such maternal effects. Second, we also expect differences between our two halves over whom to deceive in the outside world (with or without self-deception). As we shall see below, this split in us runs deep, both from early-acting genes affecting growth and consumption of parental resources to later-acting ones affecting adult behavior.

PARENT/OFFSPRING CONFLICT

Because parents typically are related to each offspring by ½ but not by 1—and vice versa—there is ample scope for conflict between the two parties. This conflict usually concerns how much parental investment the offspring receives and what its behavioral tendencies are, as these affect its relatives. The parent is selected to maximize the number of surviving offspring it produces, but the child is twice as related to itself as to its full siblings, so it is selected to try to gain more than its fair share of resources—though not so much more that it inflicts twice the cost on its siblings as the gain it enjoys itself. Deception is an important part of the child's repertoire, pretending greater need than is actually being experienced and manipulating the parent psychologically, sometimes against the parent's better instincts. The parent may be selected to minimize the appearance of available resources, the better to save some for other offspring. One critical choice the parent has is whether to impose its will, insofar as it can, or opt for a fair split with the offspring. The latter, in principle, should reduce future conflict with the offspring, especially if in response it adopts a similar posture. One danger of complete domination is the turmoil that may erupt when the offspring is as physically large as the parent and is cognizant of the parental style to which it has long been exposed.

Regarding the offspring's general behavior, it is selected to act altruistically toward a relative only when the benefit times degree of relatedness is greater than the cost to itself (B>2C for full siblings), but the parent would prefer to see altruism whenever there is a net benefit to the parent's offspring—in this example, B>C. Thus, parents are selected to mold their offspring into being better people (more altruistic, less selfish) than they are inclined to act on their own. This may take the form of punishing behavior as being generally immoral (instead of merely counter to the parent's self-interest).

CASES OF EXTREME ABUSE

The long period of parental investment in humans means that there are many opportunities for each party to respond to the other's actions. One important consequence is that a child who is receiving insufficient investment or actual abuse may be put in an awkward position where resistance is concerned. In the extreme case, resistance is likely to only make matters worse; it will provoke additional abuse and withdrawal of investment. Thus, until children reach the teen years, they may, in general, have to submit—and the more so the harsher the regime under which they live. There also are more things they need to hide from the outside world, so lack of resistance includes lack of disclosure to others, and here the evidence is clear. For abuse in general (physical, emotional, and sexual), the more the abuse is perpetrated by a close relative (or stepparent), compared to that from a more distant figure, the longer the children take to disclose the abuse, if at all. We are talking about delays of a year or more. Intervention is less likely and caregivers less supportive. And there are negative immune effects that endure into adulthood (see Chapter 6).

Here the child may be favored by natural selection to keep up a good front, which may involve self-deception, such as disassociation and selective recall. In disassociation, the mind is split into two (or more) relatively separate parts, one of which fails to recall the abuse or see it as such—perhaps the self that is usually shown to the parent. Disassociation is more common than selective recall in those who have been abused, and this disassociation compromises intellectual performance, for example, on the Stroop test (recognizing words denoting color that are written in different colors).

The notion that children completely repress memory of extreme trauma, only to recall it years later in full detail, has been shown to be unlikely in most cases, but this does not mean that amnesic factors are not at work in trauma, of which disassociation is only one example. Again, it is the closeness of the abuser that is associated with the greatest memory defects. For all forms of impairment, abuse by a caregiver induces more memory impairment than similar abuse by non-caregivers. Is this because it is inherently more offensive and in need of memory eradication or because pressure from caregivers to keep one's silence is especially strong? It may be both. We know that the tendency to share with others is less frequent when the abuse comes from a caregiver.

GENOMIC IMPRINTING

As mentioned already, one of the most striking discoveries in the past thirty years of genetics is that we are expected not to be unitary creatures with a single self-interest, but to have a paternal genetic interest and a maternal one, which may differ, with each acting to promote a view of the world from its standpoint. Biologists used to think that genes had no memory of where they came from, thus they computed the average degrees of relatedness cited earlier—half chance through Mom, half chance through Dad. In the 1980s, biologists began to discover a minority of genes whose expression level depended on which parent contributed it. Often one copy was active and one inactive. So there are paternally active genes and maternally active ones. With activity limited by parental origin, these genes can act not on average relatedness, but on exact relatedness to each parent (0 or 1) and their relatives.

The first two imprinted genes described in mice tell the whole story. *Igf2* (insulin-like growth factor 2) is a paternally active gene that activates growth in fetal life by increasing rates of cell division. A single active copy increases size at birth by 40 percent compared to no active copies. Why does this make sense? In competition over access to maternal investment, paternal genes in offspring are inevitably less related to siblings than are maternal genes. Multiple mating by a female for each litter or changing fathers between litters lowers paternal relatedness among the resulting siblings while leaving maternal relatedness unchanged. Thus, paternal genes will weigh effects on self relatively more heavily than effects on siblings (compared to unimprinted genes or maternally active ones), preferring faster fetal growth rates and relatively larger size at birth.

The proof is in the pudding. An oppositely imprinted gene has exactly opposite effects. *Igf2r* (insulin-like growth factor 2 receptor) is maternally active and, in mammals, its protein has evolved a secondary binding site to Igf2, which is carried to lysosomes and degraded. Indeed, Igf2r gets rid of 70 percent of all of the Igf2 that is produced. As a result, it lowers the fetal growth rate by about 30 percent. This is no way to build a railroad. Here are two large, costly opposing effects that virtually wipe each other out. This is not good for the individual but is exactly what you would expect if there were two opposing forces within the offspring. Evidence confirms that imprinted genes that affect early development almost always obey "Haig's rule"—paternally active genes have positive effects on growth during maternal investment, while maternally active ones have negative effects.

One final line of evidence is worth mentioning. Although mice that are artificially manipulated to have a doubly paternal or doubly maternal genome fail to develop, individuals will develop successfully if only a fraction of their cells are doubly paternal (their nuclei from two sperm cells) or doubly maternal (nuclei from two eggs) and the rest are normal cells. Such chimeras reveal a striking fact. The more doubly maternal cells, the smaller the newborn; the more doubly paternal cells, the larger the newborn, exactly as expected. But there is a surprise: the relative size of organs inside the mouse is also changed. For example, the greater the number of doubly paternal cells, the smaller the neocortex and, hence, the brain. The hypothalamus is affected in the opposite way: doubly paternal cells do well in the hypothalamus, while doubly maternal ones disappear. Let us see why.

INTERNAL CONFLICT FROM OPPOSITELY IMPRINTED GENES

Just as conflict between individuals sets the context for deception between them (including self-deception), so conflict within the individual sets the stage for deception between its competing parts—something we might call "selves-deception," which may involve different parts of the brain. The neocortex is largely the social brain, differentially involved in interactions with close relatives and other social relationships; the hypothalamus is involved in hunger and growth, much more egocentric motives. One can well imagine an argument between the two, with the (maternal) neocortex saying, "Family is important; I believe in family; I will invest in family," while the (paternal) hypothalamus replies, "I'm hungry." That is, each argues for its favored position as if arguing for the good of the entire organism ("I").

And there can be no doubt that the requisite genetic variability is available. It is a striking discovery regarding imprinted genes (in mice, at least) that more than half of them affect neural development and later adult behavior. Work is still in its infancy, but here is one striking example. In mice, *paternally* active genes in females are especially important in directing maternal behavior. A few paternally active genes in adult females mediate such important maternal activities as retrieving the pups, licking them, and huddling over them to transfer heat. Sound like a paradox? Not really. Absent inbreeding, the two kinds of genes in a female—maternal and paternal—have the same chance of showing up in her progeny, so no bias is expected on this basis. But females also invest in their sisters' progeny and other relatives, and they are more closely related to these on their maternal side, so such genes are more likely to compromise on personal reproduction, saving some investment for others, while paternal genes will emphasize investment in their offspring.

Or consider a young woman contemplating whether to enjoy a sexual adventure with her cousin (let us say, the son of her father's sister). Her paternal genes will at once see an increase in relatedness to any resulting progeny, from one-half to five-eighths (the upside to inbreeding), while maternal genes will see no increase in relatedness at all—but both sets of genes will suffer the resulting decrease in quality of the offspring due to increased genetic homogeneity (the downside to inbreeding). In short, paternal genes in her are more likely to seek out the sexual relationship and maternal ones to resist it. The first declares that "kissing cousins are cute"; the other speaks moralistically about the dangers of defective young via inbreeding. To the individual, this may be experienced as internal argumentation, without any necessary resolution and with each side tempted to overstate its case.

Imagine also a possible society-wide effect. Imagine a patri-local society, in which a woman moves into her husband's village at marriage, rarely if ever to return to her village of origin. This is common in rural India and many other parts of the world. All of her children will grow up in a world in which they are more related to most surrounding individuals on their paternal side and not their maternal one (mother and full siblings excepted). Thus, growing up in such societies, youngsters are expected to experience internal conflict between their two genetic selves over behavior affecting others. Altruistic behavior that will increase inclusive fitness of paternal genes will not necessarily do so for maternal ones, and so forth. Sons are destined to remain in this patri-local world while daughters will, like their mothers, migrate to other villages, so sons should be especially conflicted. The mother, in turn, will support the maternal genes in her sons, urging sons especially to be less kin-group oriented than the rest of his genes (and his father) might wish.

PARENTAL MANIPULATION AND IMPRINTING

Parents are selected to manipulate their offspring to serve parental interests, and offspring are selected to resist such manipulation. A key variable is the offspring's degree of altruistic and selfish tendencies, insofar as these affect other relatives. Parents will tend to encourage an equality ethic among their offspring, because the parents are equally related to all, but each offspring is more related to self than to siblings, so that a more personally biased ethic would seem more appropriate.

Of course, each parent is expected to represent its own interests and not those of both parents, so there will be maternal manipulation and paternal manipulation, and in turn possible conflict between the two representations in the offspring. What is more to the point is that maternally active genes in the offspring are expected to be receptive to maternal manipulation and vice versa for paternal genes. Thus, parental manipulation should coevolve with imprinted genes in progeny, each reinforcing the other. This strengthens the case for a "maternal voice" and a "paternal voice," each based on effects from the same-sex parent as reinforced by imprinted genes.

I must say this interaction first occurred to me when I was trying to poison the minds of my three daughters against their mother's people. Not against their mother, God forbid—I was not crazy—only against her relatives. As their faces lit up with, so far as I could tell, full agreement, I felt good, another case of successful parental manipulation in the guise of teaching. Then as they walked away, it hit me: I had been looking only at the paternal genes in them, vibrating in unison with my paternally biased argument. As soon as they were on their own, they would take a more balanced view of the matter, and what was worse, as soon as they were with their mother, the whole matter would be reversed.

Incidentally, as people age, their important categories of relatives change from those with important genetic asymmetries (parents, half-siblings, and cousins) to those without asymmetries (children and grandchildren)—in short, from relatives over whom genomic conflict is expected to occur to those in whom it is not. So perhaps we become less internally conflicted as we age because our relatedness structure to the outside world becomes more symmetrical.

THE EFFECT OF MARITAL CONFLICT ON GENETIC CONFLICT

The above line of thinking leads to a very important question: What is the genetic effect of marital strife on the psyche of the child who is both witness and actor in the drama? By logic, one would expect the child's paternal genome to accept or acquiesce in the paternal viewpoint, while the maternal genome would be biased to embrace the maternal position. With increasing strife, one can easily imagine that the two genetic sides in the child—maternal and paternal—are hyped by the escalating conflict toward excessive production of their products (proteins, small-interfering RNAs, or anti-sense RNAs, all capable of regulating other genes). Thus with greater marital strife, the intensity of the child's internal conflict may increase at the genetic level and the biochemical, as well as at the psychological. If so, this must be an important factor in intensifying the child's internal suffering.

A striking feature of children, noted in anecdotes, is how often they respond to the news of an impending half-sibling—let us say, Dad's child by his new wife—with intense hostility. Rather than gladdening their hearts at the arrival of a half-sibling, children seem instead to see the less related sibling as a threat to investment in themselves (and in their full siblings). Again, one would expect maternal genes to take the lead in such reactions, because they have *no* interest in Dad's new progeny. Thus, the genetic conflict induced by this new situation may be more intense than the direct psychological one.

IMPRINTING AND SELF-DECEPTION

The relevance of genomic imprinting to deceit and self-deception is several-fold, of which the most important is the internal fragmentation and conflict it generates. In important parts of our family lives, we are two separable people (not one) with partly divergent aims, theories of reality, and degrees of deceit and self-deception—two people who are also tempted to deceive each other. We call these two people our maternal and our paternal selves.

What difference is expected between the two sides in degree of consciousness? This depends, of course, on which personality we are most inclined to hide from others. Let us say the maternal side is more selfish in its orientation (fewer relatives to interact with). It will wish the more to hide itself from the outside world, as well as its other genetic half (the paternal). Thus, the conscious mind will show paternally oriented behavior and be unaware of its maternal biases, while the maternal side will have full opportunity to study (and exploit) the paternal, much as happens in dual personalities, where the unconscious one knows the conscious one but not vice versa. These are merely the first speculations. Inevitably, the subject of the two halves of our minds—their interactions and their differing effects on deceit and self-deception—will grow to become a major subarea regarding family.

Here is an interesting possibility. Can one half of you feel guiltier than the other? Yes. Can half of you feel ashamed and the other not? I believe so. If guilt concerns harm to other, then by logic, harm to a relative is worse than harm to a stranger, so your paternal side could feel guilty for a hurt to a paternal relative, while your maternal side scarcely notices. If shame deals with damage to the self, especially in public, then when the public includes relatives related, say, through Mom, you may feel strong shame through your maternal genes and much less or none at all through your paternal. Guilt and shame are feelings that are both produced by us and *induced* in us. Someone may try to make us feel guilty when there is no good reason to do so, and someone may also attempt to shame us. Their own relatedness asymmetries may affect their tendency to induce these feelings in us. The induction may split each of us in two, which can produce both internal conflict and confusion.

DECEPTION IN CHILDREN

At what ages do humans become capable of deception? We talk of the innocence of children, but dissembling and lying show up at very early ages—both in everyday observations and in scientific studies. Children show a wide array of deception by ages two and three, and the earliest clear signs appear at about six months. Fake crying and pretend laughing are among the earliest. Fake crying can be discerned because infants often stop to see whether anyone is listening before resuming. This shows that they are capable of moderating the deception according to the victim's behavior. By eight months, infants are capable of concealing forbidden activities and distracting parental attention. By age two, a child can bluff a threat of punishment, for example, by saying, "I don't care," about a proposed punishment when he or she clearly cares. In one study, two-thirds of children age two and a half practiced deception at least once in a two-hour period. Motives for children's lies seem broadly similar to those of adults. Lies to protect the feelings of others—so-called white lies—appear only by age five.

Temper tantrums, violent instances of rage with the child threatening even self-harm, are well known in humans, but also in chimpanzees and even pelicans. Pelican chicks will work themselves into a frenzy, swirling around violently and, in the process, chasing away their siblings, before falling prostrate at their parent's feet, in effect demanding immediate investment, which indeed they often receive. Instead of banging their head on the ground, as a child or a young chimp might do, the pelican attacks *its* most critical part and bites its own wing.

Offspring deception can be extremely subtle, as the following two anecdotes suggest. A woman with a close, loving relationship to her happy five-month-old daughter picks her daughter up at day care. The girl is playing happily with a staff member, but when she spots her mother there is a flash of joy, followed at once by collapse and tears. The mother's interpretation? The daughter is genuinely happy to see her but then immediately hides the happiness to express her suffering at not being cared for continuously by her mother, in other words, to induce guilt in Mom. In another anecdote, the same girl, now more than two years of age, uses "need" when she wants something ("I *need* . . . "), as if to stress how critical the matter is, but when she does *not* want something, she no longer speaks of need but says more gently that she does not "want" it, both asserting that she too has wants and now speaking more slowly, almost plaintively, "But, Mom, I don't *want* that." She is manipulating her mother toward greater investment in the first case, and in the second, trying to get her mother to sympathize with her as someone with her own wants.

In fact, deception in children starts even before birth. In the last trimester of pregnancy, there is a striking change in the control of the mother's major blood variables—pulse rate, blood-sugar level, and distribution of her blood. Normally these are under the control of maternal hormones, produced at very low levels. In the third trimester, control shifts to the offspring, who either produces the same chemicals or their very close mimics, but does so at one hundred to one thousand times higher concentrations. Why this shift in control to the offspring and to a grossly inefficient signaling system, at that?

Control has shifted to the fetus, to its own advantage. It acts to increase maternal blood-sugar levels and pulse rate above what the mother favors, because this will increase nutrient transfer to itself via the placenta. For the same reason, it also acts to deprive the mother's legs and arms of blood and to concentrate the blood near itself. If one assumes a coevolutionary struggle in which increases of fetal hormones are matched by increasing maternal insensitivity to them, one can easily see how hormone levels could grow over evolutionary time to many times greater than when mother alone controls her own blood. As an expert in this field put it, when there is no disagreement, a whisper will do; shouting suggests conflict.

As children mature, they become increasingly intelligent and increasingly deceptive. This is not an accident. The very maturing capacity that gives them greater general intelligence also gives them greater ability to suppress behavior and create novel behavior. There is also clear evidence that natural variation in intelligence, corrected for age, is positively correlated with deception. A child is left in a room and told not to look in a box. By the time the experimenter returns, most children have peeked. Now they are asked whether they peeked. Most say no, and the brighter the children are on simple cognitive tests, the more likely they are to lie. Even health of the child at birth (as measured by a weighted sum of multiple factors) is positively correlated with lying. Because we experience deception aimed toward ourselves as negative does not imply that as deceivers we experience it as negative, at least when undetected.

Although the critical evidence is lacking for adult humans, smarter ones, as we saw in monkeys and apes (Chapter 2), are expected to practice *more* deception, not less, and more skillfully. By theory, they are also expected to be *more* self-deceived than the less gifted. This creates special dangers—high intellectual ability combined with high self-deception—for example, a malevolent person who is good at being malevolent. It is easy for the intellectually gifted to argue otherwise, that their special talents will save them from the failings lesser mortals are prone to, but by evidence and logic, we expect the opposite. Until shown otherwise, we should assume that the intellectually gifted are often especially prone to deceit and self-deception, including in many of the academic disciplines they produce (see Chapters 10 and 13). Those who take pride in their alleged intellectual gifts or of their particular group might well contemplate whether they are also more regular liars and self-deceivers. They are expected to be better at it.

When children are told to tell white lies (for example, that they like a gift when they do not), they direct all their smiling toward the intended victim (the gift giver). When receiving a gift they actually like, they share their smiles more broadly. Like adults, children tend to suppress true facial expressions more often than they invent novel ones, and they are better at it—when inventing faces, people of all ages tend to exaggerate, while suppression is achieved more exactly.

It is interesting that more dominant five-year-old children of both sexes are better at fooling observers in laboratory experiments, but in the same experiments, dominance confers no advantage in detecting deception by others. Among adults, the same is true for men, but women's deceptive behavior is unaffected by dominance (as is their ability to spot it). As we saw in the first chapter, when people are given a "power prime," they see the emotional expressions of others *less* accurately, so if anything, we expect them to be more vulnerable to deception.

It is noteworthy that parents play "pretend" with their children at very early ages, that children play pretend with one another and with themselves, and that most of children's literature is fantasy. Consider how common (and popular) games are that involve deception—hide-and-seek, card tricks, magic, liar's dice, and so on. Hence, there seems to exist some drive to incorporate pretense into life at very early stages. It certainly stimulates imagination and learning and also prepares the child for living in a world where practicing and spotting deception are important. I have certainly never seen any signs of natural guilt in children when practicing deception. Quite the contrary, regarding their parents at least, children seem to regard deception as their first line of defense, as well they might. Their parents are bigger, stronger, more experienced, and in control of most of the resources at issue.

PARENTAL EFFECTS ON CHILDREN'S DECEPTION

Even though parents may encourage white lies in their children, they often seek to penalize, suppress, and (sometimes) harshly punish deceptive behavior (especially directed toward them). Parents have the power; they need only the facts. A common parental device is to stare into the child's face at close range and force the child to look into the parent's eyes. I have seen parents use this successfully with their twenty-year-old children. College students consistently tell me that they think their parents read their deception better than anyone else and sometimes with near-perfect accuracy. The threat of punishment in general tends to induce deception in children to avoid it, and this is true also for punishment in response to deception itself. Punishment (especially harsh varieties) may drive the deception deeper, perhaps inducing greater self-deception to hide rising fear and pain (with unknown downstream immune effects).

Parents may also have a huge effect when they themselves indulge in deceptive behavior that the offspring are then tempted to mimic. Children may learn that it is fine to deceive; it may even be a legitimate lifestyle. This may range from lying to friends to hide misdeeds ("Oh, I'm sorry I didn't pick you up; I had a medical emergency with one of my children") to more serious misrepresentations. If a parent is a drug addict and tells a lot of "stories" trying to cover up the addiction, the children may tell lies to cover up the parent's addiction and may then grow to lie to people in general. On the other hand, children are notoriously sensitive to parental contradictions and hypocrisy, especially when directed toward them. If you get caught by your child doing something you have prohibited the child from doing (throwing trash into a flower bush near your front porch), you may be in for a long afternoon of recriminations.

Psychologists have argued that a key initial stage in a child's development is whether the child has learned to trust the world around it. This is usually navigated successfully with considerable parental care, but not always: diminished care may mean that the child can't trust the world to provide the necessary care. In the extreme case, parents can so abuse the child's trust that it develops no trust and lies for fear of telling the truth. It is as if the child has learned to fear reality itself, certainly its own representation of reality. If a child can't trust its parents to act appropriately with the truth, then it may lie out of defense and distrust. This syndrome can be deep enough to endure in relations more generally. After all, parents are closely related to their children and are expected mostly to have their children's interests at heart, so that distrust engendered by them may easily extend more broadly, to individuals with less interest at stake in them.

Parents will often act to deceive their children regarding the degree of their commitment and care. "I am doing this for your own good," a child may hear while being beaten, or later, "I only have your best interests at heart," while the child's behavior is being further restricted. Really? People are expected to have their *own* best interests at heart, and these may conflict with their children's. More extreme opportunities for parental deception of children are nicely illustrated in some single-parent households. "Where is my father?" asks the child. "He left us," says the mother (in fact, it is the other way around). "He doesn't want to have anything to do with you, so get over it." Here the mother's initial behavior inflicts a cost on the child, as does its continuation—no relationship whatsoever with one's father, nor an image of the paternal half of oneself. Or says the mother, "He is dead" (in fact, he is in prison). Later the child learns the truth and is angry at the deception and its associated costs—again, no chance to develop a relationship with the father, through visits to prison, correspondence, and phone calls. Here is a particularly unfortunate example: One child reported that Mom said the man living in the house was her brother. They did sleep in separate rooms, yet the child is sure they have sex together. So which is it: They are not brother and sister and Mom is lying, or they are and she is committing incest? Family and sex could hardly be a more volatile psychological combination. To gain a deeper understanding of the family, we need, indeed, to include sex. Parent must be replaced by mother and father, offspring by son and daughter, and sibling by brother and sister. At the same time, the two sexes have meaning beyond families and attract deceit and self-deception specific to their roles. We turn to this topic next.

CHAPTER 5

Deceit, Self-Deception, and Sex

Few relationships have more potential for deceit and self-deception than those between the sexes. Two genetically unrelated individuals get together to engage in the only act that will generate a new human being—sex, an intense experience that is at best ecstatic and at worst deeply disappointing, or when forced, extremely painful and damaging. The act is often embedded in a larger relationship that will permit the two to stay together for years or even life—long enough to raise children. Opportunities for misrepresentation and outright deception are everywhere, and selection pressures are often strong. Likewise, each partner's knowledge of the other is usually detailed and intense and (absent denial) grows with time.

Sex itself is fraught with psychological and biological meaning at every depth. Are we misrepresenting our level of interest, sexual or romantic, our deeper orientation toward the other, positive or negative, or our very sexual orientation? To analyze deceit and self-deception between the sexes, we must first describe the underlying logic for the evolution of the sexes and relations between them, including sex. Then we can link this to sex differences in deceit and self-deception regarding extra-pair sex, uncertain paternity, the female monthly cycle, female sexual interest, fantasy, betrayal, and murder.

The key to the two sexes, as it is to sex, is the offspring they may produce—the very function of life. In the evolutionary context, there are only two variables we need to pay attention to—genes and parental investment. The offspring is only made up of the two. It receives its genes from both parents (roughly equally) and the investment (that is, labor and resources to build it) from both parents or, as is usually true of other species, from the mother alone. The genes it receives from each parent arrive at the same time—fertilization—but the parental investment may have started well before fertilization and will continue long afterward, split, as in the human case, between the two sexes in a complex, changing manner. But before we get into these complexities, why sex itself? Why bother?

WHY SEX?

Why sexual reproduction? Why not go the simple, efficient route and have females produce offspring without any male genetic contribution? Females typically do all the work; why not get all the genetic benefits? In other words, why males? There are, in fact, many all-female species, but they tend to be clustered in small animals (very small insects, mites, protozoa, and so on), with some notable exceptions, such as are found in some lizards and fish. And among those with larger body sizes, asexual species do not persist long over evolutionary time, they go extinct. Why these two facts?

The advantages of sex must come from the benefits of producing genetically variable offspring. Two human parents can—through the magic of everyday recombination—produce billions of genetically different offspring, while an asexual female is stuck with her own genome and the few mutations she can give each offspring. And why is it important to produce genetic variability? Logic and evidence strongly suggest that there are two important forces. By continually breaking up gene combinations, recombination permits genes to be evaluated in many different genetic combinations, instead of always being tethered to the same set of genes. This increases the rate at which beneficial genes can evolve. The major pressure for this, in turn, often comes from one's parasites, which are numerous and costly, and rapidly evolve new means of attacking you. Parasites favor in their hosts both the production of genetically variable offspring and offspring with high internal genetic diversity (heterozygosity). This underlying genetic imperative of sex has important implications for mate choice and other aspects of sex, as we shall see.

TWO SEXES—TWO COEVOLVING SPECIES

Sex has been the dominant form of reproduction in most species for hundreds of millions of years. Two partly competitive morphs, males and females (defined by whether they produce sperm cells or egg cells), are caught in a stable frequency-dependent equilibrium over huge stretches of time in which the relative increase in the numbers of one sex makes the opposite sex more valuable, thus increasing its numbers, so that many species have evolved to produce the sexes in roughly equal numbers.

The two sexes, in turn, are described by their relative parental investment. Females produce expensive eggs so that the number of eggs is strictly limited by their cost. Males produce sperm so inexpensive that 100 million typically do not weigh even a gram, and a man at rest can generate that number in less than an hour. When additional investment is added, it is usually added on the female side, so that in general female parental investment exceeds that of the male. This is true even in our own species, where male parental investment is often substantial.

For many millions of generations, male deception must primarily have concerned male genetic quality, since males offered nothing but their genes. It is generally believed that female choice has repeatedly favored signs of male quality that are reliable and hard to fake—size, symmetry, bright coloration, and complex song, to name but a few. Mating with such males usually produces genetically superior offspring. Sometimes high-quality males are temporarily in short supply and females may have been selected to advertise fertility to attract one quickly.

Of course, almost every trait is capable of being advertised or hidden. I once thought bodily symmetry was so often a marker of genetic quality (in plants, insects, birds, mammals, and so on) not only because it was a good measure but also because it was impossible to mimic. But the bluegill sunfish soon taught me otherwise. Males are brightly colored on both sides of their body and typically swim back and forth displaying both sides. But some asymmetrical males always swim showing only one side, the more colorful one. There are probably few females so dull they do not notice they are watching a "single-sider," but they still do not know how asymmetrical he is—merely that he has something to hide. Such males do not do as well as two-siders but might do even worse if they revealed both their sides.

In everyday life, the importance of this first occurred to me when I was chatting with a young student who had a remarkably attractive face, and it seemed that whenever she wanted to impress herself fully on me, she turned so that both sides were shown equally and then gave a dazzling smile. The effect was very strong. So the rest of us, unconsciously and sometimes consciously, must be altering the frequency with which the two sides are displayed, with a bias to the more attractive side and to hiding asymmetry.

I would have thought by now that scientific work would have shown a series of general differences in the sexes regarding deceit and self-deception. I would expect females to be better at seeing through males than vice versa, on grounds of social expertise and amount of time devoted to social interactions, and I would expect males to be more self-deceived than females—more opportunities for benefit through self-inflation and overconfidence. I believe that women often make a deeper study of deception in their relationships than do men—self-deception, of course, is always another matter. I will never forget the sense of vulnerability I felt when I first realized my wife of eighteen months had been catching me in a series of lies without telling me. She was building up a library of my behavior for future use. I almost felt betrayed. Being simple-minded, the first time you lie to me, I am apt to point it out to you (unless there is a dominance problem).

Whether any of my speculations here are true, I have no idea, because there is no real scientific research on this subject. There is no evidence of women's systematic ability to spot deception better or to propagate it more deftly. Nor is there a clear bias in self-deception when comparing the two sexes—except perhaps for overconfidence, where there surely appears to be a male bias.

DECEPTION AND SELF-DECEPTION AT COURTSHIP

To explore deception and self-deception between the sexes at first contact, it is helpful to know that in humans, female choice usually focuses primarily on a male's status, resources, and willingness to invest as well as signs of genetic quality (especially when she is ovulating). The latter may be revealed by physical attractiveness (for example, facial symmetry and facial masculinity). So we expect males to misrepresent their standing on these attributes upward. They appear to have more to give than they actually do, they are more likely to give it than in fact they will, and their genes are better than they really are (this last one perhaps is the hardest to fake).

Male choice focuses on physical evidence of fertility and fecundity—youth, waist/hip ratio (curvaceousness), breast size and symmetry, and evidence of genetic quality, as in degree of facial symmetry and femininity. Finally, males place a value on female sexual monogamy (never mind their own tendencies).

Given the large initial difference in parental investment—at its extreme, a sperm cell weighing one-trillionth of a gram and a nine-month pregnancy producing a seven-and-a-half-pound baby—it is hardly surprising that men (compared to women) place relatively greater emphasis on short-term mating relations than on long-term. This leads to a large and consistent psychological difference between the sexes regarding sex itself. Men all over the world show a greater preference for sexual variety than do women. Men desire more sexual partners over various time intervals, are more likely to consent to sex with an attractive stranger, have twice as many sexual fantasies per unit in time, and are more likely to seek out prostitutes and to lower their standards in choice of women for short-term relationships. Women more than men report being deceived about partner ambition, sincerity, kindness, and strength of feeling. Only in willingness to have sex are women seen as more deceptive than men—hardly surprising given men's interest in sex.

Likewise, there is selection for females to simulate orgasm, but rarely a pressure (or necessity) for males to do likewise. Women fake orgasms to massage the male ego and to bring an end to unwanted sex. Some men are completely fooled, many probably at least some of the time. The real orgasm is assumed to act positively regarding sperm movement, that is, sucking it inside. It also makes future sex with the same partner more likely.

One can, in turn, measure how much either sex is upset by particular deceptions of the opposite sex. As expected, women are more upset at male overrepresentation of resources and status than vice versa. But these are minor factors. Where women really get upset is in response to two related deceptions: men misrepresenting the depth of their feelings prior to first having sex and men failing to call or contact them after sex. That these behaviors may also involve self-deception, I have no doubt. In the early '60s, when I was a young man, I was conscious of something I called "false emotion." I would meet a woman, develop a strong attraction, wheel out my full show, feel I was in love, have sex with her two or three times, and then find the entire attraction collapsing—indeed, often turning into aversion. The false emotion of romantic love must have been generated the better to induce the sex that ended it, but I was conscious of this only after the fact. The women, of course, were bitter.

WHOSE BABY IS IT?

One of the most important issues for a man occurs nine months after a sexual act is said to produce a brand-new child, of which he is claimed to be the father. But is he? The difference is critical, related by ½ or related by 0. Sex all but guarantees maternity, of course, but it does not guarantee paternity. Men are expected to be especially likely to be concerned with problems of parentage, as indeed they are. How about our powers of perception? There appears to be no difference between the sexes in ability to recognize whether children are the offspring of a given parent, but both sexes more easily spot relatedness through the mother, and each is better at spotting relatedness when the baby is of their own sex. There is, however, a striking sex difference in *attribution* of relatedness to newborns—women and their relatives overwhelmingly comment on resemblance to the father (more so for sons than daughters) and, as expected, putative fathers may be both taken along and somewhat skeptical of the claimed resemblance. Experiments in which people's faces are morphed onto unrelated children so as to create an artificial resemblance show that men are affected by greater self-resemblance in claiming greater willingness to adopt, pay child support, forgive after something is broken, and so on—but women are not. Many societies have jokes on the subject. In Senegal: "Better to have an ugly baby who resembles you than a good-looking one who resembles your neighbor." In Jamaica, to give a man a "jacket" is to father a child he takes as his own. The better the jacket fits (resembles him), the happier he will be. To "cut a man a waistcoat" is to produce a perfect mimic, since waistcoats have to be individually tailored to fit properly.

How can you know for sure that the child a woman carries is genetically your own? Of course, you can't. Some men torture themselves over the possibilities, the hours when she was not around, phone calls with old friends, whatever. But I always believed the issue was overrated, because I did not believe it was possible for me to look at a child for long and not be able (without DNA tests) to know whether it was my own. There are enough dominant genetic markers in my lineage that the truth must be there in front of my eyes. If that is really true, then we are talking at worst about nine months or so of wasted investment, a trivial cost, put to a good social purpose. Let us put the matter behind us and go on with the rest of our lives. In other words, we need not spiral off into the fatal land of jealousy, but that is alas all too common, as the following suggests.

MALE RESPONSE TO FEMALE INFIDELITY

A man's response to signs of his partner's infidelity in an intimate relationship seems general the world over: anger and aggression, an attempt to suppress the behavior by threatening, beating, isolating, and sometimes murdering the woman. The result often is a thoroughly frightened and dominated woman, told that any attempt to flee will be met by murder, leading to a defensive form of imposed self-deception, where the woman often comes to believe her tormentor and blames herself. Genital cutting of women (to reduce desire), foot binding (to reduce mobility), and claustration (to isolate socially) all serve to reduce female choice in advance of temptation, though I doubt they are usually rationalized this way within the societies that practice them.

The law is stacked as well. It was a historic and cross-cultural universal for "unauthorized" sexual contact with a married woman to be a crime (for both the man and the woman), with the husband as the victim. In some parts of the United States, the very sight of adultery was, until very recently, considered sufficient justification for murder—of either party—by the husband. What all this means is that extramarital sex or the mere suspicion of it can be very dangerous to the woman (and the other man). Very powerful selection pressures—murder and imprisonment, for example—may be associated with deception regarding extra-pair relations. My own (very limited) experience is that one can hardly deny from consciousness other ongoing relations, so that extra-pair relations inevitably involve conscious deception, and self-deception must at best serve self-confidence in the face of possible accusation.

Consider homicide. In many American cities, sexual jealousy is the second or third leading cause of murders, and in many societies it is the first. In Detroit, one-third of all murders in 1972 were "crime specific" (as part of a robbery, for example), but of the remaining, fully one-fifth were due to sexual jealousy. The detailed breakdown is of some interest (total N = 58). Men were four times as likely as women to instigate jealous actions leading to murder. In roughly equal numbers, men killed their partner or the other male and were almost as often killed themselves by the woman (sometimes aided by a relative of hers). Two men murdered their unfaithful homosexual lovers—no problem of uncertain paternity there! When a partner was unfaithful, women were somewhat more successful, killing one of the adulterers in nine instances while being killed by the mate in only two.

In Canada, 55 percent of all wife-beating court cases involve at least some jealousy. Men respond to possible infidelity with anger, drunkenness, threats, and sexual arousal. The last is a most interesting subtlety. In many species of more or less monogamous animals, the sight of one's own female having sex is sexually arousing to the male. Even ducks being raped by groups of males are often re-raped by their mate immediately afterward, presumably to introduce sperm in competition with that just introduced. So it is a feature of male psychology that evidence or fantasies of mate involvement with others may be sexually arousing. I have never found the reverse to be true. Women respond to extra-pair copulations with tears, feigned indifference, and efforts to increase their attractiveness. Men get angry and drunk.

Men are, of course, prone to self-deception in evaluating their partner's extra-pair activities. The lower their self-image, the greater their expected suspicion, if not full-blown paranoia. The lower his intrinsic quality, the greater is her temptation. Given that he is of lower putative genetic quality, she may more easily dominate him, so that he may not dare voice his suspicions for fear of being dropped entirely. A second reason men may practice self-deception arises from their own guilt. Many times I have seen men accuse their innocent partners of exactly what they themselves are up to, another case of denial and projection, the accusations presumably serving mostly as camouflage.

DECEIT AND A WOMAN'S MONTHLY CYCLE

A woman's biology changes in very interesting ways during her monthly cycle, with many implications for deceit and self-deception. Women are more attractive at the time of ovulation—they appear to be physically more symmetrical and their waist/hip ratio is slightly more curvaceous. They also derogate the looks of other women more than at other times in the cycle. Are they (unconsciously) comparing other women to themselves and derogating other women because they themselves are relatively more attractive when ovulating, or are they adding a degree of derogation so as to accentuate their own superior appearance when it most matters? I would imagine the latter, but the evidence is not sufficient to say.

Women appear to be more sexual in general at the time of ovulation but with a distinct bias toward more genetically attractive men and extra-pair sex. In several clubs in Vienna where partners were studied over many months, a woman was less likely to show up with her partner near her time of ovulation while displaying more skin (wearing less clothing). At time of ovulation, women's preferences for men's faces shift toward those that are relatively more masculine and symmetrical, signs of genetic quality but not paternal investment. (Women's preference also shifts toward slightly darker men and less hairy ones.) If employed as a lap dancer and not on the pill, a woman earns 30 percent more per hour when ovulating than when not (excluding during menstruation, when she earns even less). If she is on the pill, then there are no differences in her earnings across the monthly cycle.

Changes across the cycle can reflect underlying subtle genetic tensions between the sexes. A particularly striking result shows that the more a woman matches her partner's genes at critical major histocompatability loci involved in defense against parasites—which is a disadvantage in that it lowers offspring survival—the less likely a woman is to have sex at ovulation, the more often she has (verbally) coerced sex, and the more often she fantasizes about sex with another man (including prior partners) while having sex. But twelve days later, when she is not ovulating, there is no effect of gene matching on her sexual behavior and fantasies (compared with women who do not match). Men show no effects of matching their partner on the major histocompatability loci at any time. They are out of the loop.

So we expect more pressure on women to act deceptively at the time of ovulation. In this case, the woman engages in a voluntary, conscious kind of self-deception—temporary fantasy—that she is unlikely to wish to share with her partner. She may start developing a private life of fantasy that recurs each month, perhaps tempting her to more overt actions at this time in the future. In any case, a private life is carved off from her partner, acting over a few critical days every month. It would be most interesting to know whether some men notice that when their partner is most attractive to them, she is least sexually interested in them. And how do they respond, if at all?

Smell is an important part of sex. A woman's sense of smell is more acute than a man's, and this is especially true at her time of ovulation, where her sensitivity to certain sex-related compounds may increase a hundredfold and her ability to discriminate men's bodily symmetry based on smell hits a peak. I am often astonished at how naive young men are regarding the olfactory dimension of life. I hear the same story from students: "I was due to meet my girlfriend later but this woman was hot for me, so I enjoyed some sex, nothing special, nothing to detract from my lady, but as soon as I saw her, it was like she knew right away something was up." I then ask these young men whether they had thought to bathe after having sex. No, hadn't occurred to them—perhaps this was their problem. They were living in one olfactory world, their partner in another. Of course, there may be no escape—a good student of your behavior may ask you why you just bathed at this odd time of day.

This difference in the olfactory dimension can be extended to many other aspects of mental life. Women are better at reading facial expressions, but men are better at picking out hostile images in a crowd. Sounds may be processed in different sections of the two brains, and it is a remarkable fact that in a variety of mental tasks, women's brains tend to act more symmetrically than men's—that is, the two hemispheres are used more equally in solving a given task. Since symmetry is so often an advantage in life and mental life in particular—for example, depth perception and location in vision and hearing both result from the use of bilateral information simultaneously—one's initial assumption must be that women thereby gain an advantage over men. The corpus callosum connecting the brain's two hemispheres in women is relatively larger than in men, meaning information is more easily shared and symmetrical functioning more likely.

MEN'S SELF-DECEIT ABOUT FEMALE INTEREST

Several lines of evidence suggest that men deceive themselves about women's sexual interest in them. Women report that men are more likely to believe that a woman has greater sexual interest in him than she really has, rather than less. By contrast, women show no bias in how they rate men's interest in them (high or low). Experimental evidence provides congruent evidence. By logic, men may gain more from such a perceptual bias than do women. They will catch more women with actual interest while making more false projections in the process. Assuming there is not much cost to the errors (the woman turns him down, he departs), the bias will give a net benefit. Of course, a reputation for overeagerness could add to the cost. This might result in a self-deceived bias toward greater interest while simultaneously thinking of oneself as "cool"—relatively restrained toward others.

There is evidence that women's behavior may heighten male illusion of female interest. When in experiments the two sexes are introduced for the first time for a ten-minute videotaped session together, female courtship behavior is higher in the first minute (e.g., nodding) but unassociated with any actual interest. Such behavior is associated with interest only in the later stages (four to ten minutes), so that women appear to display interest before they develop it. This will give men the illusion of interest before it develops and, indeed, female nodding behavior in the first minute predicts male talking in the later stages.

MALE DENIAL OF HOMOSEXUAL TENDENCIES

It has long been argued that denying one's homosexual impulses will cause one to project them onto others. It is as if we detect some homosexual content in our immediate world, and denying our own portion, we go looking for it in others. That this homosexual denial can lead to homosexual aggression is not surprising, because someone else's homosexual content may be a direct threat to our own hidden identity—do we respond, in spite of ourselves, to an attractive young man with a bouffant hairdo and a woman's perfume? We had better attack him before anyone notices our arousal. This is also sometimes called a reaction-formation. What is attractive to the self but unacceptable is disdained and denied for self but attacked aggressively when seen in others. A man thereby supports his image of heterosexuality by attacking homosexuals.

Recent work supports this kind of dynamic. In the United States, A-1 heterosexual men by Kinsey criteria—no homosexual behavior, no homosexual thoughts or feelings (or so they say)—were divided into those who were relatively homophobic, that is, upset and hostile toward homosexuals, and those who were relatively relaxed and unconcerned.

The fun part came when these men got to watch three six-minute erotic movies—a man and a woman making love, two women, and two men—while a plethysmograph attached to the base of each penis measured penile circumference very precisely. In addition, after the film, each man was asked how erect and how sexually aroused he had been. An interesting result emerged. Relatively homophobic and non-homophobic men responded similarly to the heterosexual and lesbian films, strong arousal to each, but more so for the heterosexual. It was only the male homosexual film that revealed a divergence. Non-homophobic men showed a small but insignificant increase in penis size, but homophobic men showed steady penis size growth throughout, reaching two-thirds the level seen in their response to the two women. Interviews afterward showed that everyone had an accurate view of the degree of his penile enlargement and arousal (which were highly correlated), except for the homophobic men viewing the male-homosexual scenario. They denied their tumescence and arousal. Whether they were actually conscious of this is unknown.

IS SELF-DECEPTION GOOD OR BAD FOR MARRIAGE?

There are two extreme forms of deception in a relationship where sex and love are concerned. The sex is great and you have to fake the love, or the love is real but you have to fake the sex. By the time we are thirty, we have all been in these situations. When we have to fake the sex, we often invoke fantasy, a prior partner, an imagined partner, an imagined sexual act. Whatever gets us off. Note that these relations are especially dangerous to the partner. If the partner is unaware of your own true reactions, he or she will be unprepared for the betrayal that so likely awaits. On the other side, it may be much harder to fake love when there is strong sexual interest. Low-love relationships are apt to be more volatile, open hostility coexisting with passionate sex.

The simple answer to the question about the effect of self-deception in a marriage is that it depends on the kind of self-deception. Self-deception of a positive, couple-reinforcing form appears to be beneficial, while self-deception associated with resolution of one's own cognitive dissonance in the conventional self-serving ways appears to have the opposite effect—over-affirmation versus distancing. The aphorism that you should go into marriage with both eyes open and, once in it, keep one eye shut captures part of the reality. When you are deciding whether to commit, weigh costs and benefits equally; when you have committed, try to be positive and not dwell on every little negative detail.

Consider first the positive form of self-deception. Couples last longer if they tend to overrate each other compared to the other's self-evaluation. This has an appealingly romantic ring—"I love you, darling, more than you love yourself, and thereby uplift you." Effects work on both sides. The more you overrate the other, the longer you stay together, and vice versa. Assuming long life together is a benefit, over-valuation is beneficial.

People have a bias toward seeing improvement in the relationship over time even if this is achieved by exaggerating how bad the past was (compared to evaluations of the present). Once the past is misremembered, the memory of progress is established and relationships with greater memories of improvement last longer. It is important to emphasize that we can't discriminate cause and effect. Self-deception may improve relationship satisfaction and duration, or it may accompany other factors that do. Perhaps success breeds self-deception (of the positive sort).

Evidence suggests that marital satisfaction declines linearly over time, but people have a biased memory—they remember early declines in satisfaction but more recent increases that offset the early decreases. In one study, both spouses reported steady increases in relationship satisfaction over two and a half years while none could be detected. By the end of the time, though, memories were readjusted so as to remember no improvement in the more distant past, only in the more recent.

In contrast, processes of self-justification within individuals make unity between the two more difficult so that, in the extreme, self-justification may be seen as an "assassin" of marriage. That is, active processes of self-justification appear to work against marital unity in a major way. Again, we do not know cause and effect. Is self-deception causing the disruption, or only facilitating it?

What we do know is that patterns of self-justification can be diagnostic. In trying to predict which couples would stay together three years later, scientists enjoyed surprising success based on studying the interaction between the two people during recorded sessions. Those who rewrote history in a more thoroughly negative way were predicted to break up. On this basis alone, the scientists correctly predicted all seven marital breakups, while incorrectly predicting three breakups that did not occur. They correctly predicted the other forty non-breakups, for a remarkable overall correct prediction rate of 94 percent. Though none discussed separation, some couples already talked as if they had forgotten why they married in the first place and were deep into processes of self-justification that appeared to function to reduce the dissonance of being in a bad marriage (while, of course, doing nothing to repair it). Other students of marriage claim to notice that when the ratio of positive to negative acts toward the partner drops below 5:1, the marriage is in trouble.

THE APPEAL AND DANGER OF FANTASY

Fantasy is an inviting and treacherous activity. It is deeply rooted in our biology. From our earliest years, we practice it spontaneously, with great pleasure, and it is easily encouraged by others. We create an artificial world and then choose to live in it. The fantasy typically replaces reality in a positive way—things would be better if the fantasy were true. For example, our five years of 24/7 work in the laboratory is, in fact, Nobel-quality work. As we do it, we can enjoy the return benefits sure to come our way later. Short of inducing fraud on our part, the fantasy may, in fact, improve the quality of our work. What the actual trade-off in additional fantasy-fueled labor and output is really worth, measured in other lost opportunities, is another matter, especially as the fantasy fails to pan out.

And what about the downside? Consider a romantic fantasy. That woman far away is, in fact, your wife-to-be, if not (in full delusional mode) your very soul mate. Now you can pour it on full time in the lab, certain that your romance and (future) sexual life are taken care of. You may send a portion of your earnings to your beloved every week and tell her that since you cannot show your love to her more directly, you take joy in showing it by sending her money. She *will* be pleased. She will be so pleased she may encourage you in your fantasy. In fact, she may have created it almost single-handedly in the first place.

Jamaicans have a term for this form of manipulation, called having a "boops." A boops is typically an older man who supports a young woman—her rent, electricity bill, runaround expenses, perhaps a small car—while receiving minimal sexual favors in return, only the fantasy of what soon will be his. In the optimal case, he receives no sex at all—the more fevered to keep his imagination and the more rewarding his behavior. Once caught up in his fantasy, he hardly wishes to question it. Contrary evidence that in other situations would put you immediately on guard or at least warrant some study is easily brushed aside (say, failure to receive any Christmas present at all while lavishing major ones on her). As one psychiatrist put it, "You do not want little, niggling details of reality to interfere with a good fantasy."

Now that someone else is driving your fantasy, it may carry you far from your true interests. Yes, you do wonderful lab work for six months, but if you have really bought into your fantasy, you are suffering numerous immediate costs and must someday suffer a painful de-fantasization in order to reconnect yourself with your actual interests. There can be no doubt that sexual and romantic fantasies, unfulfilled, must rank as among the most costly. Not only is a greater portion of your potential reproductive success on the line, but so is your vulnerability.

THE PAIN OF BETRAYAL

If deceit and self-deception in the family have the deepest effects on one's life, then those concerning sex are the most painful. There is nothing like sexual betrayal for pure pain—nothing like learning a loved partner is betraying you left, right, and center to split your soul in two. Deception and self-deception coming from early family life may be associated with pain akin to chronic arthritis, but with sexual betrayal, the pain is more like being hit by a truck. I believe this is true for both sexes.

There are at least three elements to this. First, the reversal in fortune can be very large—a child assumed to be your own is not, a life of love assumed to be two-sided goes in one direction only. Second, the (so-called) betrayal often rests on a bed of lies, of willful deception that may have gone on for months or years. You have played your part in all of this, by believing the lies—often with active self-deception or at the very least with failure to show due diligence.

Finally, the deceptions reach in all directions. Many lies in life are largely between you and the liar. Sexual lies inevitably encompass others, sometimes dozens of others who know a side of your life that you do not, increasing the degree of public shame. For a truly extreme example, consider the dreadful case of Elin Woods, who had to endure the knowledge that her husband, Tiger, had sex regularly with a waitress who worked across the street—at a diner they frequented—seduced the daughter of a next-door neighbor, a family she had known for several years, employed numerous people to hide his sexual life who also interacted directly with her, and then—to top it all off—let a billion people in on the secret. Arnold Schwarzenegger has now pulled off his own stunt along these lines, also available for full public enjoyment.

Why is sex so often associated with shame? One reason is that sexual activity often acts against self-interest directly—the damaged self. This includes, in principle, masturbation, bestiality, homosexuality—all sexual behavior that fails to benefit self. Unrelated individuals will have no direct self-interest but relatives will—their self-interest is directly harmed by your sexual misbehavior, as may be their reputation. So they may feel special pressure to shame you.

In principle, your inappropriate sexual behavior can upset many individuals.

Again the contrast with the family offers insight. We could have grown up under complete subjugation while being sold an ideology of equality, but usually we fall somewhere along a continuum of relative domination and misrepresentation. But infidelity (like pregnancy) is not spread along a continuum. You either are unfaithful or you are not—pregnant or not. The reversal of fortune is often absolute.

Perhaps you say to yourself, "What's the appropriate reward for someone who has lied to me, disrespected me, and plundered from me for two years?" and strangulation comes to mind. But should you not strangle yourself as well? Every deception was received and ignored by you. Your own self-deception was manipulated against you, probably both consciously and unconsciously by your partner. The two of you made that bed and lay in it.

There is often some kind of relationship between the family situation you grew up in and the one you find yourself in. Surely some of the resemblance is both genetic and through imitation. But there are also logically related effects of a different kind. Chris Rock, the American comedian, likes to joke that every woman has "a daddy problem" and you, her current partner, have to pay the price. Imagine dating a woman and taking her one day from an abusive relationship with her father. At first she will be happy, but with any hint that a man strong enough to do that could dominate her worse than her father in other ways, you have a problem on your hands.

Sexually induced pain is presumably greater the more intimate a couple have been—probably independent of the chance of propagation. Why? Imagine a sex life of relatively modest physical commitment—an embrace, a few kisses, the man climbs on top, and the two enjoy a good copulation. Contrast this with lovemaking that involves the intimate exploration of and numerous loving acts toward the body of the other person, and vice versa. After betrayal, the second is the much more painful of the two, loss in the pleasure of intimacy being the greater and also suggesting greater long-term love lost. And the greater intimacy is more painful to your imagination on both sides—he now has done those things with someone else, giving you a stabbing pain, and you also did such-and-such with him and he has gone elsewhere.

There is little doubt that pain from a relationship is among the worst of pains. With physical pain, you can almost always do something to ease it, but with emotional pain, you have to wait until it eases itself. The pain is felt on the inside and the outside—there is a social dimension that only adds to the personal. Remember that betrayal often links your partner to a web of lies involving many others—people who knew but did not speak, and so on.

Another very painful part of the interaction is that when evidence suggests that a long-term relationship is hopeless, the best strategy may be to cut the relationship in half, discard the other person, and minimize interactions, but this in itself is very painful, as if you are cutting yourself in two. Grown up between the two of you may be multiple lines of communication, now severed, so that you suffer extreme social deprivation. Two or three phone calls a day give way to oppressive silence. The sharing of joys, of minor insights, of hopes and fears, all fall by the wayside. The desire to reestablish contact—even hostile contact—is almost overwhelming. You find yourself talking to the person, and not usually in a nice way, either. If you engage in spiteful behavior or fantasize about payback time, you risk being caught in a passionate embrace, not warm but passionate, time-consuming, painful, costly, and negative.

We now have come full circle, from some of the most tender, loving, and physically exciting moments in our lives to some of the bitterest memories, as victims of lies, treachery, and even public shaming. From love to murderous impulses. This transformation is not created by self-deception but is fed by it at every stage.

CHAPTER 6

The Immunology of Self-Deception

So far we have concerned ourselves with an individual's relationship to the outside world—his or her competitors, friends, mates, and family. How does success or failure in each of these relationships involve deceit and self-deception? What kinds of self-deception are special to each realm, and what are their costs? But there is also an inner world that has strong effects on the costs and benefits of self-deceptive behavior (costs and benefits, as usual, are ultimately defined and measured by their effects on survival and reproduction). This inner world consists of a very large number of parasites (which cause disease)—invading organisms bent on eating us from the inside—and a very complex immune system of our own arrayed against them.

The importance of this world to self-deception comes primarily from the fact that the immune system is very expensive. It can act as an immense reservoir of energy and proteins and is very flexible—benefits and costs can be transferred to other functions at the flick of a molecular switch. Divert resources to attacking another male for possible immediate reproduction? Let's deal with disease later. Such decisions have very important downstream effects on health, freedom from disease, and ultimately survival and reproduction. And many of these decisions, as we shall see, involve choices between psychological states with differing degrees of self-deception. Put differently, self-deception may have strong negative or, less often, positive effects on the immune system and therefore survival and reproduction—in short, reproductive success (RS).

The inner world is populated by a series of antagonistic actors, mostly parasites—that is, species specialized to attack and devour us from the inside but also including cancer cells, mutated forms of one's own cells now replicating out of control. Parasites come in such major categories as viruses, bacteria, fungi, protozoa, and worms. They cause an enormous array of diseases: malaria, AIDS, rheumatic fever, tuberculosis, pneumonia, dysentery, smallpox, mumps, whooping cough, and elephantiasis, to name only some of the deadlier forms. Indeed, it is a sobering thought that more than half of all species on earth are parasitic on the other half—and this is a gross underestimate of the relative frequency of the two, since species of parasites are usually much smaller and harder to detect than are their host species. Most parasites have relatively mild effects, but in aggregate effects on RS, the inner world of parasites is almost as important as the outer, causing perhaps as much as 30 percent of total mortality every generation. This huge selective force has generated a very large, complex, and highly diverse system to counter the internal enemies—our immune system.

The immune system sends many cellular types to detect, disable, engulf, and kill invading organisms. One part, the innate immune system, is automatic, acts as the first line of defense, and does not rely heavily on learning. The second is based on experience and learning, the preferential production of defenses against parasites one has already encountered. This system produces as many antiparasite defenses (antibodies) as there are parasites. It has been called our "sixth sense," directed inward to spot invaders as well as cancer cells and stop them. This kind of defense, with a detailed memory of past parasitic attacks, is so important it is found even in bacteria (whose parasites are viruses).

So disease is important and we invest heavily in protecting ourselves from it—nothing surprising there. What does this have to do with deceit and self-deception? Surprisingly enough, the answer is "a lot." As we shall see, hiding one's sexual orientation (or HIV status) is costly—not just in social relations and identity but in impaired immune function and associated early death. Shame, guilt, and depression are all associated with depressed immune function, but shame has greater effects than does guilt. Sharing thoughts about a trauma—even with a private journal—is associated with improved immune function. Good marriages appear to be associated with immune benefits and bad ones with immune costs. Meditation that improves mood also improves immune function. Religiosity is associated with better immune function, as is optimism. And so on. In short, there seems to be a general rule that suppressing the truth is costly to immune function and health, as is negative affect. The key is to understand why. Why should psychological suppression of reality be associated with immune costs and sharing reality or facing it, with immune benefits? And why should an upbeat personality be associated with immune benefits, and depression with immune costs?

Perhaps the most important aspect of the immune system in this regard is its enormous cost, measured in energy and protein consumption. These resources can easily be diverted for other purposes. No one has figured out yet how to estimate the aggregate cost of the immune system, whether in energy or in other critical units, but there can be no doubt that it is large, probably on the order of the brain itself (20 percent of resting metabolic energy). We turn first to this key point.

THE IMMUNE SYSTEM IS EXPENSIVE

The beginning of wisdom about our immune system is to understand that it is extremely costly, both in energy and in the building blocks of life, proteins. It is ongoing and active twenty-four hours a day, seven days a week. To keep it running, every two weeks (roughly the maximum life span of many white blood cells), the body produces a set of cells greater in volume than two grapefruits. Some immune cells are among the most metabolically active cells in the body. Each of several thousand B cells specialized to produce antibodies grinds out about two hundred antibodies per second. Put differently, in one day's time, they generate their own weight in antibodies, the proteins that bind to parasites and disable them. Of course, they can manage this feat for only about a day and a half and must be continually replenished. Because the immune system employs a bewildering array of cell types in a very complex manner, nobody has come close to estimating its total metabolic cost, though survival costs of heightened immune activity have been measured in several bird species. Mice lacking an immune system have been created in the lab, but these animals are prone to infections of every sort and must be maintained in sterile or near-sterile conditions, where they do not thrive, in part because they are not exposed to the useful bacteria we depend upon (for digestion and skin health, for example).

Scientists have been able to show that the short-term immune response to an immediate parasite attack typically is costly in energy. Fever is often a response because it is harder on the parasite than on the host, but for every 1 degree C increase in human temperature due to fever, there is about a 15 percent increase in metabolic rate (roughly translated: the rate at which we consume energy), so the response is costly. Immunizations, which merely mimic parasite attack, commonly elevate metabolic rate by about 15 percent for several days, while real attacks impose twice the metabolic cost per unit time. This is measured not only in energy but also protein consumed—as much as 20 percent loss in total body protein in sick humans, while in some sick rats more than 40 percent of muscle protein is broken down and new synthesis is sharply reduced. Chickens reared in germ-free environments enjoy about a 25 percent gain in body weight compared to those raised in conventional environments. Of course, this reflects absence of immune costs as well as those of the parasites themselves. The metabolic requirements of mammals raised in germ-free environments drops by as much as 30 percent. Supplying antibiotics in food is associated with growth gains in birds and mammals on the order of 10 percent. The take-home message should be clear. Inside us is a system of which we are mostly unconscious that is vast, powerful, and very expensive. As we shall see, it has numerous psychological correlates, cause and effect often go in both directions, and processes of self-deception produce striking effects.

It is also striking that about one-tenth of all the proteins our cells produce are promptly degraded and their peptides recycled—a wasteful process involving largely two cell organelles specialized for this purpose (the proteosome and lysosome). Some of this involves regulating proteins that are being produced at too high numbers or are misshapen, but the rest consists of grinding up proteins made by viruses, bacteria, and cancerous cells, both to mediate their effects and to recognize them for future attack.

Thus the immune system is expensive in both energy expended and proteins consumed. But this also means that it is an energy and protein reservoir that can be drawn on for other purposes—and this is probably the key to understanding many of its behavioral and psychological correlates.

One piece of evidence for how expensive (and important) the immune system is comes from "sickness behavior"—the cost the immune system imposes on the rest of the body when it needs to repair itself. Right after the immune system has fought off a parasitic invader—let us say a virus or bacteria—it is physiologically exhausted. It has drawn down heavily on its own resources to deal with the invader, and it now needs to rebuild itself to be ready for the next one. To do this, it induces a state of torpor, apathy, and lack of interest in life in the larger organism—the "blahs." This is achieved by releasing a hormone (a particular cytokine) that acts on the brain to make the person anhedonic, that is, not taking pleasure in anything. In rats, this can be shown experimentally by releasing into healthy individuals the immune cytokine that targets the brain—the rat simply will not work as hard (on a treadmill) for sugar or other rewards.

To me, this finding was especially striking because I had always thought you felt bad after the initial attack of parasites (disease) because you were still fighting them, perhaps just mopping up operations but still enough to keep the immune system busy. Now I see that the immune system—fresh from heroic work on the barricades—merely wants to rebuild itself, and can we kindly help out by becoming inactive? To redirect energy to itself, the immune system makes other activities unrewarding so they will no longer be sought out. Internally you experience this as akin to depression. Would we suffer it better if we understood its purpose and went along with the program? Stay in bed; do not try to eat or have sex or pursue other activities that are usually fun but that make demands on the immune system and its regeneration—be satisfied with a "vacation from pleasure." Preserve your energy and be humble. Things will soon get better.

THE IMPORTANCE OF SLEEP

A profound role for sleep and immune replenishment is emerging from a variety of studies. The simple logic goes as follows—more sleep is more time for immune system regeneration (which occurs preferentially at low activity levels, such as during sleep). But self-deception often interferes with sleep. It causes internal conflict and dissatisfaction—tossing and turning mentally and physically. Since active suppression of thoughts and repression of emotions may cause a rebound effect—people may think more about what they are trying to suppress than if they didn't even try—it may directly interfere with sleep. Other things being equal, one predicts better sleep—and, therefore, better health—with less self-deception.

What the immune work shows is that there is a direct, strong, and positive relationship between sleep, immune function, and health: the more the better. Mammals generally respond more strongly to infection with increased sleep, while those rabbits that sleep more following artificial infection survive better. Meanwhile, totally sleep-deprived rats soon die from systemic bacterial infections. It is probably wise to be conscious of this connection. If you find yourself sleeping more, you may already be infected. You should probably indulge the sleep and "go with the flow."

Within a species, the more time individuals can spend sleeping, the higher are their white blood cell counts for most cell types, while red blood cells, which originate from the same tissues but are not part of the immune system, are unaffected. This correlation applies to both REM sleep (with dreaming) and non-REM sleep. Perhaps the most striking fact about the hidden benefits of sleep comes from comparing different species of mammals. Individuals from species that spend more time asleep are less likely to be infected by parasites. Mammal species range from those that sleep as little as three hours a night to those that sleep more than twenty-one. Across this range, species with ten more hours of sleep per night have rates of parasitism twenty-four times lower. In short, for long-sleeping species, life may be dull, but it sure is healthy. It is worth noting, however, that sleep and dreaming play complementary roles in consolidating memories acquired during wakefulness. Both are required for initial memory storage and then several days later, spreading the memories to the neocortex—the more social part of the brain. So for all we know, small species of mammals (with long sleep) may have superb memories.

We should also note that deliberate sleep deprivation, as practiced in various penal colonies and torture centers around the world, is expected to increase parasite attack on the victims (on top of its other negative effects).

TRADE-OFFS WITH IMMUNITY

Trade-offs appear to explain major hormonal correlates of immune activity. For example, testosterone suppresses immune function in males. Since increases in testosterone are associated with both sexual opportunities and aggressive threats, the body faced with either one appears in effect to be saying, "I will deal with my tapeworm later; right now I'll use some of those immune resources to defeat a rival male, or perhaps enjoy an extra copulation." Consistent with this, among the lowest testosterone levels are those found in men living monogamously and with children; next higher, monogamously and without children; higher still, monogamously with outside sexual activity; and highest of all, no children, no partner, in full competition. In fact, some homosexual men show the highest levels of testosterone of any, perhaps for just this reason: no parental investment, minimal marital ties, maximum male-male competition.

Health maps inversely on testosterone. Marriage tends, for example, to increase life span in men. As expected, work on monkeys, apes, and humans shows that males with higher testosterone are more likely to become infected (with such diseases as malaria) and that disease itself lowers testosterone levels—in other words, the body lowers testosterone levels to shift investment to its immune system. There is nothing magic about testosterone. It is only a signal, not a source of potency. Some of the same correlations are found in insects, in which testosterone is not involved: males have a weaker immune system than females and suffer higher parasite loads and lower survival, just as in most mammals. This difference is probably general to most animals—certainly males typically suffer higher mortality. A testosterone-associated trait—degree of fat-free muscle mass—is associated with greater self-reported sexual activity in men and earlier age at first sexual experience. The trait is also associated with higher energy consumption and lower immune function.

Likewise, corticosteroids—produced in response to stress and associated with anxiety and fear—are immune suppressors. For example, subordinate monkeys who are harassed by dominants are often high in corticosteroids and low in immune function. The immune correlation suggests that the immune system is making resources available for dealing with whatever is causing the stress and, in any case, for maintenance in the face of anxiety and fear—even if doing so temporarily increases risk of disease. (Of course, the effects of prolonged stress are another matter.) In short, whether we are pumped up on testosterone or empowered by a corticosteroid such as cortisol, we sacrifice our long-term internal defenses for short-term gains. We shall soon see that this may be yet another cost of self-deception, hyping the aggressive or the threatened, with adverse immune consequences.

The brain is also a very costly organ. Although representing only 3 percent of total body weight, the brain consumes 20 percent of all resting metabolic energy. When a person is awake, this price seems to be invariant. In the 1950s, it was shown that doing arithmetic did not require additional mental energy, a finding that now seems quaint, given that the 20 percent energy cost is known to be constant whether you are happy, depressed, schizophrenic, or on an LSD trip. The cost is slightly diminished during nondreaming sleep but slightly elevated during dreaming. Thus throughout the full twenty-four-hour cycle, the brain's resting energy cost remains virtually constant. In our species, 20 percent is the price of poker—the price to play life with a functioning brain. You must pay it or else. Indeed, not paying it for five minutes typically leads to death or, at the least, irreversible brain damage. This is just a fact of life—and an extraordinary one at that.

The invariant cost is important because one might easily imagine that different psychological functions have different energetic costs. Perhaps part of the benefit of depression is that the brain thereby saves energy. No—depression appears to have no effect on the 20 percent of energy the brain extracts. If depression lowers energy demands, it does so by lowering overall activity and metabolic rate. Likewise, if repression (suppression of truth from the conscious mind) lowers immune function, as it appears to, this is unlikely to mean that repression itself requires extra energy over and above normal function, the energy being supplied by the immune system. Instead, we must look for other changes associated with repression—which the immune system then pays for.

It has also been known for some time that the brain is the most genetically active tissue in the human body. In other words, a higher percentage of genes are active in the brain than in all other tissues, almost twice as high as in the liver and in muscle, the nearest competitors. A good one-third of all genes are so-called housekeeping genes, useful in running most kinds of cells, so they are widely shared, but the brain is unique both in the total number of genes expressed and in the number expressed there and nowhere else. By some estimates, more than half of all genes express themselves in the brain: that is, more than ten thousand genes. This means that genetic variation for mental and behavioral traits should be especially extensive and fine-grained in our species—contra decades of social science dogma. This includes, of course, such traits as degree of honesty and degree and structure of deceit and self-deception.

What we do not know is what the parallel facts are for our immune system. How much of our genes are also activated there? Are there important chemicals common to both the brain and immune system so that depletion in one system causes problems in the other? Certainly we would expect there to be, and if there are we would expect to see immune/psychological correlates we would not otherwise imagine. An analogy may help. Beginning in 1982, it was shown that female birds choose brightly colored males as a way of getting parasite-resistant genes for their offspring. This result has been documented many times since then—both that females like brightly colored males and that such males are relatively low in parasite number. It seems to be difficult to be brightly colored and sick at the same time, but why? Only in the 1990s was it shown that carotenoids—which give us orange, yellow, and red and which are not manufactured by any vertebrate but must come from their diet—play a vital role in immune function. This means that a more active immune system—for example, in response to infection—must draw carotenoids from surrounding tissues to help fight the invaders, as indeed it does. Those that are strong and healthy have color to spare, which they move to the body's exterior as an advertisement.

Are there important brain function genes that also have immune correlates? A possible example was first described in a honeybee. When the bee is given a harmless antigen to which it mounts a response, the response interferes with associative learning but not with perception or discrimination. Since it is unlikely that any of these activities increases the brain's energy budget, the explanation must lie elsewhere. In honeybees we know that associative learning depends on octopamine, a chemical that happens to be important in their immune system. In vertebrates we know that cytokines produced by the immune system can directly affect the hippocampus and reduce memory consolidation, but the functional meaning is obscure. We know that parasitic infection has a dramatic and negative effect on learning abilities. This effect must result because the activated immune system deprives the brain of other chemicals vital for learning—or has other effects, such as a decrease in sleep or dreaming, both known to be vital in consolidating learning in various species.

In birds there is clearly an intimate relationship between the immune system and the brain, one that appears to be heightened by the action of sexual selection. Two organs are intimately involved in immune function (mostly B cell production and storage)—the bursa of Fabricius of juvenile birds and the spleen of adults. The relative size of these two organs is positively associated with relative brain size across a range of species: the bigger the brain, the greater the investment in the immune system.

This may in part be due to big brains' being associated with long life span (which places a premium on parasite defense), but the correlation is especially strong when the sexes differ in brain size. That is, the bigger the relative size of the male's brain compared to the female's, the greater the relative size of the two key antiparasite organs in the species. The assumption is that males are especially likely to suffer from parasite load and its associated cognitive impairment (shown numerous times for birds), so that selection, especially in big-brained birds, will favor heavier investment in immune functions the better to protect against cognitive impairment. In this view, the two systems are complementary—the greater the investment in one (the immune system), the better the functioning of the other (the brain), presumably because the brain is especially vulnerable to parasite damage. For example, river otters that are parasitized by nematode worms show brain damage and reduction in brain size, but the effects are more prominent in males. In humans it has recently been shown that national averages in adult intellectual development are lower the greater the average parasite load.

WRITING ABOUT TRAUMA IMPROVES IMMUNE FUNCTION

In a series of important experiments from the 1980s to the 2000s, scientists showed that writing about trauma produced clear immune benefits. Although most of this writing was done in English, the same effect holds for Spanish, Italian, Dutch, and Japanese, that is, broadly. In one set of experiments, people were asked to imagine the most traumatic event in their lives. They were then split into two groups—those who spent twenty minutes each day for four consecutive days writing in a private diary about their trauma and those who wrote for twenty minutes each day on superficial topics (for example, what they had done that day). Blood was drawn before the experiment began, after the last day of writing, and six weeks later. Although those writing on their trauma said they felt worse at the end of the writing than those who wrote on innocuous topics, their immune system already showed improvement, which was still detectable six weeks later, at which time they also reported feeling better (than those who had not written about their traumas). In summary, the immediate feeling of confronting trauma is negative but the immune effects tend to be positive, and the longer-term effects on mood and immune system are both positive.

Note that the positive immune effect *precedes* the positive effect on mood, and how little writing is necessary to beget a measurable immune effect some weeks later. A recent review of about 150 studies confirms that there is a general pattern in which emotional disclosure, even in the form of occasional autobiographical writings, is often associated with consistent immune benefits.

Writing about trauma in a private journal in a lab is obviously an evolutionarily recent event, but it probably acts as a substitute for sharing this information with others. Certainly rituals of confession are common in most religions, whether public, as in many New World Amerindian religions, or private, as in the Catholic confessional. Indeed, the injunction to confess one's sins to God herself in prayer may serve a similar disclosure function. The benefits of the "talking cure," psychotherapy, may also arise in part from disclosing traumatic or shameful information that one is, in fact, hiding from others. When traveling, we will often tell secrets to complete strangers, people we have never met before and, crucially, do not expect to see again. The more that people talk in small groups, the more they claim to have learned from the group. As one psychologist drily notes, sharing our thoughts is apparently "a supremely enjoyable learning experience." For this reason, particular theories of human development—say, Freud's psychosexual stages—may be as valid as astrology, yet talking to one's analyst may provide benefits for the same reason that writing in a journal does.

One important possibility is that some of these positive correlations may in fact be caused by effects on sleep. If disclosing trauma to others results in fifteen more minutes of sleep, or at least less fitful sleep, this alone could induce the known immune benefits. A striking effect of disclosure is how quickly the benefit kicks in, as would happen if it immediately led to less troubled sleep. One final feature of the work on expressive writing is worth emphasizing. Computer-based analysis has isolated three aspects of the writing that produce beneficial effects: emotion words, cognitive words, and pronouns. The more people use positive emotion words, the more their health improves. Even writing "not happy" is better than writing "sad," perhaps because the focus in the first remains on the positive emotion. Using lots of negative emotion words and none at all are both associated with no benefit, while a moderate number is. Perhaps one is overwhelmed in the first case and in complete denial in the second. The value in taking alternative perspectives on a problem is suggested by the fact that changing back and forth from the first person ("I," "me," "my") to all other pronouns ("they," "she," "we") is associated with improvement, while remaining in one or the other perspective is not.

Conversely, there is evidence that inhibition is associated with health problems. Consistent with this, those with undisclosed childhood traumas (sexual, physical, or emotional abuse, parental death or divorce) show more illness as adults, including cancer, high blood pressure, flu, headaches, and so on. In one study, 10 percent reported sexual trauma before age seventeen, and these people had the greatest health problems of any group—fewer than half had *ever* discussed the problem. From this, one might easily imagine that a spouse's suicide, for example, would be talked about less than spousal death by other causes and would be expected to be more traumatic. But in fact suicide support groups permit more talking about these kinds of deaths, a nice example of a cultural invention that permits sufferers to come together to enjoy the benefits of sharing and disclosure.

One striking effect of writing about recent traumas is not immunological but still important: writing about job loss improves one's chance of reemployment. This sort of writing appears to be cathartic—people immediately feel better. More striking, at least in one study, is a sharply increased chance of getting a new job. After six months, 53 percent of writers had found a new job, compared with only 18 percent of nonwriters. One effect of writing is that it helps you work through your anger so it is not displaced onto a new, prospective employer or, indeed, revealed to the employer in any form. This presumably makes you more attractive to them.

HOMOSEXUALITY AND THE EFFECTS OF DENIAL

Given the global importance of HIV and AIDS, it is hardly surprising that the effects of disclosing or suppressing information have been well studied in those who are infected with HIV. Here disease progression itself can be taken as a sensitive measure of immune function, and the main findings above have been replicated almost exactly. Even relatively modest writing interventions improve apparent health status (immune chemicals per viral load). A form of "expressive" group therapy also lowers viral counts while boosting an immune measure. As has been discovered more generally, the writing/disclosure benefits tend to occur only when the writing includes increasing insight/causation and social words. Whether this is cause and effect or merely diagnostic is not known, but the correlation is strong.

Homosexuality and HIV status also turn out to be especially useful in studying deceit and self-deception because each invites a form of denial that, unlike the experimental work, occurs almost daily over a long period of time. Homosexual men often differ in the number of people to whom they reveal their sexual identity (degree to which they are "out of the closet")—from only a few heterosexual close friends, to those plus one's family, to all of those plus one's workmates, to the whole world. Likewise, it is possible to deny HIV-positive status to others and to attempt to deny it to self. All of these efforts bring negative immune and health effects, which may be substantial.

Relative to HIV-positive men who are mostly or completely out of "the closet," those who were at least half in the closet enjoyed 40 percent less time before they suffered from AIDS itself and 20 percent lower survival rate overall. Three separate studies show that denying one's HIV-positive status to others or even to self ("I am not really sick") is associated with lower immune function and/or more rapid progression of the eventually fatal HIV infection. In HIV-positive women, evidence of emotional support was not associated with immune change but evidence of psychological inhibition (use of inhibition words in daily speech) was associated—more inhibition, faster immune decay.

One study of the progression of HIV in gay men as a function of the degree the men were in the closet also controlled for unprotected sex of the dangerous kind (anal receptive). Sure enough, those in the closet practiced more of this kind of sex (being in denial, they probably prepared less for the sex likely to occur later that night). This factor had a positive effect on the rate at which their HIV progressed (probably due to the addition of competing HIV strains), but independently, being in the closet was much more harmful for resistance to HIV. At least in this respect, truth appears to be healthy for the organism expressing it: your immune system is stronger, and at the same time you are more conscious—in this case, less likely to act in obviously self-destructive ways. The US government's recent policy on service by homosexuals, "don't ask, don't tell," is an immunological disaster. You are asked to deny your sexual identity, which will invite a host of unwanted and unnecessary immune problems for you, all in order to keep everyone else relaxed.

Here is one vivid account of what it would be like to hide your heterosexual identity if this were required (as in the US military):

> Try never mentioning your spouse, your family, your home, your girlfriend or boyfriend to anyone you know or work with—just for one day. Take that photo off your desk at work, change the pronoun you use for your spouse to the opposite gender, guard everything you might say or do so that no one could know you're straight, shut the door in your office if you have a personal conversation if it might come up. Try it. Now imagine doing it for a lifetime. It's crippling; it warps your mind; it destroys your self-esteem. These men and women are voluntarily risking their lives to defend us. And we are demanding they live lives like this in order to do so.

The ill effects of concealing one's homosexual orientation are not limited to HIV-positive men. In a sample of twenty-two HIV-negative gay men studied for five years, those who concealed their homosexual identity were about two times as likely to suffer cancer and infectious diseases, such as bronchitis and sinusitis, as those who did not. These results are independent of a variety of potentially confounding factors such as age, socioeconomic status, drug use, exercise, anxiety, depression, and so on. What is especially striking is that for both cancer and infectious diseases, the effect is strictly dose-dependent—the more you are in the closet, the worse for you. Recent evidence suggests that disclosing homosexual orientation may bring correlating cardiovascular benefits as well.

Not all homosexual men are alike, of course; some are more sensitive to rejection than are others and this can have important effects. Those who are more rejection-sensitive are more likely to remain in the closet, where they avoid rejection and benefit from this immunologically. Apparently there is a general cost to remaining in the closet, but a variable benefit when one is rejection-sensitive, and this benefit can overwhelm the cost.

Have you heard of the latest twist in this saga? There are gay men who are said to be living in a glass closet. They project heterosexuality to their friends, because they believe they would be rejected if people knew about their homosexuality, but in fact the friends know about it and merely go along with the charade. It would be interesting to know where these men lie along the immune continuum. I would guess they are healthier than those in conventional closets, but not by much.

POSITIVE AFFECT AND IMMUNE FUNCTION

Direct experimental tests confirm a strong association between positive affect and immune function but are unclear regarding the correlates of negative affect. Challenging people who have never been exposed to hepatitis B with a hepatitis B vaccine shows a clear positive association between positive affect and a strong, positive immune response, no matter whether the measure of positive affect emphasizes calm, well-being, or vigor. Although negative affect has the opposite effect, this was not significant when corrected for positive affect. In general, it seems as if positive affect is not merely the absence of negative and vice versa. In some cases negative and positive affect act as independent variables and in others as only partly independent ones.

The activity of neurotransmitters such as dopamine and serotonin provides a partial explanation. Dopamine shows a phasic spike in single neurons in response to the anticipation of a reward. If the reward equals expectation, the spikes continue apace; if it exceeds, the spikes increase in rate, and if it is less than anticipated, the spikes shrink to less than the spiking baseline rate for negative rewards. Positive affect increases both dopamine and serotonin production, but negative affect has no direct effect on dopamine (though it may indirectly do so via serotonin production). Dopamine modulates immune functioning and there is an asymmetry between positive and negative affect—positive having stronger effects than negative—both on cognitive and immune function. The deeper reason for this asymmetry remains unclear.

Measures of positive affect are also associated with better survival in relatively healthy elderly people who are living independently in their communities, but curiously enough, positive affect appears to be associated with reduced survival among those already institutionalized. Likewise, those with terminal conditions, such as malignant melanoma and metastatic breast cancer, are worse off with positive affect, but in diseases with higher long-term survival, such as AIDS and non-metastatic breast cancer, positive affect is beneficial.

A possible functional explanation for these anomalies comes again from considering the rate of reward necessary to maintain positive affect and positive immune function. If your body is deteriorating quickly and you feel bad because your illness is proceeding so fast, then the expected reward of positive dopamine spikes dies down rapidly, as do the dopamine spikes. This reduces the positive cognitive and immune benefits of enhanced dopamine production. If, on the other hand, a person is caught in a long-term degenerative condition, the rate of deterioration may be slow enough that dopamine spikes and positive affect are capable of generating the cycle of positive feedback necessary to sustain improved mental and immune function.

THE EFFECTS OF MUSIC

By choosing to listen to music, people can alter their mood and their immune system. Some of the music experiments are almost too good to be true. For example, Musak (bland, peaceful music designed to calm people in a claustrophobic situation, such as an elevator) produced an increase in output of an important immune chemical by 14 percent, while jazz did so by only 7 percent. No sound had no effect, and simple noise had a 20 percent negative effect. Melodic music may suggest a happy and harmonious structure to the immediate world, while noise is cacophonous and connotes disorder, uncertainty, even danger. Music composed to match the pitch and tempo of natural monkey (tamarin) sounds but not using the monkey sounds themselves induced behavioral changes in the lab in tamarins similar to those observed in our own species. Though tamarin music based on threat vocalizations induced more anxious activity, music based on positive social interactions had positive effects: less surveillance, less sociality, and more foraging—exactly what one finds in other animals when external threat is reduced. Almost certainly there were parallel immune changes, negative to threat and positive to warm affect, so that the human response to music must have a very long past.

Two recent results stand out. Injecting about five hundred cancer cells into mice that have been stressed by exposure to noise at midnight results in much less cancerous growth if the mice then enjoy five hours of melodious music each morning. An equally dramatic example comes from humans. People undergoing bronchial physiotherapy (aspirating medicine, breathing, coughing) while listening to Bach's music (in a major key) recover much more quickly than those enjoying the therapy without music. (Minor keys show neutral or negative effects.) The point is that the right kind of music can induce positive feelings that are in turn associated with positive immune and health effects.

Certainly we know that female choice has forced a cognitive burden on males, the better to keep the females entertained. Song repertoire size in birds, which is favored by females, is controlled in males by a substantial set of neurons in the brain that completely regress during the nonbreeding season (clear evidence of the cost of running the show). We would expect pleasing male song to be both sexually arousing in females and immunologically positive. The same thing might be said for human courtship and for relations between a pair—surely there are many immunologically positive interactions possible on both sides, including good sex, and many negative ones, such as conflict, anger, suppressed feelings, and bad sex.

POSITIVITY IN OLD AGE

I suggest that an old-age positivity effect operates in a similar fashion to choosing to listen to pleasing music. By age sixty (if not earlier), a striking bias sets in toward positive social perceptions and memories. The original experiment had people looking at two faces next to each other on a screen, one with a neutral expression and one with either a positive or a negative one. After one second, the faces are removed from the screen and a dot appears where one of the faces was located. The person must hit a button as soon as the dot is perceived, one button for left side, one for right side. At ages twenty to thirty, people are equally quick to spot the dot no matter what face it was associated with. But by age sixty, a bias appears: the dot is perceived more quickly if it succeeds the positive face and more slowly if it succeeds the negative one. Study of eye movements shows that the older people spend more time inspecting faces with positive expressions than negative, and the positive ones are remembered later more often. Young people show none of these biases. These results are true among Asians, Europeans, and Americans. They appear to involve a measurable effect in the amygdala, where positive faces evoke a stronger response than negative ones in older people but not in younger people. Finally, older people tend to respond to a negative mood induced by unpleasant music by preferentially looking at positive faces, as if attempting actively to counter the negative and maintain or induce a positive mood. Young people tend, if anything, to be mood congruent—if made to feel bad, they look more at negative faces.

Why show such a positivity bias? Young people would be wise to pay attention to reality—both positive and negative—the better to make the appropriate responses later. Avoiding negative information seems risky on its face—negative events may have as big an effect on one's interests (inclusive fitness) as positive ones. By contrast, in old age it hardly matters what you learn, but greater positive affect is associated with stronger immune response, so you may be selected to trade a grasp of reality for a boost in dealing with your main problem, that of internal enemies, including cancer. A positivity bias sacrifices attention to and learning from negative stimuli the better to enjoy strong immune function now. If you haven't learned to spot an external enemy by now, chances may be low that you will learn to, and in the meantime you can enjoy a positive mood and immune response. Grandchildren may admire Gramps and Grandma because nothing seems to faze them, but Gramps and Grandma are living in positivity land—they may scarcely know the difference.

It is an interesting coincidence that although people's implicit bias in favor of youth over old age hardly changes with age (as measured by an IAT)—from twenty to seventy, they favor young over old—by our forties, our *explicit* bias in favor of youth (what we say we care about) declines until at exactly sixty, people start to say they think older is better than younger. Like everyone else, they implicitly associate youth with positive features, but they start preaching the opposite at roughly the same time they display the old-age positivity bias.

Note that the positivity effect requires no suppression of negative information or affect. The bias occurs right away. People simply do not attend to the negative information, do not look at it, and do not remember it. Thus, the possible negative immune effects of affect suppression do not need to arise. This must be a general rule—the earlier during information processing that self-deception occurs, the less its negative downstream immunological effects. At the same time, there may be greater risk of disconnect from reality, since the truth may be minimally stored or not at all.

Given what I have just said, the question arises of why old people are often perceived as being cranky or grumpy. This appears to result from an entirely independent mechanism, which sometimes cancels out or overwhelms the positivity bias. With increasing age, for reasons that are not entirely clear, people suffer greater deficits in their inhibitory abilities, that is, their ability to stop behavior under way that they may wish to stop. Since people often wish to inhibit behavior that will be seen as socially inappropriate, it is not surprising that with increasing age comes exactly that, increasingly socially inappropriate behavior. This includes discussion of private material in public, more frequent overt expressions of prejudice and stereotype, greater difficulty taking the perspective of another, and more off-target verbosity ("Don't get me started!"). Perhaps many of these traits are later described or rationalized by saying that Gramps sure is "cranky" today.

AN IMMUNOLOGICAL THEORY OF HAPPINESS

All of this work is consistent with an immunological theory of human happiness in which a finely tuned immune system purring along at near-peak efficiency with hardly a target in sight would be experienced internally as a highly enjoyable state. Even such variables as absence of food (hunger) or water (thirst) must be at least partly aversive because of their negative effects on the immune system. At the very least, it must be true that as the brain looks outward and acts to increase inclusive fitness in part by increasing happiness, then surely the same must be true when looking inwardly.

According to this view, the brain is split between outward-directed and inward-directed activity. In the outside world, many features are stationary and predictable—the shape of your bedroom, the location of food in your refrigerator, the way to work, etc. Within this world, of course, there is important variation: a predator appears, a food source, a possible mating opportunity, a hole in the street, for all of which you are selected to make appropriate responses. You have an internal reward/punishment system that goads you in appropriate directions.

Now imagine the whole thing all over for the internal system. Your brain looks inward and sees many constant features—feet and hands farther from it than the trunk, a particular circulatory system through which almost all chemicals must ultimately pass, including those produced by the brain to regulate downstream chemical activity. But in this world also live (in principle) hundreds and even thousands of species of parasites, at the moment just a few, perhaps, but taking particular configurations that need to be countered. The brain may receive or note signals that a major infection is under way in the lower left abdomen but miss the fact that a core of parasitic cells resides in the right big toe and are capable of generating the primary attack.

One important distinction concerns consciousness. We are highly conscious of interactions outside our bodies but highly unconscious of interactions within the body. Why? Part of it is that many signals to self need no consciousness, but one wonders why we are so unconscious of parasitic interactions—for example, failing to appreciate the meaning of "sickness behavior" or the value of more sleep.

Despite its importance, almost no attention has been directed toward measuring the correlates of immune function with such major components of individual fitness—or reproductive success—as survival, fecundity, physical attractiveness, and so on. The comparative work has all been done in birds. Here the pattern is clear. A greater natural immune response to some kind of challenge is positively associated with survival in nature and in the lab, and the effect size is relatively large—18 percent of variation in survival is explained by immune variation, while the closest competitor, degree of bodily symmetry, explains only 6 percent of variance in survival.

How is optimism related to immune function? A number of studies have shown a positive correlation between optimism and health outcomes, immune function, and survival. A recent study is especially striking. Law students were assayed five times throughout the year both for optimism regarding their studies and for a major immune parameter.

Within a student's year, high optimism was associated with high immune function, but when comparing students, there was no effect; that is, optimistic students were not more likely to have stronger immune systems. Although psychologists almost uniformly assume that mood affects immune system, the reverse is equally plausible. With your immune system at near-top efficiency, you should feel happy, positive, and optimistic.

The psychological and immune systems are deeply intertwined, cause and effect go in either direction, and it is hardly possible for one system to react without affecting the other. For reasons that are not always obvious, self-deception appears to have strong immune effects, usually according to the rule more self-deception, lower immune strength, but occasionally, more self-deception, better immune function.

This field is still in its infancy. Some interesting things are known, but much more remains to be found out. Which levels of information suppression are associated with what immune effects? And what chemicals are common to the brain and the immune system, leading to important trade-offs between the two? And what questions do we not even know enough to ask?

CHAPTER 7

The Psychology of Self-Deception

How do we achieve our various self-deceptions? If not in precise mechanistic terms, then in psychological ones, what are the psychological processes that help us achieve self-deception? We both seek out information and act to destroy it, but when do we do which and how do we do it? To give an answer to this, we need to trace the flow of information from the moment it arrives until the moment it leaves, that is, is represented to others. From the "rooter to the tooter," as we say for pigs. At every single stage—from its biased arrival, to its biased encoding, to organizing it around false logic, to misremembering and then misrepresenting it to others, the mind continually acts to distort information flow in favor of the usual good goal of appearing better than one really is—beneffective to others, for example. Misrepresentation of self to others is believed to be the primary force behind misrepresentation of self to self. This is way beyond simple computational error, the problems of subsampling from larger samples, or valid systems of logic that occasionally go awry. This is self-deception, a series of biasing procedures that affect every aspect of information acquisition and analysis. It is systematic deformation of the truth at each stage of the psychological process. This is why psychology is both the study of information acquisition and analysis and also the study of its continual degradation and destruction.

One important fact is worth stressing at the outset. Self-deception does not require that the truth and falsehood regarding something be simultaneously stored—as in our example of voice recognition (Chapter 3). Falsehood alone may be stored. As we saw for the old-age positivity bias (Chapter 6), the earlier the information is shunted aside—or indeed entirely avoided—the less storage of truth occurs and the less need there will be for (potentially costly) suppression later on. At the same time, since less information is stored, there are greater potential costs associated with complete ignorance. As time after acquisition increases, the choice between suppressing and retaining the truth should be more subtle and complex. The study of exactly how these conflicting forces have played out over time is a completely open field whose exploration will be most revealing.

In what follows, I begin with a review of some of the biasing that takes place during information processing. This is by no means an exhaustive look but more an impressionistic one of the ways in which various psychological processes support a deceptive function. This may include biases in predicting future feelings. Especially important are the roles of denial, projection, and cognitive dissonance in molding deceit and self-deception.

AVOIDING SOME INFORMATION AND SEEKING OUT OTHER

However much we champion freedom of thought, we actually spend much of our time censoring input. We seek out publications that mirror or support our prior views and largely avoid those that don't. If I see yet another article suggesting the medical benefits of marijuana, you can trust me to give it a careful read; an article on its health hazards is worth at best a quick glance. Regarding tobacco, I couldn't care less. The scientific facts were established decades ago and it has been years since my last cigarette. So this bias in my attention span is both directly adaptive—I smoke marijuana, so I am interested in its effects—and serves self-deception, because I hype the positive and neglect the negative, the better to defend the behavior from my own inspection and that of others.

A lab experiment measured this kind of bias precisely by confronting people with the chance that they might have a tendency toward a serious medical condition and telling them a simple test would suggest whether they were vulnerable. If they applied their saliva to a strip of material and it changed color, this indicated either vulnerability or not (depending on experimental group). People led to believe that a color change was good looked at the strip 60 percent longer than did those who thought it would be bad (actually the strip never changed color). In another experiment, people listened to a tape describing the dangers of smoking, while being asked to pay attention to content. Meanwhile, there was some background static and the subjects had the option of decreasing its volume. Smokers chose not to decrease the static, while nonsmokers lowered the level, the better to hear what was being said.

Some people avoid taking HIV and other diagnostic tests, the better not to hear bad news. "What I don't know can't hurt me." As expected, this is especially likely when little or nothing can be done either way. It is also not surprising that those who feel more secure about themselves are more willing to consider negative information. In short, we actively avoid learning negative information about ourselves, especially when it can't lead to any useful counteraction and when we feel otherwise insecure about ourselves. Self-deception is here acting in service of maintaining and projecting a positive self-view.

In many situations, we can choose what to concentrate on. At a cocktail party, we could overhear two conversations. Depending on which views we wish to hear, we may attend to one conversation instead of the other. We are likely to be aware of the general tenor of the information we are avoiding but none of its details, so here again biased processes of information-gathering may work early enough to leave no information at all that may later need to be hidden. In one experiment, people were convinced that they were likely—or highly unlikely—to be chosen for a prospective date. If yes, they spent slightly more time studying the positive rather than negative attributes of the prospective date, but if no, they spent more time looking at the negative, as if already rationalizing their pending disappointment.

BIASED ENCODING AND INTERPRETATION OF INFORMATION

Assuming we do attend to incoming information, we can still do so in a biased way. One experiment invited people to look at a figure that could be either a capital B or the number 13 (or a horse or a seal) and were told the stimulus could be either a letter or a number (or a farm animal or ocean animal). Having been provided differential food reward for the general categories ahead of time, people quickly developed a sharp perceptual bias in the appropriate direction on items presented for only four hundred milliseconds, that is, ones just reaching consciousness. Eye tracking showed that the first look was usually toward the preferred category (about 60 percent). These studies suggest that the impact of motivation on processing information extends to preconscious processing of visual stimuli and thus guides what the visual system presents to conscious awareness. Similar work has now been done using colors.

The point is that our perceptual systems are set up to orient very quickly toward preferred information—in this case, shapes associated with food rewards. This itself has nothing to do with deceit and self-deception—it will often give direct benefits. But the same quick-biasing procedure is available to us when the information is preferred because it boosts our self-esteem, or our ability to fool others. There are few more powerful forces in the service of self-deception than personal fantasies, so when these are aroused, selective attention is expected to be especially intense.

The related effect was shown sixty years ago: hungrier children, when asked to draw a coin, draw coins larger. Instruments to gain satisfaction (money buys food) are more attractive and are perceived as being larger. Recent confirmation shows that a glass appears larger to you when you are thirstier, especially if attention is called to your thirst, and even garden implements appear larger if gardening has subliminally been linked to suggestions that it is fun.

Our initial biases may have surprisingly strong effects. In one experiment, people were preselected for strong attitudes for and against capital punishment. They were then presented with a mixed bag of facts supporting both positions. Instead of leading to group cohesion, this action split the group more sharply. Those who were already against capital punishment now had a new set of arguments at hand, and vice versa. Biased interpretation ran the process. Those in favor of capital punishment accepted pro arguments as sound and rejected anti arguments as unsound. As before, self-affirming thoughts were negatively associated with this behavior—think better of yourself, you practice less self-deception. One important implication is that self-deception is a force that often drives people apart—certainly friends, lovers, neighbors—although under common group aims, such as war, shared self-deceptions are also uniquely powerful in binding people together.

BIASED MEMORY

There are also many processes of memory that can be biased to produce welcome results. We more easily remember positive information about ourselves and either forget the negative or, with time, transmute it to be neutral or even positive. Differential rehearsal, as in telling others, can itself produce the effect, an example of self-deception at the end of the process (the "tooter") affecting earlier processes. Complementary memory biases may actively work in the same direction. When given a "skills class," people remember their skills prior to the class as being worse than they rated them at the time, probably to create an illusion of progress. They then later misremember their actual performance after the class as being better than it was, presumably in service of the same delusion. What we are doing here is producing a consistent set of biases in our own favor by a series of biased memories.

Memories are continually distorted in self-serving ways. Men and women both remember having fewer sexual partners, and more sex with each partner, than was actually true. People likewise remember voting in elections they did not and giving to charity when they did not. If they did vote, they remember supporting the winning candidate rather than the one they actually voted for. They remember their children as being more precocious and talented than they were. And so on.

Although people often think of memory as a photo whose sharpness gradually degrades with time, we know that memory is both reconstructive and easily manipulated. That is, people continually re-create their own memories, and it is relatively easy to affect this process in another person. If a police officer asks a witness about a nonexistent red sports car right near an accident, the officer will often learn about the red sports car in subsequent questioning—it can sometimes end up as one of the most vividly remembered details of the accident itself. As mentioned, differential rehearsal of material after the fact can produce reliable biases in memory.

Take another example. Health information can easily be distorted in memory even when it is presented in a clear and memorable fashion. People were given a cholesterol screening and then one, three, and six months later tested for the memory of the result. Respondents usually (89 percent) recalled their risk category accurately and their memory did not decay with time, but more than twice as many people remembered their cholesterol level as lower rather than higher than it actually was. This same kind of memory bias is true of daily experiences in which people recall their good behavior more easily than bad but show no such bias in recalling the behavior of others.

Or we can invent completely fictitious memories. As has been said, "My memory is so good I can remember things that never happened." One case is memorable in my own life. For many years I told the story of how in 1968 I went deep into the bowels of Harvard's Widener Library to find a book coauthored by my father in 1948, published by the State Department, which laid out the de-Nazification procedures for all Nazis too unimportant to be hung at Nuremberg. It was a complex system of graded steps. If you were a member of SA, two slaps on the wrist; if SS, you lost your job for five years—that kind of thing. Yet almost none of this is true. No such book exists. Yes, the trip to the bowels took place and a book on Nazis was located with my father as a coauthor and it was published by the State Department. Only it was published in 1943 and is a minor piece on the structure of Nazi organizations in Nazi-occupied territories. Hardly the basis for the reinvention of Germany, but is this not the point of false memory—to improve things, especially appearances? I added nice little touches along the way. I liked to say that I trusted no one, including myself, and thus went to the bowels to see whether this family story was true. But this added to the falsehood, since there really was no "family story" about this minor 1943 work, and is a general feature of false-memory construction—new details are added that support the general argument and then become part of memory.

One can even reverse exactly who is saying what to whom. Gore Vidal remembers an interview with Tom Brokaw on NBC's morning *Today Show* in which Brokaw began by asking about Vidal's writings on bisexuality, to which Vidal replied that he was there to talk politics. Brokaw persisted with bisexuality; Vidal stood firm until they concentrated on politics. Yet years later, when Brokaw was asked what his most difficult interview had been, he cited his interview with Vidal. Reason: Vidal kept insisting they talk about bisexuality when all he wanted to discuss was politics. Positions exactly reversed—and, as expected, in the service of self-improvement: Brokaw looks better being interested in politics than in bisexuality.

In arguments with other people, lab work shows that we naturally tend to remember the good arguments on our side and the poor ones on the other, and to forget those that turn out badly for us and good for the other. This bolsters our own side and image, of course, which presumably is its function. Memory distortions are more powerful the more they are motivated to maintain our self-esteem, to excuse failures or bad decisions, and to push into the deeper past causes of current problems. Thus, most people maintain the illusion of improvement, where such mistakes as must be acknowledged can at least be attributed to the failings of an earlier version of oneself.

RATIONALIZATION AND BIASED REPORTING

We reconstruct internal motives and narratives to rationalize otherwise bad or questionable behavior. We can attribute behavior to external contingencies rather than internal, thereby helping defend ourselves. So a general belief that cheating is not bad—or is unintentional or occurs in a world without free will—will all serve to rationalize our cheating, as indeed they do.

Biases show up in unexpected places, even when there are no clear benefits or costs at issue. The classic experiment in this domain was beautifully designed to put people in an awkward situation with one of two escapes. People were offered the chance to sit next to a crippled person or one who was not. Each was watching a television set in front of him or her. Sometimes the two TVs had the same show, sometimes different ones. When it was the same show, people preferentially chose to sit next to the handicapped person, as if demonstrating their lack of bias, but if the two TVs had different shows, people chose to sit away from the crippled person, as if now having a justification (more interesting show) for an otherwise arbitrary choice. Similarly, a meta-analysis of many studies shows that white Americans choose to help black Americans more or less equally (compared to helping whites) but not when they can rationalize less helping on grounds such as distance or risk. Here people are not denying or misremembering their behavior—rather, they are denying the underlying intention and rationalizing it as the product of external forces. This has the advantage of reducing their responsibility for behavior performed.

A belief in determinism can provide a ready excuse for misbehavior, just as can unconsciousness: the "I had no choice" defense. Relatively deterministic views of human behavior may provide some cover for socially malevolent behavior. Experimentally inducing a deterministic view (reading an essay on how genes and environment together determine human behavior) increases cheating on a computer-based task that permits cryptic cheating. What this work shows is that by manipulating a variable that reduces personal responsibility, we easily induce immoral behavior in ourselves (at least as viewed by others).

PREDICTING FUTURE FEELINGS

It is an interesting fact that we show systematic biases in our ability to predict our own future feelings. We make systematic errors in the process, under the general rule that what we are feeling at the present will extend into the future. When imagining a good outcome, we overestimate our future happiness, and vice versa for a bad one. It is as if we assay our current feelings and then project them into the future. We do not imagine that we will "regress to the mean," that is, return naturally to the average value of happiness. We do not assume we will be less happy in the future than our current state of happiness or happier in the future if we are currently down. Thus, one week after the 2004 US elections, Kerry supporters were less dejected than they thought they would be and Bush supporters, less ecstatic.

There is evidence that we make similar mistakes when trying to predict the feelings of others, whether friends or strangers. We overestimate the effect of an emotional event on their future feelings, much as we do for ourselves. Indeed, our forecasting of them is positively correlated with their own, but neither is very predictive of the future.

The problem is in the interpretation. Some see this as a form of self-deception in which we are unconscious of the degree to which our system of self-deception will readjust our thinking in the future. I doubt this. We project easily into the future because it expresses our current emotional state. Verbal predictions regarding our future mental states may be a relatively recent invention with limited selective effects. The relevant trade-offs are already built into our behavior whatever our verbal predictions.

Certain exceptions to this rule also stand out. I remember "courting" a Nigerian beauty at a very safe distance at a club in Amsterdam for three hours without ever having the courage to approach her. When she left, she threw me a look of withering contempt that burned right into my soul. If a social psychologist had been there to measure my "affective forecasting," I doubt I would have guessed that twenty-five years later, the memory still sears in my consciousness. I believe I would have predicted that within a year or two the whole evening would have been completely forgotten.

ARE ALL BIASES DUE TO SELF-DECEPTION?

A hallmark of self-deception is bias. Mere computational error is not enough. Such error is often randomly distributed around the truth and shows no particular pattern. Self-deception produces biases, patterns where the data point in one direction—usually that of self-enhancement or self-justification. Are there biases that are real but not driven by self-deception? Of course there are.

Consider the following. Sounds that are coming toward us are perceived as closer and louder than they really are, while the opposite is true for receding sounds. This is a bias and it has a perfectly good explanation. Approaching objects are inherently more dangerous than are receding ones—hence the value of earlier and more acute detection. Perhaps the organism is measuring distances in Darwinian units rather than Newtonian ones. From that viewpoint, there is no bias.

Or consider another example. From the top of a tree, the drop to the ground looks much farther than does the same distance viewed from the ground up. There is no social component to these biases. You are directly saving yourself—not trying to manipulate the opinions of others. Many other errors have similarly innocent explanations. Some are simple optical illusions, holes in our sensory system that produce startling biases under particular conditions. Others are general rules that work well in most situations but fail badly in some.

Of course the errors we make are very numerous. In the words of one psychologist, we can fall short, overreach, skitter off the edge, miss by a mile, take our eyes off the prize, or throw the baby out with the bathwater. And we can exaggerate our accomplishments, diminish our defects, and act vice versa regarding those of others. Many of these may serve self-deceptive functions but not all. Sometimes when we take our eyes off the prize, we have only been momentarily distracted; sometimes when we miss by a mile we have only (badly) miscalculated. At other times, it is precisely our intention to throw out the baby with the bathwater or to miss by a mile, so in principle we have to scrutinize our biases to see which ones serve the usual goal of self-enhancement or, in some other fashion, deception of others, and which ones subserve the function of rational calculation in our direct self-interest.

DENIAL AND PROJECTION

Denial and projection are fundamental psychological processes—the deletion (or negation) of reality and the creation of new reality. The one virtually requires the other. Projecting reality may require deleting some, while denial tends to create a hole in reality that needs to be filled. For example, denial of personal malfeasance may by necessity require projection onto someone else. Once years ago while driving I took a corner too sharply and my one-year-old baby fell over in the backseat and started to cry. I heard myself harshly berating her nine-year-old sister (my stepdaughter) for not supporting her—as if she should know by now that I like to take my corners on two wheels. The very harshness of my voice served to signal that something was amiss. Surely the child's responsibility in this misdemeanor was, at most, 10 percent, the remaining 90 percent lying with me, but since I was denying my own portion, she had to endure a tenfold increase in hers. It is as if there is a "responsibility equation" such that decrease of one portion must necessarily be matched by an increase elsewhere.

A rather more serious example of denial and projection concerns 9/11. Any major disaster has multiple causes and multiple responsible parties. There's nothing wrong with assigning the lion's share of cause and responsibility to Osama bin Laden and his men, but what about creating a larger picture that looks back over time and includes us (US citizens) in the model, not so much directly causing it as failing to prevent it? If we were capable of self-criticism, what would we admit to? How did we, however indirectly, contribute to this disaster? Surely through repeated inattention to airline safety (see Chapter 9) but also in our foreign policy.

This final admission is often hardest to make and is almost never made publicly, but sensible societies sometimes guide behavior after the fact in a useful way. It is easy for personal biases to affect one's answer here, but I will set out what seem to me to be obvious questions. To wit, are there no legitimate grievances against the United States and its reckless and sometimes genocidal (Cambodia, Central America) foreign policy in the past fifty years? Is there any chance that our blind backing of Israel—like all our "client states," right or wrong, you're our boys—has unleashed some legitimate anger elsewhere, among, say, Palestinians, Lebanese, Syrians, and those who identify with them or with justice itself? In other words, is 9/11 a signal to us that perhaps we should look at our foreign policy more critically and from the viewpoint of multiple others, not just the usual favored few? One need not mention this in public but can start to make small adjustments in private. Again, the larger message is that exterminating one's enemies is not the only useful counterresponse to their actions, but becomes so if one's own responsibility is completely denied and self-criticism aborted.

DENIAL IS SELF-REINFORCING

Denial is also self-reinforcing—once you make that first denial, you tend to commit to it: you will deny, deny the denial, deny that, and so on. In the voice-recognition experiments, not only do deniers deny their own voice, they also deny the denial. A person decides that an article on which he is a coauthor is not fraudulent. To do so, he must deny the first wave of incoming evidence, as he duly does. Then comes the second wave. Cave in? Admit fault and cut his losses? Not too likely. Not when he can deny once more and perhaps cite new evidence in support of denial—evidence to which he becomes attached in the next round. He is doubling down at each turn—double or nothing—and as nothing is what he would have gotten at the very beginning, with no cost, he is tempted to justify each prior mistake by doubling down again. Denial leads to denial, with potential costs mounting at each turn.

In trading stock, the three most important rules are "cut your losses, cut your losses, and cut your losses." This is difficult to do because there is natural resistance. Benefits are nice; we like to enjoy them. But to do so, we must sell a stock after it has risen in value; then we can enjoy the profit. By the same token, we are risk averse. Loss feels bad and is to be avoided. One way to avoid a cost is to hold the stock after it has fallen—loss is only on paper and the stock may soon rebound. Of course, as it sinks lower, one may wish to hold it longer. This style of trading eventually puts one in a most unenviable position, holding a portfolio of losers. Indeed, this is exactly what happens. People trading on their own tend to sell good stocks, buy less good ones, and hold on to their bad ones. Instead, "cut your losses, cut your losses, cut your losses."

YOUR AGGRESSION, MY SELF-DEFENSE

One of the most common cases of denial coupled with projection concerns aggression—who is responsible for the fight? By adding one earlier action by the other party, we can always push causality back one link, and memory is notoriously weak when it comes to chronological order.

An analogy can be found in animal species that have evolved to create the illusion that they are oriented 180 degrees in the opposite direction and are moving backward instead of forward. For example, a beetle has its very long antennae slung underneath its body so they protrude out the back end, creating the illusion of a head. When attacked, usually at the apparent "head" end (that is, the tail) it rushes straight forward—exactly the opposite of what is expected, helping it to escape. Likewise, there are fish with two large, false eyespots on the rear end of their body, creating the illusion that the head is located there. The fish feed at the bottom, moving slowly backward, but again, when attacked at the apparent "head" end, take off rapidly in the opposite direction. What is notable here is that the opposite of the truth (180 degrees) is more plausible than a smaller deviation from the truth (say, a 20-degree difference in angle of motion). And so also in human arguments. Is this an unprovoked attack or a defensive response to an unprovoked attack? Is causation going in this direction, or 180 degrees opposite? "Mommy, he started it." "Mommy, she did."

COGNITIVE DISSONANCE AND SELF-JUSTIFICATION

Cognitive dissonance refers to an internal psychological contradiction that is experienced as a state of tension or discomfort ranging from minor pangs to deep anguish. Thus, people will often act to reduce cognitive dissonance. The individual is seen to hold two cognitions—ideas, attitudes, or beliefs—that are inconsistent: "Smoking will kill you, and I smoke two packs a day." The contradiction could be resolved by stopping cigarettes or by rationalizing their use: "They relax me, and they prevent weight gain." Most people jump to the latter task and start generating self-justification in the face of a much more difficult (if healthier) choice. But sometimes there is only one choice, because the cost has already been suffered: you can rationalize it or live with the truth.

Take a classic case. Subjects were split into two groups, one comprising people who would endure a painful or embarrassing test to join a group and the other people who would pay a modest fee. Then each was asked to evaluate the group based on a tape of a group discussion arranged to be as dull and near-incoherent as possible. Those who suffered the higher cost evaluated the group more positively than did those who paid the small entry fee. And the effect is strong. The low-cost people rated the discussion as dull and worthless and the people as unappealing and boring. This is roughly how the tape was designed to appear. By contrast, those who paid the high cost (reading sexually explicit material aloud in an embarrassing situation) claimed to find the discussion interesting and exciting and the people attractive and sharp.

How does *that* make sense? According to the prevailing orthodoxy, less pain, more gain, and the mind should measure accordingly. What we find is: more pain, more post-hoc rationalization to increase the apparent benefit of the pain. The cost is already gone, and you cannot get it back, but you can create an illusion that the cost was not so great or the return benefit greater. You can choose, in effect, to get that cost back psychologically, and that is exactly what most people do. This particular experiment has been replicated many times with the same result. But it is still not quite clear why this makes sense. Certainly it works in the service of consistency—since you suffered a larger cost, it must have been for a larger benefit. People can be surprisingly unconscious of this effect in their own behavior. Even when the experiment is fully explained and the evidence of individual bias demonstrated, people see that the general result is true but claim that it does not apply to them. They take an internal view of their own behavior, in which lack of consciousness of the manipulating factor means it is not a manipulating factor.

The need to reduce cognitive dissonance also strongly affects our reaction to new information. We like our biases confirmed and we are willing to manipulate and ignore incoming information to bring about that blessed state. This is so regular and strong as to have a name—the confirmation bias. In the words of one British politician, "I will look at any additional evidence to confirm the opinion to which I have already reached."

So powerful is our tendency to rationalize that negative evidence is often immediately greeted with criticism, distortion, and dismissal so that not much dissonance need be suffered, nor change of opinion required. President Franklin Roosevelt uprooted hundreds of thousands of Japanese-American citizens and interned them for the remainder of World War II, all based on anticipation of possible disloyalty for which no evidence was ever produced except the following classic from a US general: "The very fact that no sabotage has taken place is a disturbing and confirming indication that such action *will* be taken."

Supplying a balanced set of information to those with divergent views on a subject, as we saw earlier in the case of capital punishment, does not necessarily bring the two sides closer together; quite the contrary. Facts counter to one's biases have a way of arousing one's biases. This can lead to those with strong biases being both the least informed and the most certain in their ignorance. In one experiment, people were fed politically congenial misinformation and an immediate correction. Most people believed the evidence more strongly after the refutation.

One important factor affecting the need for cognitive dissonance reduction is post-hoc rationalization of decisions that can no longer be changed. When women are asked to rank a set of household appliances in terms of attractiveness and then offered a choice between two appliances they have ranked equally attractive, they later rank the one they chose as more attractive than the one they rejected, apparently solely based on ownership. A very simple study showing how people value items more strongly after they have committed to them focused on people buying tickets at a racetrack. Right after they bought their ticket, they were much more confident that it was a good choice than while waiting in line with the intention of buying the same ticket. One upshot of this effect is that people like items more when they cannot return them than when they can, despite the fact that they say they like the option to return items.

A bizarre and extreme case of cognitive dissonance reduction occurs in men sentenced to life imprisonment without the possibility of parole for a crime—let us say a spousal murder, using a knife repeatedly. Surprisingly few will admit that the initial act was a mistake. Quite the contrary: they may be aggressive in its defense. "I would do it again in a second; she deserved everything she got." It is difficult for them to resist reliving the crime, fantasizing again about the victim's terror, pain, unanswered screams for help, and so on. They are justifying something with horribly negative consequences (for themselves as well now) that they cannot change. Their fate is instead to relive the pleasures of the original mistake, over and over again.

SOCIAL EFFECTS OF COGNITIVE DISSONANCE REDUCTION

The tendency of cognitive dissonance resolution to drive different individuals apart has been described in terms of a pyramid. Two individuals can begin very close on a subject—at the top of a pyramid, so to speak—but as contradictory forces of cognitive dissonance come into play and self-justification ensues, they may slide down the pyramid in different directions, emerging far apart at the bottom. As two experts on the subject put it:

> We make an early, apparently inconsequential decision and then we justify it to reduce the ambiguity of the approach. This starts a process of entrapment—action, justification, further action—that increases our intensity and commitment and may take us far from our original intentions or principles.

As we saw in Chapter 5, this process may be an important force driving married couples toward divorce rather than reconciliation. What determines the degree to which any given individual is prone to move down the pyramid when given the choice is a very important (unanswered) question.

A novel implication of cognitive dissonance concerns the best way to turn a possible foe into a friend. One might think that giving a gift to another would be the best way to start a relationship of mutual giving and cooperation. But it is the other way around—getting the other person to give you a gift is often the better way of inducing positive feelings toward you, if for no other reason than to justify the initial gift. This has been shown experimentally where subjects cajoled into giving a person a gift later rate that person more highly than those not so cajoled. The following folk expression from more than two hundred years ago captures the counterintuitive form of the argument (given reciprocal altruism):

> *He that has once done you a kindness*
> *will be more ready to do you another*
> *than he whom you yourself have obliged.*

COGNITIVE DISSONANCE IN MONKEYS AND YOUNG CHILDREN

It is of some interest to know whether animals show cognitive dissonance and at what age children show such effects. Birds often show the human bias of preferring items for which the birds work harder (in their case, food) over identical items achieved through less work. The same is true sometimes of rats.

A more novel set of experiments shows that when a monkey is forced to choose between two items it is equally fond of (say, a blue M&M instead of a red one), it will then prefer another color (say, a yellow M&M) over the one it just rejected (red), as if needing consistency. That is, having rejected red once, to remain consistent it must do so again. But if the initial choice is made by the human experimenter (blue over red), this either has no effect on the monkey's subsequent choice or the monkey then chooses the one the human kept for itself, as if this must be the better one.

Nearly identical experiments run on four-year-olds produce nearly identical results. When the children are forced to choose between two equivalent objects, they continue to reject the one they rejected the first time, as if staying true to themselves. That is, having rejected one, the child acts as if there must have been a good reason and rejects it again. This occurs even if the child does not see which item it chose until after having made its choice. Once again, as with the monkeys, when the experimenter makes the choice instead of the child, this either has no effect on how the child chooses or it chooses the one the experimenter kept for itself, as if this must be the better one.

In short, though there are only a few studies of cognitive dissonance in other animals and in children, they tend to give similar results: each party acts as if it is rationalizing its prior choice as having been based on sound logic and hence worth repeating when given the same opportunity. Given the theory advanced in this book, it is tempting to argue that the children and the monkeys may be projecting a general illusion of consistency to impress others.

By now we have laid the foundations for an understanding of the evolution, biology, and psychology of self-deception. We can now apply our logic to everyday life, including airplane crashes, historical narratives, warfare, religion, other intellectual systems, and our own lives. The applications extend in all directions.

CHAPTER 8

Self-Deception in Everyday Life

The logic we have been developing applies with full force to everyday life—so much so that its validity can, in part, be tested there. How much does our system of thought help us understand our lives? What interesting facts of everyday life are completely hidden from us until research or logic reveals them? Some biases in our thinking have been studied in surprising detail, and others are known only from anecdotes. I begin with the study of the stock market and what it reveals about sex differences in overconfidence, as well as the unconscious use of language to hype the upside of the market—that is, to encourage trading.

SEX DIFFERENCES IN OVERCONFIDENCE

Overconfidence must—in competitive situations—sometimes give an advantage, but insofar as it induces risky and ultimately unprofitable behavior, it must also have costs. Clearly our confidence in ourselves is an important variable in many situations affecting and predicting our behavior. Others would do well to attend to our self-confidence—that is, if they can measure it accurately. After all, they may just have met you, but you have known yourself all your life. So we expect overconfidence on deceptive grounds alone (see Chapter 1). In general, across many species, including our own, males are more likely to profit from overconfidence than are females. Certainly their potential reproductive success is usually higher (because males usually invest less per offspring), so the payoff for successful overconfidence is likely to be higher as well (see Chapter 5).

Stock trading by amateurs (via computer-placed options) provides a nice situation from daily life to study the bias. Competitive interactions are at a minimum—your overconfidence is not directly affecting any of the other investors you are competing against, none of whom knows you—so with no benefits from overconfidence, costs are expected to dominate. Under perfect information, stock prices are at their true value, so that trading produces random effects. Under mildly imperfect information, prices are close to true values, so that trading produces near-random direct effects. But trading is costly, as you pay a fee for every trade. Given these facts, it is clear that there is substantial overtrading in the US stock markets. Nearly 100 percent of stocks change hands every month and five billion are traded per day (2007). Given the cost of each trade, the net effect of this level of trading is negative. To cite but one example, in the general population, males trade stocks more often than do females (45 percent more in one sample), and they suffer accordingly: 2.7 percent annual loss in returns compared to 1.7 percent loss for females. The sex difference probably reflects the possibility of greater reproductive returns for males of financial success than females, an upside bias that is expected in many male activities given their greater chance in general to achieve especially high reproduction.

One work was notable for studying multiple kinds of overconfidence as possible correlates of trading volume. The key correlate to overconfidence turned out to be the good old "above-average effect." The average investor rated him- or herself above average in ability and past performance. And the more an individual did so, the more he or she traded, even though there was no correlation with actual past performance. This resulted in more trading with no average gain and an average loss due entirely to the transaction costs. Believing that there is more information than in fact there was, that is, underestimating the variance of the signal, was not correlated with trading activity, only overestimation of self.

Overconfidence in currency markets provides a nice contrast. Here transaction costs are negligible (about one-hundredth of the 3 percent stock cost), so there is no immediate downside to overtrading. There is a widespread tendency for professional traders to overestimate their success and their ability to forecast correctly. Overconfidence has no effect on profitability (expected given negligible transaction costs), but there are positive social correlates. Overconfidence is positively associated with individual rank and trading experience. Here cause and effect are by no means certain, since in other domains it is well known that people of superior rank and age show higher confidence, with no superiority in actual performance.

Greater male than female overconfidence has been detected in studies of arithmetic contests. An individual can either be paid piecemeal (50 cents per correct answer adding sets of five numbers for five minutes) or in competition with three others, winner take all: $2 per correct answer for the highest scorer, nothing for the other three. Under perfect information about one's relative skill, the top one-fourth should choose to compete and the rest should choose to work piecemeal. This is far from what happens: 35 percent of women choose to compete, close to expected, but fully 75 percent of men choose to compete in a task in which, on average, only 25 percent can win. Overall, when matched for ability, women have a 38 percent lower chance of deciding to compete. This means that on the upper end of ability, women undercompete, and on the lower end, men greatly overcompete. This is yet another example of a degree of self-deception—here in the form of overconfidence—having a positive effect under certain circumstances and negative under others, the net effect being negative.

Another cause of misbehavior in stock trading is a tendency toward thrill seeking. Like those who are overconfident, those who have a special need for thrills tend to trade more often to their own disadvantage, and this is independent of overconfidence. Men, in turn, are vastly overrepresented among thrill seekers, at least as measured by speeding tickets, drug use, gambling, and participation in dangerous sports (such as hang gliding). In Finland, those with more speeding tickets trade more often to their own disadvantage. What the advantage is to the thrill seeking has not been measured, but it probably has to do with showing off—the stunt properly executed may be a sight worth recounting.

METAPHORS IN THE STOCK MARKET

A nice example of unconscious persuasion concerns metaphors about the stock market taken from daily news broadcasts. The stock market moves up or down in response to a great range of variables, about most of which we are completely ignorant. The movement mirrors a random walk, with no particular pattern. And yet at the end of the day, its movements are described by the media in two kinds of language (agent or object) that are often used for movement more generally. The average listener will be completely unconscious of the metaphors being used. The key distinction is whether an agent controls the movement of something or it is an object moved by outside forces (such as gravity). Here are examples of the agent metaphor for stock movements: "the NASDAQ climbed higher," "the Dow fought its way upward," "the S&P dove like a hawk." The object metaphors sound more like: "the NASDAQ dropped off a cliff," "the S&P bounced back."

Agent metaphors tempt us to think that a trend will continue; object ones do not. The interesting point is that there is a systematic bias in the use of the language—up trends are more the action of agents, while down trends are externally caused. Both of these metaphors are stronger for movement that is consistent, and the bias exists whether reporting is occurring after a long up market or a long down market. Even experimental student commentators unconsciously adopt the appropriate bias: agent for the up trends, external factors for the down. Now here is the average upward bias. The more a market moves up during a day, the more it is given an agent metaphor that, in turn, (unconsciously) suggests continued upward movement. Since the opposite is true for down days—less agent metaphor, less expectation of continued downward movement—the net effect is positive. Investment information should lead to more investment, on average. Surely the effect of this bias in media language is to encourage investment overall, just as supplying information about the day's trends instead of merely reporting them (up or down) gives a greater expectation of a trend and hence greater trading after up movement, at greater net loss (there is a cost to trading and no benefit during a random walk). Perhaps the function of the financial commentators in the first place (from the standpoint of those who employ them) is to hype interest in the market.

MANIPULATIVE METAPHORS IN LIFE

The use of metaphor is a key part of language, structuring meaning by embedding more abstract concepts in day-to-day events—such as moving into new spaces at a given rate, and so on. Metaphor often flies just below radar and may have important unconscious effects. Euphemisms, for example, may not just soften meaning but invert it. "Waterboarding" sounds like something you would like to do with your children on a Mediterranean vacation, and "stress positions" the perfect way to end a workout, while all of us could benefit from some good "sleep management." But each of these, in fact, refers to a form of torture—repeated near-drowning, long-term painful bending and stretching, wholesale sleep deprivation. In the same vein are terms such as "collateral damage" (civilians killed during military operations), "extraordinary rendition" (kidnapping followed by torture), "enhanced interrogation" (torture), "friendly fire" (death at the hands of your own soldiers), and the "final solution" (genocide of European Jews).

There is also something that has been aptly called the euphemism treadmill, in which each new euphemism soon becomes tainted by what it refers to so that a new euphemism must be invented to take its place. "Garbage collection" becomes "sanitation work," which morphs into "environmental services." "Toilet" turns into "bathroom" (so you are washing in there), which turns into "restroom" (so you are taking a nap in there). "Slum" to "ghetto" to "inner city," with "ghetto" making a modest comeback lately as a synonym for lower-class black culture—"he is so 'ghetto.'" It seems as if we are running from the negative connotations of words, with no net progress. The association is soon reestablished, so we have to keep running.

We all know of examples. In my younger days, "retarded" went to "disabled" to "mentally challenged," and is now a person with "special needs." "School security guard" is now a "school safety agent." The other day a phone "operator" told me he was an "information assistant." Not quite sure how much elevation he thereby achieved, but notice that the euphemism is longer than what it replaces, as often happens—in other words, this enterprise is trending in the wrong direction, at least where efficiency is concerned.

The euphemism treadmill has several important implications. For one thing, it means that concepts are in charge, not words, contradicting entire disciplines (see Chapter 13 for cultural anthropology). That is, the words keep changing, but not, so far as we can see, the underlying concept. It also means that we are expected to be vigilant about the various changes introduced—otherwise, why make them? But any advantage tends to be strictly temporary.

The treadmill also suggests the novel notion that we will finally have relaxed about some of our distinctions—racial, sexual, whatever—when the treadmill stops. Some of the running has deeper meaning than simply running from negativity. Yes, "Negro" is Spanish for black, but it is uncomfortably close to the common "white" mispronunciation of "Nigrah," itself rather too close to the racially insulting "n-word." The initial attempt to fight back is to overstate the case. Hence, "black" is chosen not just to achieve parity with "white" but to frighten anti-black people with their worst racial nightmares, the black man unfettered—Black Panthers—invisible at night except for their yellow eyes. Incidentally, "colored people" was the genteel acknowledgment of intermixing (without taking any responsibility for it), so it was condescending. When you are in a time of revolutionary mind change, you push for racial solidarity—"all of us brothers and sisters are 'black.'" But then you want to move to the next stage, defined not by some other group but by your own roots. All other people do it: Italian Americans, Chinese Americans, Japanese Americans, etc. What is a person supposed to say, "oppressed black slave American"?

So there was a natural turn to "African American"—at least it says where most of the genes came from. In this case, then, linguistic change seems to match logically the stages through which a particular group passes.

There is also something one could call the malphemism treadmill, where a word is forced to take on negative connotations. Thus, "tendentious" originally meant strongly stated minority views apt to provoke a response. In the UK and Australia, this is still its meaning, but in the United States, a negative connotation has been added—being of the minority, the views are likely to be wrong—so it is incorrect views that arouse natural resistance. Perhaps the fact that "tendentious" rhymes with "pretentious" makes this shift in meaning easier. Criticism of Israel is often said to be tendentious, which in the United States is often literally true; such criticism is a strongly stated minority opinion likely to provoke disagreement. That it is thereby false is another matter. The larger tendency to produce malphemisms in the press is suggested by the following double whammy: "the tragedy of the vitamin D deficiency epidemic," probably referring to a small increase in D-deficient individuals with negligible overall health effects.

An extraordinary verbal one-step has been spearheaded in multiple disciples in the past fifty years—the switch from "sex" to "gender" as words to denote the two sexes. From time immemorial (at least a thousand years), sex referred to whether an individual was a male (sperm producer) or a female (egg producer). In the past hundred years, the word was extended to "having sex." "Gender" was strictly a linguistic term. It referred to the fact that in various languages, *words* may be feminine, masculine, or neuter, apparently in almost random ways. "Sun" is feminine in German, masculine in Spanish, and neuter in Russian, but "moon" is feminine in Spanish and Russian, and masculine in German. In German, a person's mouth, neck, bosom, elbows, fingers, nails, feet, and body are masculine, while noses, lips, shoulders, breasts, hands, and toes are feminine and hair, ears, eyes, chin, legs, knees, and the heart are neuter. Pronouns are assigned by gender, so you can say about a turnip, "He is in the kitchen." You tell me. I have been a biologist for forty-five years and I can see no rhyme or reason to this system. It seems completely arbitrary, and this is perhaps the point. Since grammatical gender is arbitrary and meaningless, so also are biological sex differences if they can be rendered in the language of gender.

In a remarkable burst of activity, in fewer than forty years, "gender" took over entirely in many disciplines as the word for sex. Thus a person's gender is male or female—not the ending on the word itself, but the person's actual sex. And likewise, for cows and everyone else, "gender" has replaced sex. The pressure for all of this was twofold: to disassociate sex differences from sexual behavior and to minimize the apparent biological differences between the sexes in favor of differences imposed by verbiage itself ("culture")—symbolized by the gender of words. The more arbitrary the gender of words, the more arbitrary the assignment of sex differences.

THE NAME-LETTER EFFECT

What about linguistic effects at a much smaller level—biases in favor of the initials of one's own name, for example? People prefer letters that are found in their own first and last names. That is, when choosing between two letters based on attractiveness (asked to do so quickly and with no thought), people consistently choose letters contained within their own names. This is especially true for the first initials of their first and last names, but in fact it is true throughout each name. The effect is robust to various forms of measurement and occurs, so far as can be seen, completely outside of consciousness—nobody appears to be aware they are choosing letters on the basis of self-similarity. The effect is found in every language examined: eleven European languages using the Roman alphabet, as well as Greek and Japanese. A similar effect is found for one's own birth dates—a preference for these numbers against a random set of numbers. The effect appears in children as young as eight and in university students, demonstrating that the effect remains strong despite the person's having been exposed by then to millions of letters and many, many numbers.

The simplest explanation would be that the name-letter bias is due solely to familiarity of one's own name, since familiarity can increase attractiveness, but there is good reason to believe that more than familiarity is involved. Young Japanese women show a strong preference for their first-name letters and a weak one for those of their last name, which they will soon change, while the opposite is true for Japanese men. This suggests that it is the personal significance of the name that produces the effect, not the frequency with which it has been encountered. Nor does the overfrequency of letters have much to do with their popularity, at least at the top end: the most frequent letters are not the most popular. At the bottom end, it is true that many letters that are rarely encountered—W, X, Y, Z, and Q—are also often unattractive, but when encountered often, as W is among the Walloons of Belgium, the letter fails to rise in popularity. More to the point, the name-letter effect is enhanced by such variables as positive parenting style (see below) that are associated with self-esteem, but not obviously with word usage. In short, the name-letter effect appears to be primarily narcissistic: with a minor frequency effect, we love the initials of our names above those of others, because they are our own.

For one brief shining moment, it appeared as if the name-letter effect had widespread important effects on our behavior of which we were completely unconscious. Too many Larry and Laura lawyers, too many Geoffreys publishing in the geosciences. Too many people's last names (first four letters) match those of towns or streets or states where they live. People appeared to be making major life decisions based on trivial egoistic coincidences. Causality was strongly implied by evidence that people tend to migrate to states that match their own last names. Fortunately, perhaps, the entire edifice collapsed when a very careful reanalysis replicated all the original findings and then showed that every single one was due to hidden biases in procedure or logic. For example, forty years ago, there was a wave of enthusiasm for naming babies Geoffrey, Laura, or Larry. Hence, they are overrepresented in a variety of enterprises today besides the geosciences and law. Likewise, place of birth for the migration study was often noted as place of residence several years later (when the child was first given a social security number) and the subjects may already have migrated away. Since people have a strong tendency to return to where they were born, this alone would create a spurious correlation as, indeed, it did.

What we do know about the costs or benefits associated with the name-letter effect are nonetheless surprising. Preference for one's own first initials can lead to a real cost, that is, lower performance when one's own initials are associated with signs of lower performance (though the reverse is not true). Self-love in this context gives a cost but not a benefit. In schools in the United States, Cs and Ds are low grades and As and Bs high. People with a C or a D at the beginning of either their first or last names show lower academic performance (grade-point average) than do those with As, Bs, or other letters, apparently because lower grades (Cs and Ds) are (unconsciously) less aversive to them. It is notable that self-love does not benefit those with initials of A or B—they score just like those with other initials—but self-love harms those with C or D. If your name is Charles Darwin, you will tend to do slightly less well academically than everyone around you. And these biases have ramifying effects in life. When law schools are ranked in terms of quality, students with first initials in their names of either C or D are preferentially located in inferior schools.

For academic performance, one could argue that teachers unconsciously downgrade students with the initials C and D, but direct experiments prove that self-initiated failure works just fine. When given the choice—after trying to solve ten difficult anagrams (of which two are impossible)—people will choose to push a button associated with failure (and a lower possible prize) if it matches their own initials, but they will not show an upward bias. Once again, self-love is associated with failure but not success. Is it possible that some among us tend not to respond to such arbitrary biases and thus succeed more often while seeing life more objectively?

How do these implicit self-biases come about? There is some evidence that early parenting style, both as remembered by individuals and, separately, by their mothers, is associated with the degree of name-letter bias and (in some cases) birth-date bias according to the following rules: warm and positive parenting produces a stronger positive self bias, while being controlling or overprotective has the opposite effect. The variables had similar effects on explicit self-esteem, as measured by asking people to rate themselves on a series of traits, such as "I feel that I have a number of good qualities" (1 to 7—completely true to completely untrue), but the implicit effect is still significant when corrected for explicit self-esteem. Recent work even suggests that daily events can affect one's name-letter bias, but only among those with low explicit self-esteem; a greater number of negative events in the previous twenty-four hours lowers implicit self-esteem, that is, preference for one's own name letters.

DECEIVING DOWN AND DUMMYING UP

As we have seen, we usually think of deception where self-image is concerned as involving inflation of self—you are bigger, brighter, better-looking than you really are. But there is a second kind of deception—deceiving down—in which the organism is selected to make itself appear smaller, stupider, and perhaps even uglier, thereby gaining an advantage. In herring gulls and various other seabirds, offspring actively diminish their apparent size and degree of aggressiveness as fledglings, to be permitted to remain near their parents, thereby consuming more parental investment. In many species of fish, frogs, and insects (see Chapter 2), males diminish apparent size, color, and aggressiveness to resemble females and steal paternity of eggs. These findings indicate that deceiving down has often been a viable strategy in other species, and thus is likely to be one in humans as well, which should lead to self-deceptive self-diminishment.

For example, appearing less threatening may permit you to approach more closely. This is a minority strategy that probably owes some of its success to the fact that most people are doing the opposite, so our guard is not as well developed in this direction. I remember students whose approach was so low-key, so noninvasive, you would never imagine that they would end up consuming far more of your time (to less effect) than many of their more talented counterparts who were representing themselves honestly or with an upward bias. Whether they were *self*-deceiving downward is, of course, difficult to say.

The most memorable version of deceiving down that I know of is referred to in African-American culture as "dummying up." This can refer to a specific situation in which you pretend not to know anything—for example, complete failure to witness a crime at which you were present or complete ignorance of a hidden relationship. But it can also refer to a general style. You can represent yourself as being less intelligent or less conscious than you really are, often the better to minimize the work you have to do. Thus an employee may dummy up to avoid doing more difficult tasks. I have often watched Spanish-speaking people in Panama and sometimes in the United States represent themselves as understanding much less English than in fact they do, all to gain benefits from English-speaking Americans who readily believe the dummying up—another example of being victimized by one's own prejudices.

I once asked Huey Newton how he dealt with dummying up directed at him, a problem he must have faced often as head of a major organization (the Black Panther Party). In reply, he imagined a situation in which a waiter always managed to avoid seeing you when you were calling him and otherwise appeared to be working while not actually doing anything. Here is how Huey would dress him down: "Oh, so you are so dumb that you happen to be looking the other way whenever I am trying to get your attention? And you are so dumb that when you know I am watching you, you decide to polish silverware that needs no polishing? And you are so dumb that you are always walking toward the pantry without ever reaching it? Well, you're not *that* damn dumb!"—followed by verbal or physical assault. Perhaps the ultimate in dummying up is that alleged of chimpanzees by several African peoples living near them—that the chimps can easily understand human speech but pretend not to in order to avoid being put to work!

FACE-ISM

It has been argued that visual depictions of the face that show more of the face relative to the rest of the body—that is, the face appears closer to you and is higher in "face-ism" —will give the impression of higher dominance, and people do indeed rate such faces as being more dominant. The word "face," after all, can be used to imply confrontation, as in "face-off," "face-to-face," "in your face," "loss of face," and so on. In short, the more I project my face on you, the more dominant I appear.

Consistent with this, the faces of a discriminated-against minority in the United States, African Americans, show lower face-ism than do those of European Americans in a variety of American and European periodicals, American portrait paintings, and US stamps. The difference shows up even when relative status is controlled for. Only when the artist is an African American is there an exception—there is no ethnic difference, with all face-ism ratings being on the high side. The degree of consciousness of the artists about these effects is, of course, unknown, but I would guess that many of the presenters of stimuli are unconscious of the effect, as are almost all of the recipients.

Similar findings have emerged for the two sexes in a wide range of US periodicals (such as *Time* and *Ms.*), in 3,500 media photos from eleven countries (including Kenya, Mexico, India, and France), in portraits and self-portraits dating back to the fifteenth century, and in amateur drawings of the faces of the two sexes. In all of these samples, men score higher in face-ism than do women. That is, relatively more of their face is presented in the picture—especially surprising since women have slightly larger heads for a given body size. On the other hand, women have breasts, and this may lead to a bias toward showing less head and more body. In any case, the correlation is true for every single country studied and every century from the seventeenth onward. The general face-ism effect appears to be all but universal, showing up in children's books, Fortune 500 websites, and prime-time television, among other places. *Ms.* magazine (feminist) is only slightly less biased in the usual direction than the rest of US publications.

There are some weak associations between higher face-ism and higher perceived intelligence, but no evidence that this affects the between-sex or ethnic comparisons, with one small exception. In photos from a variety of US periodicals, men shown in relatively intellectual professions had higher face-ism scores than similar women, and the effect was reversed for more physical professions.

Even politicians' self-presentations—that is, the photos they choose to post on their websites—show the usual bias, at least in the United States, Canada, Australia, and Norway. The bias remains the same whether twice as many women per men are serving in the legislature or one-tenth as many (compare Norway and the United States). Once again, though, in the United States, African-American politicians are an exception, showing the highest face-ism index for any ethnic group. Again, this suggests awareness among them that higher face-ism equals higher perceived dominance (and perhaps intelligence). Among female politicians in the United States, the more a woman's votes are interpreted as "pro-women," the more she emphasizes her face in photos of herself.

The degree to which people are conscious of face-ism is unknown, and so is its mechanism. Does a white photo selector see a black face and say "subordinate," then search for a relatively low face-ism picture? Or does he or she find black faces somewhat aversive, and so prefer them when they are smaller? And do black people viewing the photos find black pictures attractive and therefore easily tolerated up close, or are they saying "equally dominant" or "I wish myself and people like me to appear equally dominant"?

There is a curious result concerning George W. Bush's head. Someone thought to analyze his face-ism index in cartoons rendered 78 days before and 134 days after the start of each of his two wars. The authors of this study predicted that, dominant leader that he was, his face-ism index would increase with the outbreak of war. In fact, it decreased in both cases. Because in every major recent US war the president has made sure to appear as if he were forced into it, after every concession and reasonable effort, the authors argued that this lowered his apparent dominance. Or perhaps cartoonists knew something the rest of us did not about how each war would turn out. More likely still, the cartoonists were unconsciously reflecting the bias toward inflating one's own country (and leaders) prior to war, so as to impress adversaries, but not continuing once war was under way.

SPAM AGAINST ANTI-SPAM

There is an analogy between the coevolutionary struggle in nature and struggles in human life over deception in which (over a period of months or years) each move by a deceiver is matched by a countermove from the deceived and vice versa. The advantage lies with the deceiver, who usually has the first move. This is true even of situations in which the very best minds are enlisted in fighting the deception. Consider the ubiquitous invasive "species" of spam, unwanted computer messages. They offer a variety of services to induce a transfer of funds, however small, directly or from third parties. In some cases, companies will send out spam to lure the unsuspecting viewer to their websites, whose visits garner them more pay from the advertising company employing them. When spam first became a problem, computer software engineers leaped in on the side of prevention and protection, devising means of spotting incoming spam and blocking it. This led Bill Gates, in a burst of enthusiasm in 2004, to proclaim that the problem of junk e-mail "will be solved by 2006." Gates saw that defenses could easily be erected against the set of spamming devices then in use, but he could not imagine that these defenses could quickly be bypassed at little cost and that newer forms of spamming would easily be invented. By 2006, the amount of spam was higher than ever, having doubled in the previous year alone. Spam, of course, is a human invention for human purposes, with the computer and the Internet serving as the tools of replication.

After an initially successful counterattack by the anti-spam forces that resulted in a decrease in spam, the protective measures introduced could all be circumvented so that by the end of 2006, roughly nine out of every ten e-mail messages were junk. The initial attack against spam blended three filtering strategies. Software scanned each incoming message and looked at where the message was from, what words it contained, and which website it was connected to. The first was bypassed in spectacular fashion by devising programs that infected other computers with viruses that sent out the spam instead. In late 2006, an estimated quarter-million computers were unknowingly conscripted to send out spam every day. This achieved two aims at once: no sender's address that could be screened, and no additional cost to send.

The second screening device searched statistically for word usages suggestive of spam, but this maneuver was overcome by embedding the words in pictures whose extra expense was offset by the first device, the use of pirated computers. Efforts to spot and analyze images were, in turn, offset by "speckling" the images with polka dots and background bouquets of color that interfered with the computer scanners. To block detection of multiple copies of the same message, programs were written that automatically changed a few pixels in each picture. It was as if an individual could change successive fingerprints by minute amounts to evade detection, reminiscent of the ability of octopuses (see Chapter 2) to rapidly spin out a random series of cryptic patterns, again to avoid targeting. The HIV virus uses the same trick, mutating its coat proteins at a high rate to prevent the immune system from concentrating on it. As for the problem of linked sites, some scams do not require any. Spam can hype so-called penny stocks (inexpensive stocks in obscure companies) that may give a quick 5 percent profit in a matter of days, when enough people invest to raise the value, after which the spammer sells his or her interest in the stock and it collapses.

The point is that each move is matched by a countermove and a new move is always possible, so deceiver leads and deceived responds with costs potentially mounting by the year on both sides with no net gain. Intellectual powers among programmers increasingly will be required on both sides. One inevitable cost in this context is the destruction of true information by spam detectors that are too stringent, thus excluding some true information. This, as we saw in Chapter 2, is a universal problem in animal discrimination. Greater powers of discrimination will inevitably increase so-called false negatives—rejecting something as false that is in fact true. So as we act to exclude more spam, we inevitably delete more true messages. And now there is something more dangerous, called malware—special infiltrating codes that download proprietary information and ship it to one's enemies. As with newly appearing natural parasites (such as living viruses), malware is increasing at a more rapid rate than defenses against it.

HUMOR, LAUGHTER, AND SELF-DECEPTION

One striking discovery is that humor and laughter appear to be positively associated with immune benefits. Humor in turn can be seen as anti-self-deception. Humor is often directed at drawing attention to the contradictions that deceit and self-deception may be hiding. These are seen as humorous. Reversals of fortune associated with showing off—usually entrained by self-deception—are often comical to onlookers. A staple of silent films is the man strutting down the street, dressed to the nines, showing off, with head held high—so that he does not see the banana peel underneath him, producing an almost perfect visual metaphor for self-deception. The organism is directing its behavior toward others, with an upward gaze that causes him to pay no attention to the surface on which he is actually walking. Result: cartwheel and complete loss of bodily control, of the strut, of the head held high, and of the well-presented clothes—the whole show destroyed by a single contradiction.

Those who are low in self-deception (as judged by a classic paper-and-pencil test) appreciate humor more (as measured by actual facial movements in response to comedic material) than do those high in self-deception. At the same time, those with greater implicit biases toward black people or toward traditional sex roles laugh more in response to racially and sexually charged humor than do those with less implicit biases. Is it possible that the greater internal contradiction in them is released by appropriate humor on the subject, resulting in greater laughter? Laughter is an ancient mammalian trait, found in rats as well as chimpanzees. Tickling a rat will produce laughter-like sounds, and the rats will seek out the pleasure of being tickled. Chimpanzees will pant-laugh when being chased, an action that signals that the chase is not aggressive or aversive.

Humor permits discussion of taboo topics and the views of disempowered groups. Also, people know self-deception is negative and costly but necessary, so humor permits us to bring out this truth for enjoyment and consumption—we are all self-deceivers. Humor permits a kind of societal-level criticism in which no one need be threatened—it is all just a joke.

DRUGS AND SELF-DECEPTION

Recreational drugs and self-deception are obviously intimately connected. For one thing, drug use is often, to varying degrees at least, harmful and addiction almost invariably so. I am speaking of a wide range of both legal and illegal chemicals with effects from mild to severe: marijuana, alcohol, tobacco, uppers, downers, cocaine, heroin, and so on. Hence, this cost must be rationalized to the mind and, through the mind, to others. Thus, self-deception is a virtual requirement of drug use. I remember the first time I tried cocaine, I said to myself, "Why, this drug will pay for itself! I am so much more clear-headed and will get so much more work done while using it." Of course, in reality the drug was very expensive and entirely counterproductive where work was concerned. Huey Newton and I used to joke that we could practice drug abuse without self-deception, thus reducing or wiping out the cost, but it was a lie. Even the pleasant joke served to minimize the problem.

A second effect of drug use is often to separate our daily life into an up phase while using the drug and a down phase while recovering from it. This tends to split our personalities into two parts that then may be in conflict. The hungover self may remonstrate with the drunken self of the night before (and more generally), but the drunken state will usually forget all of this as soon as its time comes. It is tempting to imagine that the hungover self is more conscious of the two selves than is the drunken self. The latter is into enjoyment and would wish to suppress information from the other self that might cut into the pleasure. But in the hungover state, you are very aware of what went on the night before. Perhaps when you are drunk, your hungover self watches with dismay and attempts to call out—and sometimes (thank God) some information gets through.

My reason for imagining that the hungover self is the more conscious of the two rides partly on an analogy to split personalities. Many years ago, it was shown that among those rare people with two personalities, the second personality usually emerged in early adulthood and may have been strikingly different from the first. The first could be a shy and retiring British gentleman, the second a flamboyant Spanish fellow with a taste for flamenco. Typically the first personality knew nothing about the second, while the second had been watching the first for many years. Thus, therapy to unite such an individual into a single personality usually focuses on the second personality as the primary one. By analogy, then, the drunken self is like the first personality: it does not know that there is a second personality watching it.

A third factor of some importance is that the cost of drug use/abuse is often experienced as physiological pain, which you are then tempted to add to the pain of a given social interaction and to project it onto those around you. So the pain of arguments is that much greater, but, denying your own responsibility for that portion of the pain due to your drug use, you project your full anger onto the other person. Abusive drunks—surely we have all met one or two by now, if not in the mirror—fit the mold. So drug addicts tend to be irritable and morally righteous about it at the same time.

Finally, let us not forget that decisions made while high—while feeling an unnatural affinity for those close by, while feeling especially good about the future—are expected often to be biased away from one's true interests, just as the drug boosts us from our natural states. It would be nice to know the answer to the question: Are relatively more self-deceived individuals relatively more likely to be drug addicts? One expects the answer to be yes, but I do not know of any evidence. Certainly it is commonly claimed that con artists and thieves end up ensnared by a hard drug—and I have seen several such cases myself—but for the rest of us people, semi-addicted to milder stuff, I do not know.

Another problem that baffles me is whence the anti-pleasure bias? It is often said by opponents of medical marijuana that we already have legal drugs that promote appetite or suppress pain, so why should we give in to illegal ones? Yet the latter also give pleasure, so that you survive with good appetite and feeling better, so why is the latter not a virtue but an impediment? In fact, I now believe the ideal medicine for a root canal is, in fact, cocaine, and not its chemical analogs (procaine) that numb the pain but don't make you feel good.

VULNERABILITY TO MANIPULATION BY OTHERS

Socially, a potential cost of self-deception is greater manipulation (and deception) by others. If you are unconscious of your actions and others are conscious, they may manipulate your behavior without your being aware of it. Consider the story of a man who insisted, "You can't make a town man drunk." This occurred in rural Jamaica some thirty-five years ago, when a man from Kingston ("town") was passing through and bragging at a bar. Of course, we locals resisted his view and for a while there was a spirited argument. Then one local had a bright idea: he switched sides. He agreed with the town man—you can't make a town man drunk—and bought him a drink. Soon we all caught on, switched sides, and bought the man a drink. The town man was now in drunkard's paradise: everyone agreed with his opinions and everyone was buying him drinks. He got drunker and drunker, finally swaying on his chair, then falling to the ground, then vomiting, then slipping and falling in his own vomit. I say this not with pride but to describe the truth: we doubled over with laughter—as he sunk each step lower, we howled the more in pleasure. As Huey Newton was fond of saying, we owned him. We could have robbed him, killed him—he no longer had any control over his destiny. This is a terrible danger in self-deception—not that he was truly deluded into thinking it was impossible to make a Kingstonian drunk but that he had entered into fantasy land, selling one and then believing the fantasy had been bought by others. He was completely unaware of what was going on and he could have died from this as certain as from a heart attack.

This must be a very general and important cost of self-deception. You are trying to deceive others socially by being unconscious of a critical part of social reality. What if others are conscious of that very part while you are not? Your entire environment may be oriented against you, all with superior knowledge, while you peer out, ignorant and hobbled by self-deception. In the town man's case, it was his sense of superiority that served as a resource mined by those surrounding him.

PROFESSIONAL CON ARTISTS

Bless Bernie Madoff. He has brought con artists back to public attention and given them the attention they deserve, almost as much as when Ponzi swindled thousands of people out of hundreds of thousands of dollars in a pyramid scheme—where early investors are paid high returns, not out of actual earnings but out of the donations of others joining the scheme. As word of mouth spreads about the high returns, more and more want to join the fun. By definition, such an operation can't continue indefinitely. Typically those who invest early and depart early earn a nice return, as does the swindler himself, though he also may suffer later prison time. Everyone else loses—most people, everything they invested. Madoff stole a staggering $50 billion. He was a classic swindler; smooth and attractive in style, he made you pursue him. Many times he told people "the books are closed" on investment with him, only later to relent and permit them to lose their money with him. As always, some people did not buy in and a few spotted the scheme for what it was. This is what we have expected all along: an evolutionary game, with multiple actors, caught in a frequency-dependent interaction such that most actors will not be forced out of the game anytime soon, and new strategies are always appearing. Incidentally, one of Madoff's victims had just published a book on gullibility when he learned that it applied to himself: he lost $400,000. In self-defense, he said he was only trying to buy a safe investment with modest returns (more than 10 percent annually) for his family. Modest? What positive feature in the universe increases by more than 10 percent annually, year after year?

Most con artists operate on a much smaller scale. They are professional thieves whose art consists of extracting money voluntarily from others, as did Madoff, just on a much smaller scale. They often survive on the unconsciousness, including self-deception, of their victims, as did Madoff. Here it is useful to distinguish between the "long con" and the "short con." The long con may run for several days, may result in tens of thousands of dollars lost at the end, and often involves activating the victim's system of self-deception, while the short con is usually over in a matter of minutes for a few dollars and typically involves lulling the victim into temporary unconsciousness regarding a key variable. During long cons, the victim is often put into a trance-like state of mind, as one of his or her weaknesses, often greed, is amplified by the con artist. Because the same illegal or "special situation" can, in principle, be repeated indefinitely, there is no upward limit to the victim's fantasies, an easily exploitable resource to help overcome contradictions should they arise. Victims in this state are said to "glow" and to be easily spotted by other con artists. Getting the victim into that state is called "putting him under the ether"—presumably into a deep state of self-deception.

As it looks to the victim: "You're experiencing the ride singing 'yo ho ho it's a pirate's life for me' but you never see any of the trappings of the ride itself." The con artist induces an internal ride in the victim that is very satisfying but is hard to view sideways so as to see where, in fact, the ride is taking you. Once we have taken the bait, we stop asking questions, much as people do in the instrumental phase of any activity, that is, when they are carrying out a project. In the memorable phrase of a great con artist of the street, "I plucked his dreams right out of his head and then sold them back to him—and at a good price, too!"

Incidentally, con artists demonstrate again the importance of frequency-dependent effects. At low frequency they do well, at high frequency not so well. A shopkeeper may be fooled once by a short-change game but usually not twice. The con artist must always be on the move to fresh victims. Here the density-dependent effect occurs directly through learning (and also passing this information on to others), while in other systems it is genetic and may require several generations of selection to show an effect.

A medium-length con (about two hours and netting $40) was run against me years ago in Jamaica. I was leaving Kingston one Saturday morning when a short, wiry man hitched a ride. When I asked him where he was going, he said Caymanas Racecourse, the local horse-racing track. He was a jockey—in fact running in the day's third race, as he proved to me, pointing to his name on the racing form, a name he had introduced at the very beginning of our relationship. He had recently lost his car in an accident, which had also left him broke. After further discussion, it was proposed that I invest in a gambling scheme—betting, as is perfectly legal, on the day's races, based on his insider knowledge. I remember my thought processes well. As a seasoned virtual Jamaican, I knew that the races were entirely fixed ahead of time, the general public betting not on horses but on how the race would be thrown. The very fact that this man was proposing such a financially advantageous scheme to me (I provide the cash for betting based on his special knowledge, proceeds to be split evenly) was a testament to my fluency in Jamaican culture—my general likability, if you will, augmented by my cultural competence. Why else had we hit it off so quickly? And it was a scheme that was foolproof as far as his stealing from me was concerned: we would buy matching sets of tickets. Our payoffs were yoked. And now that I had made the key breakthrough, it could be repeated ad libitum, $2,000 won this time, $20,000 the next, and so on.

I do remember one feature of his style that was off-putting: he called me "boss" more than once. This is something I have never liked but in this situation it jarred with my self-image as a fellow Jamaican: someone able to get this opportunity in part because I was not a boss. At one point I asked him not to call me "boss," as if to say, "please, don't interfere with my fantasy."

We bought $80 worth of matching bets, many coupled with each other, so that should multiple horses come in, the winnings would be very large, but if a single horse failed, we would win nothing. *No problem for me*, I thought. *This is as near to a sure thing as I have seen in my lifetime. Let's maximize gains!* The first horse did come in, as my friend crouched down on the imaginary winner and whipped it home—in a bar where we were now drinking. Didn't he have to run in the third race? Again, this caused some small internal unease because of the obvious contradiction—not only did he risk being late for his own race, but he also risked arriving drunk—but I was willing to suppress the truth to maintain the fantasy. I dropped him at the track and continued on my way. Within four races, all of my bets were busted. Rounding a corner too quickly, now half drunk, I struck a rock and had to change a tire. Outside in the broiling-hot Jamaican sun, the truth had plenty of time to sink in. The man knew nothing about the track, was certainly not a jockey, and could no more predict the future than I but he was only too happy to have a series of risky bets bought for him by a complete stranger who, as an additional bonus, would deliver him to the track.

The whole experience now seems to be a metaphor for self-deception itself: the smooth and seamless takeoff, the intoxicating heights, the occasional doubts easily brushed aside, followed by reality itself and an appreciation of the growing costs: no longer just the monetary losses but also an inability to deal with moment-to-moment reality. The upside is temporary and psychological, while the downside is real and enduring.

LIE-DETECTOR TESTS

Given the importance of perceiving deception, for example, in spotting an intended "terrorist," there is a great demand for anyone who can scientifically uncover a lie—hence, the vaunted lie-detector test and a series of new ones, accessing deeper regions of our brains. The classical test measures three variables: heart rate, breathing amplitude, and galvanic skin response (GSR), a measure of physiological arousal. A series of innocuous questions are interspersed with incriminating ones, and systematic deviations in the underlying three measures are recorded. Especially significant, it is argued, are contrasts between key lies ("did you kill Betty Sue?"), to which only the perpetrator is guilty, and much more minor infractions, to which most people are probably guilty ("did you ever steal from your office?"). The guilty are presumed to respond more to the main question and the guiltless to the harmless lie. But these hard and fast rules rarely work so well in real life, and some people appear nearly completely unresponsive to variation in these questions.

The only question that gives truly reliable results is called the "guilty knowledge test." Among otherwise innocuous questions is one interspersed that refers to a fact that only the criminal could know—the victim was lying on a *red* satin sheet before she met her demise. *Any* deviation from the background responses is evidence of deception—high arousal, low arousal, anything different from the responses to questions about which the person is ignorant.

I once inadvertently experienced the benefits of the guilty knowledge test when I was trying to counsel a youngster (thirteen years old) about his unfortunate tendency to steal his neighbors' bicycles, an escalation of his previous petty larceny. I told him, "Don't steal; don't steal your neighbors' tools; don't steal your neighbors' toys." At first his eyes showed alarm as I talked about stealing, but as I ran down my boring list, he visibly relaxed and looked me in the eye. Then I added "and don't steal your neighbors' bicycle." Suddenly his eyes darted up, down, and around, until I continued droning through my list and he relaxed again. Guilty knowledge.

There is now a raft of new lie-detector tests coming out of neurophysiology and heavily funded by "antiterror" money coursing through the US government. Each test tends to claim high success, but this is usually based on modeling neurophysiological data after the fact against known honest and deceptive responses in a study population to gain the tightest fit. The tightness of the fit is then highlighted, but this is an illusion. The key is whether your method applied to a fresh set of subjects gives any fit at all, much less the high one claimed.

Another weakness of this line of work is the tendency to believe that lying per se gives off cues—not a particular kind of lie in a particular kind of situation. Contrast two kinds of lies. A little recorded lie you have waiting and ready for an expected question—where have you been the past two hours? This lie should light up memory areas of the brain, among others. By contrast, a simple denial, in which you suppress the truth and assert a falsehood, should light up areas involved in cognitive control. And so on. But at this time we are nowhere near devising a neurologically valid lie-detector test.

CHAPTER 9

Self-Deception in Aviation and Space Disasters

Disasters are always studied in retrospect. We will not have an experimental science of the subject anytime soon. Disasters range from the personal—your wife tells you she is leaving you for the mailman—to the global—your country invades the wrong nation, with catastrophic effects all around. Disasters, of course, are expected to be closely linked to self-deception. There is nothing like being unconscious of reality to make it intrude upon you in unexpected and painful ways. In this chapter we will concentrate on one kind of disaster—airplane and space crashes—because they typically are subject to intensive investigation immediately afterward to figure out the causes and avoid repetition. For our purposes, these accidents help us study the cost of self-deception in-depth under highly controlled circumstances. The disasters produce a very detailed and well-analyzed body of information on their causes, and they form a well-defined category. As we shall see, there are repeated ties to self-deception at various levels: the individual, pairs of individuals (pilot and copilot), institutions (NASA), and even countries (Egypt).

But there is one striking difference between space and aviation disasters. In the United States, aviation disasters are immediately and intensively studied by teams of experts on twenty-four-hour notice in an institution designed to be insulated from outside interference, the National Transportation Safety Board. The NTSB generally does a superb job and publicizes its findings quickly. It almost always discerns key causes and then makes appropriate recommendations, which appear to have helped reduce the accident rate steadily for some thirty years, so that flying is, by far, the safest form of travel. I know of only one case of a delayed report (about three years) and this was because of interference on the international level, when Egypt fought the truth to the bitter end.

By contrast, NASA's accidents are investigated by a committee appointed to study only a specific disaster, with no particular expertise, and sometimes with a preordained and expressed goal to exonerate NASA. Study of one disaster does not prevent another, even when it has many of the same causes identified in the first case. Of course, safety corners can more easily be cut when only the lives of a few astronauts are at stake, instead of the great flying public, including airline personnel.

Aviation disasters usually result from multiple causes, one of which may be self-deception on the part of one or more key actors. When the actors number more than one, we can also study processes of group self-deception. A relatively simple example of this is the crash of Air Florida Flight 90 in 1982, in which both pilot and copilot appear to have unconsciously "conspired" to produce the disaster.

AIR FLORIDA FLIGHT 90—DOOMED BY SELF-DECEPTION?

On the afternoon of January 13, 1982, Air Florida Flight 90 took off from Washington, D.C.'s National Airport in a blinding snowstorm on its way to Tampa, Florida. It never made it out of D.C., instead slamming into a bridge and landing in the Potomac River—seventy-four people died, and five survivors were fished out of the back of the plane. Perhaps because one of those who died was an old friend of mine from Harvard (Robert Silberglied), I was listening with unusual interest when soon thereafter the evening news played the audiotape of the cockpit conversation during takeoff. The copilot was flying the plane, and you could hear the fear in his voice as he also performed the role the pilot should have been playing, namely reading the instrument panel. Here is how it went:

Ten seconds after starting down the runway, the copilot responds to instrument readings that suggest the plane is traveling faster than it really is: "God, look at that thing!" Four seconds later: "That doesn't seem right, does it?" Three seconds later: "Ah, that's not right." Two seconds later: "Well . . ."

Then the pilot, in a confident voice, offers a rationalization for the false reading: "Yes, it is, there's 80," apparently referring to an airspeed of 80 knots. This fails to satisfy the copilot, who says, "Naw, I don't think that's right." Nine seconds later, he wavers: "Ah, maybe it is." That is the last we hear from the copilot until a second before the crash when he says, "Larry, we're going down, Larry," and Larry says, "I know."

And what was Larry doing all this time? Except for the rationalization mentioned above, he only started talking once the mistake had been made and the plane was past the point of no return—indeed when the device warning of a stall started to sound. He then appeared to be talking to the plane ("Forward, forward." Three seconds later: "We only want five hundred." Two seconds later: "Come on, forward." Three seconds: "Forward." Two seconds: "Just barely climb."). Within three more seconds, they were both dead.

What is striking here is that moments before we have a human disaster that will claim seventy-four human lives, including both primary actors, we have an apparent pattern of reality evasion on the part of one key actor (the pilot) and insufficient resistance on the part of the other. On top of this, typical roles were reversed, each playing the other's: pilot (ostensibly) as copilot and vice versa. Why was the copilot reading the contradictory panel readings while the pilot was only offering a rationalization? Why did the copilot speak while it mattered, but the pilot started talking only when it was too late?

The first thing to find out is whether these differences are specific to the final moments or we can find evidence of similar behavior in the past. The answer is clear. In the final forty-five minutes of discussion between the two prior to takeoff, a clear dichotomy emerges. The copilot is reality-oriented; the pilot is not. Consider their discussion of snow on the wings, a critical variable. Pilot: "I got a little on mine." Copilot: "This one's got about a quarter to half inch on it all the way." There were equal amounts of snow on both wings but the pilot gave an imprecise and diminutive estimate, while the copilot gave an exact description.

And here is perhaps the most important exchange of all, one that occurred seven minutes before takeoff. Copilot: "Boy, this is a losing battle here on trying to de-ice those things. It gives you a false sense of security is all that it does" (!!). Pilot: "This, ah, satisfies the Feds." Copilot: "Yeah—as good and crisp as the air is and no heavier than we are, I'd . . ." Here is the critical moment in which the copilot timidly advanced his takeoff strategy, which presumably was to floor it—exactly the right strategy—but the pilot cut him off midsentence and said, "Right there is where the icing truck, they oughta have two of them, pull right." The pilot and copilot then explored a fantasy together on how the plane should be deiced just before takeoff.

Note that the copilot began with a true statement—they had a false sense of security based on a de-icing that did not work. The pilot noted that this satisfies the higher-ups but then switched the discussion to the way the system *should* work. Though not without its long-term value, this rather distracts from the problem at hand—and at exactly the moment when the copilot suggests his countermove. But he tried again. Copilot: "Slushy runway, do you want me to do anything special for this or just go for it?" Pilot: "Unless you got something special you would like to do." No help at all.

The transcript suggests how easily the disaster could have been averted. Imagine the earlier conversation about snow on the wings and slushy conditions underfoot had induced a spirit of caution in both parties. How easy it would have been for the pilot to say that they should go all-out but be prepared to abort if they felt their speed was insufficient.

A famous geologist once surveyed this story and commented: "You correctly blame the pilot for the crash, but maybe you do not bring out clearly enough that it was the complete insensitivity to the copilot's doubts, and to his veiled and timid pleas for help, that was the root of all this trouble. The pilot, with much more experience, just sat there completely unaware and without any realization that the copilot was desperately asking for friendly advice and professional help. Even if he (the pilot) had gruffly grunted, 'If you can't handle it, turn it over to me,' such a response would have probably shot enough adrenaline into the copilot so that he either would have flown the mission successfully or aborted it without incident." It is this dreadful, veiled indecision that seems to seal the disaster: the copilot tentative, uncertain, questioning, as indeed he should be, yet trying to hide it, and ending up dead in the Potomac.

The geologist went on to say that in his limited experience in mountain rescue work and in abandoned mines, the people who lead others into trouble are the hale and hearty, insensitive jocks trying to show off. "They cannot perceive that a companion is so terrified he is about to 'freeze' to the side of the cliff—and for very good reasons!" They in turn freeze and are often the most difficult to rescue. In the case of Flight 90, it was not just the wings that froze, but the copilot as well, and then so did the pilot, who ended up talking to the airplane.

Earlier decisions infused with similar effects contributed to the disaster. The pilot authorized "reverse thrust" to power the airplane out of its departure place. It was ineffective in this role but apparently pushed the ice and snow to the forward edge of the wing, where they would do the most damage, and at the same time blocked a key filter that would now register a higher ground speed than was in fact obtained. The pilot has been separately described as overconfident and inattentive to safety details. The presumed benefit in daily life of his style is the appearance of greater self-confidence and the success that this sometimes brings, especially in interactions with others.

It is interesting that the pilot/copilot configuration in Flight 90 (copilot at the helm) is actually the safer of the two. Even though on average the pilot is flying about half the time, more than 80 percent of all accidents occur when he is doing so (in the United States, 1978–1990). Likewise, many more accidents occur when the pilot and copilot are flying for the first time together (45 percent of all accidents, while safe flights have this degree of unfamiliarity only 5 percent of the time). The notion is that the copilot is even less likely to challenge mistakes of the pilot than vice versa, and especially if the two are unfamiliar with each other. In our case, the pilot is completely unconscious, so he is not challenging anyone. The copilot is actually challenging himself but, getting no encouragement from the pilot, he lapses back into ineptitude.

Consider now an interesting case from a different culture. Fatal accident rates for Korea Airlines between 1988 and 1998 were about seventeen times higher than for a typical US carrier, so high that Delta and Air France suspended their flying partnership with Korea Air, the US Army forbade its troops from flying with the airline, and Canada considered denying it landing rights. An outside group of consultants was brought in to evaluate the problem and concluded, among other factors, that Korea, a society relatively high in hierarchy and power dominance, was not preparing its copilots to act assertively enough. Several accidents could have been averted if the relatively conscious copilot had felt able to communicate effectively with the pilot to correct his errors. The culture in the cockpit was perhaps symbolized when a pilot backhanded a copilot across the face for a minor error, a climate that does not readily invite copilots to take strong stands against pilot mistakes. The consultants argued for emphasizing copilot independence and assertion. Even the insistence on better mastery of English—itself critical to communicating with ground control—improved equality in the cockpit since English lacked in-built hierarchical biases to which Koreans responded readily when speaking Korean. In any case, since intervention, Korea Air has had a spotless safety record. The key point is that hierarchy may impede information flow—two are in the cockpit, but with sufficient dominance, it is actually only one.

A similar problem was uncovered in hospitals where patients contract new infections during surgery, many of which turn out to be fatal and could be prevented by simply insisting that the surgeon wash his (or occasionally, her) hands. A steep hierarchy—with the surgeon unchallenged at the top and the nurses carrying out orders at the bottom—was found to be the key factor. The surgeon practiced self-deception, denied the danger of not washing his hands, and used his seniority to silence any voices raised in protest. The solution was very simple. Empower nurses to halt an operation if the surgeon had not washed his hands properly (until then, 65 percent had failed to do so). Rates of death from newly contracted infections have plummeted wherever this has been introduced.

DISASTER 37,000 FEET ABOVE THE AMAZON

Another striking case of pilot error occurred high above the Amazon in Brazil at 5:01 p.m. on September 26, 2006. A small private jet flying at the wrong altitude clipped a Boeing 737 (Gol Flight 1907) from underneath, sending it into a horrifying forty-two-second nosedive to the jungle below, killing all 154 people aboard. The small American executive jet, though damaged, landed safely at a nearby airport with its nine people alive. Again, the pilot of the small jet seemed less conscious than his copilot when the disaster was upon them, but neither was paying attention when the fatal error was made, nor for a long time afterward.

The key facts are not in doubt. The large commercial jet was doing everything it was supposed to do. It was flying at the correct altitude and orientation (on autopilot); its Brazilian pilots were awake, alert, and in regular contact with their flight controllers. In addition, they were fully familiar with the plane they were flying and spoke the local language. The only mistake these pilots made was getting out of bed that morning. By contrast, the American crew was flying a plane of this kind for the first time. They were using the flight itself to master flying the craft by trial and error as they went along. Although they had had limited simulation training on this kind of airplane, they did not know how to read the instrument panel and, as they put it while in flight, were "still working out the kinks" on handling the flight management system. When attempting to do so, they could not compute time until arrival or weather ahead, much less notice whether their transponder was turned off, as soon enough it was. They tried to master the airplane display systems, toyed with a new digital camera, and planned the next day's flight departure.

They chatted with passengers wandering in and out of their cockpit. They did everything but pay attention to the task at hand—flying safely through airspace occupied by other airplanes.

They were, in fact, flying at the wrong altitude, contradicting both normal convention (even numbers in their direction) and the flight plan they had submitted (36,000 feet for the Brasilia–Manaus leg of their trip). But their own error was compounded by that of the Brasilia controller who okayed their incorrect orientation. They had managed to turn off their transponder (or it had done so on its own), so they were flying invisible to other planes and were blind themselves—a transponder warns both oncoming craft of your presence and you of theirs—yet they were completely unaware of this. They were barely in contact with the flight controllers, and when they were, the pilots showed little evidence of language comprehension or of interest in verifying what they thought the controllers were saying ("I have no idea what the hell he said"). They had spoken disparagingly of Brazilians and of the tasks asked of them, such as landing at Manaus.

Their flight plan was simplicity itself. They were to take off from near Sao Paolo on a direct leg to Brasilia at 37,000 feet; then they were to turn northwest toward Manaus at 36,000, since planes flying in the opposite direction would be coming at 37,000 feet. They then were to land at Manaus. Automatic pilots would attend to everything, and there was only one key step in the whole procedure: go down 1,000 feet when they made their turn high over Brasilia. This is precisely what the flight plan they submitted said they would do, it was the universal rule for flights in that direction, and it was assumed to be true by the flight bearing down on them from Manaus.

It was not, however, what they did. Instead, as they made their turn, they were at that moment busying themselves with more distant matters—trying to calculate the landing distance at Manaus and their takeoff duties the next day. This was part of their larger absorption in trying to master a new plane and its technology. For the next twenty minutes, the mistake was not noticed by either the pilots or the Brazilian air controller who had okayed it, but by then the plane's transponder was turned off and there was no longer clear evidence to ground control of who and where they were. There is no evidence of deception, only of joking around as if jockeying for status while being completely oblivious to the real problem at hand. This is a recurring theme in self-deception and human disasters: overconfidence and its companion, unconsciousness. Incidentally, it was the copilot who seems first to have realized what may have happened, and he took over flight of the plane, later apologizing repeatedly to the pilot for this act of self-assertion. He was also the first to deny the cause of the accident on arrival and provide a cover-up.

In the example of Air Florida Flight 90, the pilot's self-deception—and copilot's insufficient strength in the face of it—cost them their lives. In the case of Gol Flight 1907, both pilots who caused the tragedy survived their gross carelessness while 154 innocents perished. This is a distressing feature of self-deception and large-scale disasters more generally: the perpetrators may not experience strong, nor indeed any, adverse selection. As we shall see, it was not mistakes by astronauts or their own self-deception that caused the *Challenger* and *Columbia* disasters but rather self-deception and mistakes by men and women with no direct survival consequences from their decisions. The same can be said for wars launched by those who will suffer no ill effects on their own immediate inclusive fitness (long-term may be another matter), whatever the outcome, even though their actions may unleash mortality a thousand times more intense in various unpredictable directions.

ELDAR TAKES COMMAND—AEROFLOT FLIGHT 593

It is hard to know how to classify the 1994 crash of Aeroflot Flight 593 from Moscow to Seoul, Korea, so absurd that its truth was covered up in Russia for months. The pilot was showing his children the cockpit and, against regulations, allowed each to sit in a seat and pretend to control the plane, which was actually on autopilot. His eleven-year-old daughter enjoyed the fantasy, but when his sixteen-year-old son, Eldar, took the controls, the teen promptly applied enough force to the steering wheel to deactivate most of the autopilot, allowing the plane to swerve at his whim.

Deactivation of the autopilot turned on a cockpit light (which was missed by the pilots), but more important, the pilot was trapped in a fantasy world in which he encouraged his children to turn the wheel this way and that and then to believe that this had an effect, while in fact the plane was (supposed to be) on autopilot. When his son actually controlled movements, the pilot was slow to realize this was no fantasy; indeed, his son was the first to point out that the plane was actually turning on its own (due to forces unleashed by Eldar's turning motions), but the plane then quickly banked at such an angle as to force everyone against their seats and the wall so that the pilot could not wrest control of the plane from his son. After a harrowing vertical ascent, the copilot and Eldar managed to get the plane in a nosedive, which permitted control to be reestablished, but alas it was too late. The plane hurtled to the ground, losing all seventy-five aboard. Besides disobeying all standard rules for cockpit behavior, the pilot appeared blissfully unaware that he was doing this high in the air and was becoming trapped in the very fantasy he had created for his children. Of course, it is easy for adults to underestimate the special ability of children to seize control of electromechanical devices.

SIMPLE PILOT ERROR—OR PILOT FATIGUE?

We now turn to self-deception at higher levels of organization—within corporations or society at large—that impede airline safety. That is, pilot error is compounded by higher-level error. For example, the major cause of fatal airline crashes is said to be pilot error—about 80 percent of all accidents in both 2004 and 2005. This is surely an overestimate, as airlines benefit from high ones. Still, evidence of pilot error is hardly lacking and is usually one of several factors in crashes. We do not know how much of this error is entrained by self-deception, but a common factor in pilot error is one we have already identified: overconfidence combined with unconsciousness of the danger at hand. Certainly this combination appears to have doomed John F. Kennedy Jr. (and his two companions) when he set out on a flight his experienced copilot was unwilling to take—into the gray, dangerous northeastern fog in which a pilot can easily become disoriented, mistake up for down, lose control, and enter a death spiral.

Consider a commercial example, documented by the flight recorder. On a cloudy day in October 2004 at 7:37 in the evening, a twin-engine turboprop approaching the airport at Kirksville, Missouri, was descending too low, too fast, though the pilots could not see the runway lights until they were below three hundred feet and soon were on top of trees. Both pilots and eleven of the thirteen passengers died in the crash. Below ten thousand feet, FAA rules require a so-called sterile cockpit, in which only pertinent communication is permitted, yet both pilots were sharing jokes and cursing frequently below this altitude. They discussed coworkers they did not like and how nice it would be to eat a Philly cheesesteak, but they did not attend to the usual rules regarding rate and timing of descent or to the plane's warning system alerting them to the rapidly approaching ground below.

Of course, the usual human bias toward self-enhancement makes this negligence more likely: "rules that apply to the average pilot do not apply to better ones, such as me." The pilot, whose job in this situation was to watch the instruments, said it was all right to descend because he could see the ground. The copilot—whose job was to look for the runway—said he could not see a thing, but he did not challenge the pilot, as rules required him to do. The pilot kept descending as if he could see the runway when he probably saw nothing at all until finally he spotted the landing lights and then immediately the tops of trees. Here we see familiar themes from the crash of Air Florida Flight 90: the irrelevant and distracting talk during takeoff in the first case and landing in this one, pilot overconfidence prevailing over the more reality-oriented but deferential copilot, and the pilot's failure to read instruments, as was his duty.

It should be mentioned that the pilots could be heard yawning during their descent, and they had spent fourteen hours on the job, after modest sleep. This was their sixth landing that day. Had they followed proper procedure, they still should have been able to land safely, but surely fatigue contributed to their failure to follow procedure, as well as to their degree of unconscious neglect of the risks they were taking.

Now here comes the intervention of self-deception at the next level. In response to this crash, the NTSB recommended that the FAA tighten its work rules for pilots by requiring more rest time, the second time it had done so in twelve years, because the FAA did not act on the first recommendation. In response to this crash, the airline industry, represented by its lobbying organization, the Air Transport Association, argued that this was an isolated incident that did not require change in FAA rules. (If accidents were not isolated incidents, we would not get on airplanes.) "The current FAA rules . . . ensure a safe environment for our crews and the flying public." Of course, they do no such thing: they save the airlines money by requiring fewer flight crews. And note the cute form of the wording "our crews" comes first—we would hardly subject our own people to something dangerous—followed by reducing everyone else to "the flying public." But neither management nor lobbyists are part of the flight crew, and predictably, the Airline Pilots Association backed the rule change. True to form, in March 2009, seven airlines sued in federal court to overturn a recent FAA rule that imposed forty-eight-hour rest periods between twenty-hour flights (e.g., Newark to Hong Kong), a decision that followed earlier pioneering work by Delta Airlines to institute the rule and to provide proper sleeping quarters for the pilots during their nearly daylong flight. The fiction is that the FAA represents the so-called flying public; the truth is that it represents the financial interests of the airlines and represents the general public only reluctantly and in response to repeated failures.

ICE OVERPOWERS THE PILOTS; AIRLINES OVERPOWER THE FAA

Ice poses a special problem for airplanes. Ice buildup on the wings increases the plane's weight while changing the pattern of airflow over both the main wings and the small rear control wings. This reduces lift and in some cases results in rapid loss of control, signaled by a sudden pitch and a sharp roll to one side. The controls move on their own, sometimes overpowering counterefforts by the pilots. Commuter planes are especially vulnerable because they commonly fly at lower altitudes, such as ten thousand feet, at which drizzling ice is more common. When icing results in loss of control, the plane turns over and heads straight to the ground.

To take an example, on October 31, 1994, American Eagle Flight 4184 from Indianapolis had been holding at ten thousand feet in a cold drizzle for thirty-two minutes with its de-icing boot raised (to break some of the ice above it), when it was cleared by Chicago air traffic controllers to descend to eight thousand feet in preparation for landing. Unknown to the pilots, a dangerous ridge of ice had built up on the wings, probably just behind the de-icing boot, so that as the pilots dipped down, they almost immediately lost control. The plane's controls moved on their own but on the right wing only, immediately tilting the plane almost perpendicular to the ground. The pilots managed to partly reverse the roll before the (top-heavy) plane flipped upside down and hit the ground at a 45-degree angle in a violent impact that left few recognizable pieces, including any of the sixty-eight people aboard.

This was an accident that did not need to happen. This kind of airplane (ATR 42 or 72 turboprops) had a long history of alarming behavior under icing conditions, including twenty near-fatal losses of control under icing conditions and one crash in the Alps in 1987 that killed thirty-seven people. Yet the problem kept recurring because safety recommendations were met by strong resistance from the airlines—which would have to pay for the necessary design changes—and the FAA ended up acting like a biased referee, approving relatively inexpensive patches that probably reduced (at least slightly) the chance of another crash but did not deal with the problem directly. As one expert put it, "Until the blood gets deep enough, there is a tendency to ignore a problem or live with it." To wait until after a crash to institute even modest safety improvements is known as tombstone technology. The regulators and airline executives are, in effect, conscious of the personal cost—immediate cost to the airlines in mandated repairs and bureaucratic cost to any regulator seen as unfriendly to the airlines—while being unconscious of the cost to passengers.

In the United States, the NTSB analyzes the causes of an airline disaster, relying on a series of objective data, cockpit and flight recorders, damage to aircraft, etc., to determine cause and then makes obvious recommendations. The theory is that this relatively modest investment in safety will pay for itself in future airplane design and pilot training to minimize accidents. In reality, everything works fine until the recommendation stage, when economic interests intervene to thwart the process. This is well demonstrated by the FAA's inability to respond appropriately to the problem of ice buildup on smaller, commuter airplanes, a problem well known for more than twenty years yet claiming a new set of lives about every eight years, most recently on February 13, 2009, in Buffalo, New York, leaving fifty dead.

A deeper problem within the FAA was its unwillingness to reconsider basic standards for flying under icing conditions, as indeed had been requested by the pilots' union. The FAA based its position on work done in the 1940s that had concluded that the chief problem was tiny droplets, not freezing rain (larger droplets), but science did not stop in the '40s, and there was now plenty of evidence that freezing rain could be a serious problem. But this is one of the most difficult changes to make: to change one's underlying system of analysis and logic. This could lead to wholesale redesign at considerable cost to—whom?—the airlines. So it was patchwork all the way around. There is also an analogy here to the individual. The deeper changes are the more threatening because they are more costly. They require more of our internal anatomy, behavior, and logic to be changed, which surely requires resources, may be experienced as painful, and comes at a cost.

The very symbol of a patch-up approach to safety is the fix the FAA approved for the well-proven habit of these planes in freezing ice to start to flip over. The fix was a credit-card-size piece of metal to be attached to each wing of a several-ton airplane (not counting passengers—or ice). This tiny piece of metal allegedly would alter airflow over the wings so as to give extra stability. No wonder the pilots' union (representing those at greatest risk) characterized this as a Band-Aid fix and pointed out (correctly) that the FAA had "not gone far enough in assuring that the aircrafts can be operated safely under all conditions." The union went on to say that the ATR airplanes had an "unorthodox, ill-conceived and inadequately designed" de-icing system. This was brushed aside by the FAA, a full six years before the Indiana crash, in which the airplane was fully outfitted with the FAA-approved credit-card-size stabilizers.

By the way, to outfit the entire US fleet of commuter turboprops with ice boots twice as large as before the Indiana crash would cost about $2 million. To appreciate how absurdly low this cost is, imagine simply dividing it by the number of paying customers on the ill-fated Indianapolis-to-Chicago trip and asking each customer in midair, "Would you be willing to spend $50,000 to outfit the entire American fleet of similar planes with the larger boot, or would you rather die within the next hour?" But this is not how the public-goods game works. The passengers on the Chicago flight do not know it is their flight out of 100,000 that will go down. Rather, the passengers know they have a 0.99999 chance of being perfectly safe even if they do nothing. Let someone else pay. Even so, I bet everyone would get busy figuring out how to raise the full amount. I certainly would. Of course, if each passenger only had to help install the boots on his or her own plane, about $300 per passenger would suffice. The point is that for trivial sums of money, the airlines routinely put passengers at risk. Of course, they can't put it this way, so they generate assertions and "evidence" by the bushel to argue that all is well, indeed that every reasonable safety precaution is being taken. Six years before this crash, British scientists measured airflow over icy wings and warned that it tended to put the craft at risk, but these findings were vehemently derided as being wholly unscientific, even though they were confirmed exactly by the NTSB analysis of the Indianapolis–Chicago crash.

Finally, a series of trivial devices were installed in the cockpit and new procedures were mandated for pilot behavior. For example, a device giving earlier warning of icing was installed and pilots were told not to fly with autopilot when this light is on, precisely to avoid being surprised by a sudden roll to one side as the autopilot disengages. But of course this does not address the problem of loss of control under icing. From the very first Italian crash over the Alps, when one of the pilots lashed out at the control system that failed to respond to his efforts with an ancient curse on the system's designers and their ancestors, it has been known that conscious effort to maintain control is not sufficient. And of course, pilots may make matters worse for themselves in a bad situation. In the Buffalo crash, the pilots apparently made a couple of errors, including keeping the plane on autopilot when they lowered their landing gear and deployed the flaps that increase lift. Suddenly there was a severe pitch and roll, suggestive of ice, which in fact had built up on both the wings and the windshield, blocking sight. Although the NTSB attributed the crash to pilot error, the fact that ice had built up, followed by the familiar pitch and roll, suggests a poorly designed airplane as well.

In short, a system has developed in which the pilot may make no errors—and yet the plane can still spin out of control. It is ironic, to say the least, that a basic design problem that deprives a pilot of control of the airplane is being solved by repeatedly refining the pilot's behavior in response to this fatal design flaw. A pilot's failure to do any of the required moves, for example, disengage autopilot, will then be cited as the cause. No problem with the airplane; it's the pilot! But is this not a general point regarding self-deception? In pursuing a path of denial and minimization, the FAA traps itself in a world in which each successive recommendation concerns more and more pilot behavior than actual aircraft design changes. Thus does self-deception lay the foundations for disaster.

Consider an international example.

THE US APPROACH TO SAFETY HELPS CAUSE 9/11

The tragedy of 9/11 had many fathers. But few have been as consistent in this role as the airlines themselves, at least in preventing the actual aircraft takeovers on which the disaster was based. This is typical of US industrial policy: any proposed safety change comes with an immediate threat of bankruptcy. Thus, the automobile industry claimed that seat belts would bankrupt them, followed by airbags, then child-safety door latches, and whatnot. The airline's lobbying organization, the Air Transport Association, has a long and distinguished record of opposing almost all improvements in security, especially if the airlines have to pay for them. From 1996 to 2000 alone, the association spent $70 million opposing a variety of sensible (and inexpensive) measures, such as matching passengers with bags (routine in Europe at the time) or improving security checks of airline workers. They opposed reinforced cabin doors and even the presence of occasional marshals (since the marshals would occupy nonpaying seats). It was common knowledge that the vital role of airport screening was performed poorly by people paid at McDonald's wages—but without their training—yet airlines spent millions fighting any change in the security status quo. Of course, a calamity such as 9/11 could have severe economic effects as people en masse avoided a manifestly dangerous mode of travel, but the airlines merely turned around and beseeched the government for emergency aid, which they got.

It seems likely that much of this is done "in good conscience," that is, the lobbyists and airline executives easily convince themselves that safety is not being compromised to any measurable degree, because otherwise they would have to live with the knowledge that they were willing to kill other people in the pursuit of profit. From an outsider's viewpoint this is, of course, exactly what they are doing. The key fact is that there is an economic incentive to obscure the truth from others—and simultaneously from self.

Only four years after 9/11, the airlines were loudly protesting legislation that would increase a federal security fee from $2.50 to $5.50, despite numerous surveys showing that people would happily pay $3 more per flight to enhance security. Here the airlines did not pay directly but feared only the indirect adverse effects of this trivial price increase. Note that corporate titans appear to slightly increase their own chances of death to hoard money, but with the increasing use of corporate jets, even this is not certain.

We see again patterns of deceit and self-deception at the institutional and group levels that presumably also entrain individual self-deception within the groups. Powerful economic interests—the airlines—prevent safety improvements of vital importance to a larger economic unit, the "flying public," but this unit is not acting as a unit. The pilots have their own organization and so of course do the (individually) powerful airlines, but the flying public exerts its effects one by one, in choice of airline, class of travel, destination, and so on—not in the relative safety of the flight, about which the public typically knows nothing. The theory is that the government will act on their behalf. Of course, as we have seen, it does not. Individuals within two entities should be tempted to self-deception—within the airlines that argue strenuously for continuation of their defective products and within the FAA, which, lacking a direct economic self-interest, is co-opted by the superior power of the airlines and acts as their rationalizing agent. In the case of NASA, those who sell space capsules to the public and to themselves never actually ride in them.

Regarding the specific event of 9/11 itself, although the United States already had a general history of inattention to safety, the George W. Bush administration even more dramatically dropped the ball in the months leading up to 9/11—first downgrading Richard Clarke, the internal authority on possible terrorist attacks, including specifically those from Osama bin Laden. The administration stated they were interested in a more aggressive approach than merely "swatting at flies" (bin Laden here being, I think, the fly). Bush himself joked about the August 2001 memo saying that bin Laden was planning an attack within the United States. Indeed, he denigrated the CIA officer who had relentlessly pressed (amid code-red terrorist chatter) to give the president the briefing at his Texas home. "All right," Bush said when the man finished. "You've covered your ass now," as indeed he had, but Bush left his own exposed. So his administration had a particular interest in focusing only on the enemy, not on any kind of missed signals or failure to exercise due caution. Absence of self-criticism converts attention from defense to offense.

THE *CHALLENGER* DISASTER

On January 28, 1986, the *Challenger* space vehicle took off from Florida's Kennedy Space Center and seventy-three seconds later exploded over the Atlantic Ocean, killing all seven astronauts aboard. The disaster was subject to a brilliant analysis by the famous physicist Richard Feynman, who had been placed on the board that investigated and reported on the crash. He was known for his propensity to think everything through for himself and hence was relatively immune to conventional wisdom. It took him little more than a week (with the help of an air force general) to locate the defective part (the O-ring, a simple part of the rocket), and he spent the rest of his time trying to figure out how an organization as large, well funded, and (apparently) sophisticated as NASA could produce such a shoddy product.

Feynman concluded that the key was NASA's deceptive posture toward the United States as a whole. This had bred self-deception within the organization. When NASA was given the assignment and the funds to travel to the moon in the 1960s, the society, for better or worse, gave full support to the objective: beat the Russians to the moon. As a result, NASA could design the space vehicle in a rational way, from the bottom up—with multiple alternatives tried at each step—giving maximum flexibility, should problems arise, as the spacecraft was developed. Once the United States reached the moon, NASA was a $5 billion bureaucracy in need of employment. Its subsequent history, Feynman argued, was dictated by the need to create employment, and this generated an artificial system for justifying space travel—a system that inevitably compromised safety. Put more generally, when an organization practices deception toward the larger society, this may induce self-deception within the organization, just as deception between individuals induces individual self-deception.

The space program, Feynman argued, was dominated by a need to generate funds, and critical design features, such as manned flight versus unmanned flight, were chosen precisely because they were costly. The very concept of a reusable vehicle—the so-called shuttle—was designed to appear inexpensive but was in fact just the opposite (more expensive, it turned out, than using brand-new capsules each time). In addition, manned flight had glamour appeal, which might generate enthusiasm for the expenses. But since there was very little scientific work to do in space (that wasn't better done by machines or on Earth), most was make-do work, showing how plants grow absent gravity (gravity-free zones can be produced on Earth at a fraction of the cost) and so on. This was a little self-propelled balloon with unfortunate downstream effects. Since it was necessary to sell this project to Congress and the American people, the requisite dishonesty led inevitably to internal self-deception. Means and concepts were chosen for their ability to generate cash flow and the apparatus was then designed top-down. This had the unfortunate effect that when a problem surfaced, such as the fragile O-rings, there was little parallel exploration and knowledge to solve the problem. Thus NASA chose to minimize the problem and the NASA unit assigned to deal with safety became an agent of rationalization and denial, instead of careful study of safety factors. Presumably it functioned to supply higher-ups with talking points in their sales pitches to others and to themselves.

Some of the most extraordinary mental gyrations in service of institutional self-deception took place within the safety unit. Seven of twenty-three *Challenger* flights had shown O-ring damage. If you merely plot chance of damage as a function of temperature at time of takeoff, you get a significant negative relationship: lower temperature meant higher chance of O-ring damage. For example, all four flights below 65 degrees F showed some O-ring damage. To prevent themselves—or others—from seeing this, the safety unit performed the following mental operation. They said that sixteen flights showed no damage and were thus irrelevant and could be excluded from further analysis. This is extraordinary in itself—one never wishes to throw away data, especially when it is so hard to come by. Since some of the damage occurred during high-temperature takeoffs, temperature at takeoff could be ruled out as a cause. This example is now taught in elementary statistics texts as an example of how not to do statistics. It is also taught in courses on optimal (or suboptimal) data presentation since, even while arguing against a flight, the engineers at Thiokol, the company that built the O-ring, presented their evidence in such a way as to invite rebuttal. The relevance of the mistake itself could hardly be clearer since the temperature during the *Challenger* takeoff (below freezing) was more than 20 degrees below the previous lowest takeoff temperature.

On the previous coldest flight (at a balmy 54 degrees), an O-ring had been eaten one-third of the way through. Had it been eaten all the way through, the flight would have blown up, as did the *Challenger*. But NASA cited this case of one-third damage as a virtue, claiming to have built in a "threefold safety factor." This is a most unusual use of language. By law, you must build an elevator strong enough that the cable can support a full load and run up and down a number of times without any damage. Then you must make it eleven times stronger. This is called an elevenfold safety factor. NASA has the elevator hanging by a thread and calls it a virtue. They even used circular arguments with a remarkably small radius: since manned flight had to be much safer than unmanned flight, it perforce was. In short, in service of the larger institutional deceit and self-deception, the safety unit was thoroughly corrupted to serve propaganda ends, that is, to create the appearance of safety where none existed. This must have aided top management in their self-deception: less conscious of safety problems, less internal conflict while selling the story.

There is thus a close analogy between self-deception within an individual and self-deception within an organization—both serving to deceive others. In neither case is information completely destroyed (all twelve Thiokol engineers had voted against flight that morning, and one was vomiting in his bathroom in fear shortly before takeoff). The truth is merely relegated to portions of the person or the organization that are inaccessible to consciousness (we can think of the people running NASA as the conscious part of the organization). In both cases, the entity's relationship to others determines its internal information structure. In a non-deceitful relationship, information can be stored logically and coherently. In a deceitful relationship, information will be stored in a biased manner the better to fool others—but with serious potential costs. However, note here that it is the astronauts who suffer the ultimate cost, while the upper echelons of NASA—indeed, the entire organization minus the dead—may enjoy a net benefit (in employment, for example) from this casual and self-deceived approach to safety. Feynman imagined the kinds of within-organization conversations that would bias information flow in the appropriate direction. You, as a working engineer, might take your safety concern to your boss and get one of two responses. He or she might say, "Tell me more" or "Have you tried such-and-such?" But if he or she replied, "Well, see what you can do about it" once or twice, you might very well decide, "To hell with it." These are the kinds of interactions—individual on individual (or cell on cell)—that can produce within-unit self-deception. And have no fear, the pressures from overhead are backed up with power, deviation is punished, and employment is put at risk. When the head of the engineers told upper management that he and other engineers were voting against the flight, he was told to "take off your engineering hat and put on your management hat." Without even producing a hat, this did the trick and he switched his vote.

There was one striking success of the safety unit. When asked to guess the chance of a disaster occurring, they estimated one in seventy. They were then asked to provide a new estimate and they answered one in ninety. Upper management then reclassified this arbitrarily as one in two hundred, and after a couple of additional flights, as one in ten thousand, using each new flight to lower the overall chance of disaster into an acceptable range. As Feynman noted, this is like playing Russian roulette and feeling safer after each pull of the trigger fails to kill you. In any case, the number produced by this logic was utterly fanciful: you could fly one of these contraptions every day for thirty years and expect only one failure? The original estimate turned out to be almost exactly on target. By the time of the *Columbia* disaster, there had been 126 flights with two disasters for a rate of one in sixty-three. Note that if we tolerated this level of error in our commercial flights, three hundred planes would fall out of the sky every day across the United States alone. One wonders whether astronauts would have been so eager for the ride if they had actually understood their real odds. It is interesting that the safety unit's reasoning should often have been so deficient, yet the overall estimate exactly on the mark. This suggests that much of the ad hoc "reasoning" was produced under pressure from the upper ranks after the unit had surmised correctly. There is an analogy here to individual self-deception, in which the initial, spontaneous evaluation (for example, of fairness) is unbiased, after which higher-level mental processes introduce the bias.

There is an additional irony to the *Challenger* disaster. This was an all-American crew, an African American, a Japanese American, and two women—one an elementary schoolteacher who was to teach a class to fifth graders across the nation from space, a stunt of marginal educational value. Yet the stunt helped entrain the flight, since if the flight was postponed, the next possible date was in the summer, when children would no longer be in school to receive their lesson. Thus was NASA hoisted on its own petard. Or as has been noted, the space program shares with gothic cathedrals the fact that each is designed to defy gravity for no useful purpose except to aggrandize humans. Although many would say that the primary purpose of cathedrals was to glorify God, many such individuals were often self-aggrandizing. One wonders how many more people died building cathedrals than flying space machines.

THE *COLUMBIA* DISASTER

It is extraordinary that seventeen years later, the *Challenger* disaster would be repeated, with many elements unchanged, in the *Columbia* disaster. Substitute "foam" for "O-ring" and the story is largely the same. In both cases, NASA denied they had a problem, and in both cases it proved fatal. In both cases, the flight itself had little in the way of useful purpose but was done for publicity purposes: to generate funding and/or meet congressionally mandated flight targets. As before, the crew was a multicultural dream: another African American, two more women (one of whom was Indian), and an Israeli who busied himself on the flight collecting dust over (where else?) the Middle East. Experiments designed by children in six countries on spiders, silkworms, and weightlessness were duly performed. In short, as before, there was no serious purpose to the flight; it was a publicity show.

The *Columbia* spacecraft took off on January 15, 2003 (another relatively cold date), for a seventeen-day mission in space. Eighty-two seconds after launch, a 1.7-pound chunk of insulating foam broke off from the rocket, striking the leading edge of the left wing of the space capsule, and (as was later determined) apparently punching a hole in it about a foot in diameter. The insulating foam was meant to protect the rocket from cold during takeoff, and there was a long history of foam breaking off during flight and striking the capsule. Indeed, on average thirty small pieces struck on every flight. Only this time the piece of foam was one hundred times larger than any previously seen. On the *Atlantis* flight in December 1988, 707 small particles of foam hit the capsule, which, in turn, was inspected during orbit with a camera attached to a robotic arm. The capsule looked as though it had been blasted with a shotgun. It had lost a heat-protective tile but was saved by an aluminum plate underneath. As before, rather than seeing this degree of damage as alarming, the fact that the capsule survived reentry was taken as evidence that foam was not a safety problem. But NASA did more. Two flights before the *Columbia* disaster, a piece of foam had broken off from the bipod ramp and dented one of the rockets, but shuttle managers formally decided not to classify it as an "in-flight anomaly," though all similar events from the bipod ramp had been so classified. The reason for this change was to avoid a delay in the next flight, and NASA was under special pressure from its new head to make sure flights were frequent. This is similar to the artificial pressure for the *Challenger* to fly to meet an external schedule.

The day after takeoff, low-level engineers assigned to review film of the launch were alarmed at the size and speed of the foam that had struck the shuttle. They compiled the relevant footage and e-mailed it to various superiors, engineers, and managers in charge of the shuttle program itself. Anticipating that their grainy photos would need to be replaced by much more accurate and up-to-date footage, they presumed on their own to contact the Department of Defense and ask that satellite or high-resolution ground cameras be used to photograph the shuttle in orbit.

Within days the Air Force said it would be happy to oblige and made the first moves to satisfy this request. Then an extraordinary thing happened. Word reached a higher-level manager who normally would have cleared such a request with the Air Force. At once, she asked her superiors whether they wanted to know the requested information. They said no. Armed with this, she told the Air Force they no longer needed to provide the requested information and that the only problem was underlings who failed to go through proper channels! On such nonsense, life-and-death decisions may turn.

This is vintage self-deception: having failed to deal with the problem over the long term, having failed to prepare for a contingency in which astronauts are alive in a disabled capsule unable to return to Earth, the NASA higher-ups then decide to do nothing at all except avert their eyes and hope for the best. With fast, well-thought-out action, there was just barely time to launch a flight that might reach the astronauts before their oxygen expired. It would have required a lot of luck, with few or no hitches during countdown, so it was unlikely. An alternative was for the astronauts to attempt crude patches on the damaged wing itself. But why face reality at this point? They had made no preparation for this contingency, and they would be making life-and-death decisions with all the world watching. Why not make it with no one watching, including themselves? Why not cross their fingers and go with the program? Denial got them where they were, so why not ride it all the way home?

The pattern of instrument failure before disintegration and the wreckage itself made it abundantly clear that the foam strike filmed during takeoff must have brought down the *Columbia*, but people at NASA still resisted, denying that it was even *possible* for a foam strike to have done such damage and deriding those who thought otherwise as "foamologists." For this reason, the investigating commission decided to put the matter to a direct test. They fired foam pieces of the correct weight at different angles to the left sides of mock-ups of the spacecraft. Even this NASA resisted, insisting that the test use only the small pieces of foam that NASA had modeled! The key shot was the one that mimicked most closely the actual strike, and it blew a hole in the capsule big enough to put your head through. That was the end of that: even NASA folded its tent. But note that denial (of the problem ahead of time) entrained denial (of the ongoing problem), which entrained denial (after the fact). As we have noted in other contexts, this is a characteristic feature of denial: it is self-reinforcing.

The new safety office created in response to the *Challenger* explosion was also a fraud, as described by the head of the commission that later investigated the *Columbia* disaster, with no "people, money, engineering experience, [or] analysis." Two years after the *Columbia* crash, the so-called broken safety culture (twenty years and counting) at NASA still had not been changed, at least according to a safety expert and former astronaut (James Wetherbee). Under pressure to stick to budget and flight schedules, managers continue to suppress safety concerns from engineers and others close to reality. Administrators ask what degree of risk is acceptable, when it should be what degree is necessary and how to eliminate that which is unnecessary. A recent poll showed the usual split: 40 percent of managers in the safety office thought the safety culture was improving while only 8 percent of workers saw it that way. NASA's latest contributions to safety are a round table in the conference room instead of a rectangular one, meetings allowed to last more than half an hour, and an anonymous suggestion box. These hardly seem to go to the heart of the problem.

That the safety unit should have been such a weak force within the organization is part of a larger problem of organizational self-criticism. It has been argued that organizations often evaluate their behavior and beliefs poorly because the organizations turn against their evaluation units, attacking, destroying, or co-opting them. Promoting change can threaten jobs and status, and those who are threatened are often more powerful than the evaluators, leading to timid and ineffective self-criticism and inertia within the organization. As we have seen, such pressures have kept the safety evaluation units in NASA crippled for twenty years, despite disaster after disaster. This is also a reason that corporations often hire outsiders, at considerable expense, to come in and make the evaluation for them, analogous perhaps to individuals who at considerable expense consult psychotherapists and the like. Even grosser and more costly failures of self-criticism occur at the national level, and we will refer to some of these when we discuss war (see Chapter 11).

EGYPT AND EGYPTAIR DENY ALL

A most unusual accident occurred on October 31, 1999, when EgyptAir Flight 990 took off from New York's JFK Airport bound for Cairo. It climbed to 33,000 feet, flew normally for about half an hour on a calm night, and then suddenly (in about two minutes) plummeted to the ocean below, killing all 217 aboard. Later work by the NTSB proved beyond a reasonable doubt that the plane was deliberately brought down by its second copilot. (Long flights have two crews: one for takeoff and landing and one for flying some of the routine work in between.) The copilot used a little deception to achieve his aim, but there is no evidence of self-deception involved in the disaster (beyond whatever may have been going on in the head of the suicidal copilot). But afterward there was a furious, long-lasting effort by EgyptAir and the Egyptian government to deny the cause of the crash, a denial that entrained self-deception and that continues to this day. Nearly every conceivable counterargument was advanced, including small bombs near the cabin or the rear, active Israeli agents nearby taking out the thirty-four Egyptian army generals aboard, and so on. This is a case of a post hoc attempt to create a false narrative to protect oneself, and we can readily appreciate Egyptian sensitivities. Because the copilot was soon reported to have murmured a standard Muslim prayer (and in Arabic at that!) before he put the plane into its dive, EgyptAir was about to acquire the unenviable reputation of being "unsafe at any speed" due to internal terrorism. If you can't trust the flight crew to try to stay alive, how on earth can you trust the flight?

Given nearly daily Egyptian resistance to NTSB findings for more than a year from well-trained aviation engineers, with the Egyptian government proposing numerous, sometimes very sophisticated alternatives, this crash is unusually well studied. Yet the basic facts were clear very early. There was no evidence of a bomb at all—fore, aft, or anywhere else.

Bombs typically leave at least three kinds of evidence: instrument readings on the flight recorder, voices and sounds on the cockpit recorder, and a certain debris pattern at the bottom of the ocean. Instead, twenty minutes into the flight, the second copilot (fifty-nine years old) maneuvered out the first one (thirty-six years old) by bullying him. He first suggested that the other, who was flying the airplane, take a break and get some rest. When the first one angrily replied that this should have been agreed upon at the start of the flight, he said, "You mean you're not going to get up? You will get up. Go and get some rest and come back." In a few moments, the man got up and left the cockpit. The second copilot then buckled himself in next to the pilot (age fifty-seven). After eight minutes of pleasant banter between two old friends, the second copilot found the pen of the first, or more likely, pretended to: "Look, here's the new first officer's pen. Give it to him, please. God spare him," he says to the captain, "to make sure it doesn't get lost." Pilot: "Excuse me, Jimmy, while I take a quick trip to the toilet." Copilot: "Go ahead, please." Pilot, exiting: "Before it gets crowded, while they are eating, and I will be back to you." As easy as that, the second copilot had the airplane to himself.

About twenty seconds later, the copilot said (in Arabic, his native tongue), "I rely on God," and the autopilot disengaged. Four seconds later, another "I rely on God," and two things happened: the throttles moved from fast to minimum idle and the massive rear elevators dropped, raising the tail and pointing the nose down. The copilot apparently choked the power and pushed the control yoke forward. The airplane dived steeply, and six times in quick succession, the copilot said calmly, "I rely on God." As the nose continued to pitch downward, the inside of the plane changed from no gravity to negative gravity, with objects hitting the ceiling.

Somehow, sixteen seconds into the dive, the pilot managed to return and yelled, "What's happening? What's happening?" He got no answer other than "I rely on God." Then the two evidently fought for control of the airplane. The pilot tried to move the nose up and the copilot held it down, so that the elevators split, one down, one up, a most unusual configuration. (That they split is a design feature allowing either pilot to overcome a mechanical jam and fly the airplane with only one elevator.) The plane descended at a maximum rate of 630 feet per second and at a downward angle of almost 40 degrees. Somewhere along the line, the copilot turns off the engines while the pilot shouts incredulously. The plane hits about 550 miles per hour at 16,000 feet, when the pilot's efforts seem to have reversed the dive. The plane then soars steeply back up to 24,000 feet, loses its left engine, and dives at high speed into the ocean. This must have been the most horrifying roller-coaster ride of a lifetime, lasting as it did for two minutes.

The NTSB did a voice-stress analysis that showed a sharp contrast between the copilot and the pilot as they fought for control of the airplane. The pilot's voice rose steadily in pitch and intensity, as one would expect from a person under growing stress and panic. But the copilot's never changed. Through a total of twelve utterances of "I rely on God," his voice never betrayed any stress or fear. He intended what he did and he was calm in his intention.

The only part of this story we do not know is why the copilot brought the plane down. Was it the presence of those thirty-four generals aboard? He was not known to be politically active. Was it the fact that he had been warned only a few days before by a very senior pilot (himself riding as a passenger on this flight) to grow up before he caused himself serious problems? He indeed had a reputation for inappropriate behavior at the hotel in New York where the airline personnel stayed: following women (uninvited) toward their rooms, for example, and similar behavior, nothing dangerous, perhaps more on the pathetic side. He was carrying items in the plane for use back in Egypt, a part for his car and so on, so perhaps the decision was made shortly before he enacted it. If he was acting vindictively toward EgyptAir, the presence of all those generals may have made the suicide more dramatic and costly. We will never know, because part of Egyptian denial was either never to investigate the copilot's possible motives or to hide whatever they found out. If NTSB investigations in the United States were typically run this way, we would have no objective data on the causes of airplane crashes. This is a case of self-deception intruding at yet a higher level—the international one—to impede the truth at an international cost. Surely we all benefit from reducing civilian crashes.

Of course, Egypt is far from alone. For example, in the United States hardly anyone is conscious of—much less concerned by—the fact that the US economic embargo has prevented Iran from directly acquiring replacement parts for its aging airplanes. The country imposing the embargo is the same one that sold the planes, so here is a country acting in gross violation of international public safety for purely petty reasons when it alone has a legal obligation (original contracts signed) to provide replacement parts. This is a form of economic warfare, perhaps with the meta-message, "Go screw yourself and may your airplanes crash, too."

SAVED BY *LACK* OF SELF-DECEPTION?

Perhaps we can end on a more positive note with the reverse of what we have described so far: the celebrated safe landing of a plane in the Hudson River shortly after takeoff from La Guardia Airport in New York on January 17, 2009, saving all 155 lives. The plane was headed for Charlotte, North Carolina, and apparently struck a column of geese at three thousand feet, disabling both engines simultaneously. The captain (fifty-eight years old)—who was not at the time flying the plane—immediately made a series of decisions, none of which was brilliant or exceptional, but all of which showed immediate rational calibration toward a serious danger, one for which the pilot had long prepared. The first thing he did was to take control of the plane. "My airplane," he announced to his first officer, the standard procedure for a takeover. "Your aircraft," the officer (age forty-nine) responded.

The pilot first decided against landing at two possible airports and chose the large Hudson Bay instead. He cut speed by lowering wing flaps and made sure that the nose was raised on landing. It was experienced as a "hard landing" by the crew in the rear of the plane, with utensils flying around, but there was no damage to anyone in the craft, beyond a flight attendant's broken leg. In the captain's own words:

> Losing thrust on both engines, at a low speed, at a low altitude, over one of the most densely populated areas on the planet—yes, I knew it was a very challenging situation. I needed to touch down with the wings exactly level . . . with the nose slightly up . . . at a rate that was survivable . . . and just above our minimum flying speed, not below it. And I needed to make all these things happen simultaneously.

The pilot had several advantages. He was very experienced and competent, having been the top Air Force Academy cadet in his class in flying ability and having flown military jets before becoming a commercial pilot. He was trained as a glider pilot, precisely what was required in this situation, the key being to keep both wings out of the water while landing on it. He had taught courses on risk management and catastrophes. He remembered from training that when in a forced landing in water, you should look to land near a boat. Within moments of landing, there were so many boats nearby, large and small, as to risk swamping the already rapidly sinking aircraft. Children as young as eight months and eighteen months emerged alive. Two women who ended up in the frigid waters were rapidly rescued.

The remarkable scene was a source of excitement for several days. The key is that the captain was highly conscious throughout, very well prepared, and ended up doing everything right. When asked whether he prayed, he said he was concentrating too hard: "I would imagine someone in back was taking care of that for me while I was flying the airplane."

CHAPTER 10

False Historical Narratives

False historical narratives are lies we tell one another about our past. The usual goals are self-glorification and self-justification. Not only are we special, so are our actions and those of our ancestors. We do not act immorally, so we owe nothing to anyone. False historical narratives act like self-deceptions at the group level, insofar as many people believe the same falsehood. If a great majority of the population can be raised on the same false narrative, you have a powerful force available to achieve group unity. Of course, leaders can easily exploit this resource by coupling marching orders with the relevant illusion: German people have long been denied their rightful space, so *Dass Deutsche Volk muss Lebensraum haben!* (German people must have room in which to live!)—neighbors beware. Or the Jewish people have a divine right to Palestine because ancestors living in the general area some two thousand years ago wrote a book about it—non-Jewish occupants and neighbors better beware. Most people are unconscious of the deception that went into constructing the narrative they now accept as true. Nor are they usually aware of the emotional power of such narratives or that these may entrain long-term effects.

There is a deep contradiction within the study of history between ferreting out the truth regarding the past and constructing a false historical narrative about it. As we have seen in this book, we make up false narratives all the time, about our own behavior, about our relationships, about our larger groups. Creating one for one's larger religion or nation only extends the canvas. Usually a few brave historians in every society try to tell the truth about the past—that the Japanese army ran a vast, forced system of sexual slavery in World War II, that the United States committed wholesale slaughter against Koreans during the Korean War and against Vietnamese, Cambodians, and Laotians in the Vietnam War, that the Turkish government committed genocide against its successful subgroup of Armenians, that the Zionist conquerors of Palestine committed ethnic cleansing against some 700,000 Palestinians, that the United States has waged a long campaign of genocide and murder against American Indians, from the nation's founding to the murder by proxy of more than a half million in the 1980s alone, not counting before or after, and it has sought through military means to determine the fate of the entire New World for well over a century. But most historians will tell only some version of the conventional, self-aggrandizing story, and most people in the relevant countries will not have heard of (or believed) the factual assertions I just made.

One noteworthy fact is that the younger the recipient of the knowledge, the greater the pressure to tell a false story. So we are apt to tell our children a heroic version of our past and reserve for our university students a more nuanced view. This of course strengthens the bias, since views learned early have special power and not everyone attends college, or studies history if they do. Fortunately, the young often appear naturally to resist parental and adult nonsense, so there is at least some tendency to resist and upgrade. Just the same, there are strong pressures on professional historians to come up with a positive story, in part to undergird what is taught more widely.

Make no mistake about it. People feel strongly about these matters. One person's false historical narrative is another's deeply personal group identity—and what right do you have commenting on my identity in the first place? Many Turkish people may well feel that I have slandered their country regarding its Armenian genocide, while I believe I have merely told the truth. The same may be true (though less strongly) for some Japanese people regarding their country's practice of sexual slavery during World War II. Most Americans could hardly care less. So we wiped out the Amerindians—so what? So we repeatedly waged aggressive war on Mexico and stole nearly half their country. They probably deserved it. And, yes, since then we have fought a staggering series of wars ourselves and by proxies—even recently supporting genocide in such diverse places as Central America, Vietnam, Cambodia, and even East Timor, while blocking international action against it in Rwanda—but so the hell what? Only a left-wing nutcase would dwell on such minor details. Isn't that what great powers do, and aren't we the greatest?

Israel is no different from any other country or group in having its own false historical narrative, and Israel's is especially important because it exacerbates a set of troubled international and intergroup relations. The narrative is also one that is accepted almost wholesale in the United States, the most powerful military nation in the world. As the old joke goes, why doesn't Israel become the fifty-first state? Because then it would have only two senators. Again, feelings run high. Some regard as anti-Semitic any attack on the behavior of Israel (or its underlying narrative). I regard this as nonsense and follow instead what seem to me to be the best Israeli (and Arab) historians—and their (largely Jewish) American counterparts—in describing a false historical narrative used to expand Israel at a cost to its neighbors by waging regular war on them to seize land and water (with near-constant US support), all in the name of fighting terrorism, while using state terrorism as the chief weapon. The narrative inverts reality: Israel wants only peace with its Arab neighbors (from as early as 1928), who to this very day reject peace at every turn and seek the total destruction of Israel and its Jewish population.

But what are we to do? Yes, feelings run high, but false historical narratives are a critical part of self-deception at the group level, often with horrendous effects on others—if not on those practicing them. To discuss the subject, we need examples. Are we to leave out this important topic because on any given example feelings are easily bruised and controversy aroused? I see no sense in this. A theory of self-deception is not of much use if it can't be applied to cases of actual human importance. Of course, I am more likely to be personally biased on these topics than on, say, the immunology of self-deception, but for me the risk of appearing foolish, indeed self-deluded, is preferable to the cowardice of not taking a position.

The point of this chapter is to paint a few false historical narratives in enough detail to see clearly some of the lies we tell about our histories, how they were constructed and maintained, and the purposes they may serve. We will also consider the costs. It has famously been said that those who do not know history are destined to repeat it, or as Harry Truman put it: "The only thing new under the sun is the history you do not know."

THE US FALSE HISTORICAL NARRATIVE

The false US narrative can be summarized with a few key facts, their rationalization, and the function of the rationalizations. The key fact is the slaughter and dispossession of an entire people (or set of peoples) to make room for Europeans (and their African slaves), a feat also accomplished by treaties not kept. Too late did Amerindians learn never to sign a treaty with a white man. For the latter, treaties were merely temporary agreements to be abrogated as soon as it was advantageous to do so.

It is fully apt that Christopher Columbus should have been elevated to historical status for discovering the Americas. On the one hand, he did no such thing. There were more than 100 million people awaiting him when he arrived. And ships had also recently arrived from Africa, Polynesia, Phoenicia, and even other European countries. On the other hand, Columbus was unique in that he combined exploration with an explicit plan for subjugating the locals and extracting their wealth and labor. This of course is not what he is celebrated for.

His first visit in 1492 merely allowed him to look around, and this is the one preserved in historical memory. The arrival of his three cute little ships—the *Nina*, the *Pinta*, and the *Santa Maria*—connoted a naive, peaceful arrival to "discover" a brand-new land. The second time (in 1493), he arrived more fully prepared: seventeen ships, at least twelve hundred men, cannons, crossbows, guns, cavalry, and attack dogs trained to rend human flesh. Yet this second visit is lost from historical memory entirely. It is the key visit, but no one mentions it.

In Hispaniola, he and his men immediately demanded food, gold, spun cotton, and access to the local women. Indians were put to work mining gold, raising Spanish food, and even carrying the Spaniards around wherever they went. Minor offenses by Indians were punished by mutilation—an ear, a nose, both hands. Failing to find gold, Columbus started slave capture and transmission on a large scale, returning to Spain with five hundred Indians (almost half dying on the way) and leaving five hundred slaves behind. He launched a reign of sadistic terror: newborns given to dogs as food or smashed against rocks in front of their screaming mothers, twenty thousand killed in Hispaniola alone, with more to come on nearby islands. Mass suicide and regular infanticide were common responses by the Indians to the horrors they were experiencing. To make a long story shorter, a mere twenty-five years later when Columbus and his immediate heirs were done with Hispaniola, its Indian population had been reduced from an estimated five million people to fewer than fifty thousand. This was a story to be repeated in North, Central, and South America except that in the mainland tropics you could never exterminate everyone, especially those living deeper or higher in the forest. Neither the invention of ships nor means of navigation allowed this conquest and holocaust to take place; it was the invention of large guns, which could be attached to sturdy ships and supported by an array of smaller guns and aggressive weapons. It was the invention of high-tech war across the sea that brought about the new wave of colonization and genocide.

The point is that our retrospective re-creation of the "founding of the Americas" minimizes the sordid details of murder, slavery, sexual exploitation, and degradation with which it began. Instead it exalts simple exploration and discovery. Thus do we deny the motives and the reality of the territorial takeover. The benefit is self-glorification and continuation of the same kind of behavior; the cost is much more long-term, depending partly on the reaction of the survivors to this kind of behavior.

The holocaust was repeated up and down the Americas: one part introduced diseases to which the local people had little or no resistance, and one part heartless slaughter—women, children, the elderly, all members of village after village after village put to the sword—in what has been described as the longest-running genocide in the world. No longer in the United States, where Amerindians were long ago wiped out with a few remnants held on "reservations," but throughout Central and South America the slaughter of indigenous peoples continues apace. In Guatemala the renewed attacks coincided with a US-supported coup in 1953. For the next fifty years, hundreds of thousands of Amerindians were killed in generalized anticommunist warfare. During the great Spanish-imposed holocaust of the 1500s and immediately afterward, local populations were more than decimated (to 5 percent or fewer of their original numbers) due to both introduced diseases and genocidal behavior on a large scale.

An important difference between what became the United States and countries north and south of it is that the pre–United States consisted of prime temperate-zone land, with neither the cold of the Arctic nor the overwhelming biological competition in the tropics, which chiefly comes from antagonistic life forms such as diseases, both human and crop. Thus removal of the original population from this space resulted in huge opportunities for rapid growth of the new powerful European industrial system. Stealing nearly half of Mexico greatly increased the available space.

And the rationale for the genocide? Manifest destiny. Very simple. A religious and racial concept: you were destined by God to do exactly what you did. "Might makes right," but with a more exalted ring. And the value of the rationale? Keep on doing what you are doing. Today the intellectuals rationalizing American misbehavior along these lines are fond of speaking about "American exceptionalism." Somehow America is exempt from the usual laws of history and reality. We are the exceptional case and permitted—no, required—to act appropriately. We are the new chosen people of the Bible, as we have seen ourselves now for more than two hundred years (see the following section "Christian Zionism").

How many of us Americans know that the Founding Fathers we venerate explicitly urged the eradication of Amerindians—genocide—by any means necessary: terror, starvation, inebriation, deliberate infection with smallpox, and outright slaughter?

- *President George Washington (stated at the time of open warfare):* The immediate objectives are the total destruction and devastation of their settlements. It will be essential to ruin their crops in the ground and prevent their planting more.
- *President Thomas Jefferson:* This unfortunate race, whom we had been taking so much pains to save and to civilize, have by their unexpected desertion and ferocious barbarities justified extermination and now await our decision on their fate.
- *President Andrew Jackson:* They have neither the intelligence, the industry, the moral habits, nor the desire of improvement which are essential to any favorable change in their condition. Established in the midst of another and a superior race, and without appreciating the causes of their inferiority or seeking to control them, they must necessarily yield to the force of circumstances and ere long disappear.
- *Chief Justice John Marshall:* The tribes of Indians inhabiting the country were savages. . . . Discovery [of America by Europeans] gave an exclusive right to extinguish the Indian title of occupancy, either by purchase or by conquest.
- *President William Henry Harrison:* Is one of the fairest portions of the globe to remain in a state of nature, the haunt of a few wretched savages, when it seems destined by the Creator to give support to a large population and to be the seat of civilization?
- *President Theodore Roosevelt:* The settler and pioneer have at bottom, had justice on their side; this great continent could not have been kept as nothing but a game preserve for squalid savages.

No one seems self-conscious in the slightest about the links between explicit racism, claims of divine design, and calls for "extirpation" of entire peoples—all to the advantage of one's own people.

CONTROL THROUGH SMALL WARS AND INSTALLED PROXIES

Most Americans have no idea how often the United States has gone to war, that is, invaded another country with its troops. For nearby countries, such visits are a regular occurrence. To take but World War I, when the United States was engaged in a major war against Germany and its allies in Europe, it still managed to invade the Dominican Republic, Haiti, Cuba, Panama, and Mexico (multiple times) while permanently stationing troops in Nicaragua. Surely this is an admirable achievement. The usual rationale was instability threatening Americans and American property, but the actual function was typically to subvert local democracy in favor of American business interests. Presidents were replaced, assemblies dissolved, new and biased constitutions rushed through rigged plebiscites, and so on.

After World War I, in Guatemala, El Salvador, Colombia, Nicaragua, Cuba, Brazil, Argentina, Chile, and Panama, the Monroe Doctrine—the notion that the United States reigns supreme in the New World—was enforced (or, in Cuba's case, was attempted) through armed invasions, local militias, and internal subversion. Most invasions set the stage for a series of dictators serving US interests: Batista, Trujillo, Duvalier, and Samosa. In Franklin Roosevelt's famous words (about Samosa), "He may be a son of a bitch, but he's our son of a bitch." Of course, such a person is much more useful to you (in the short term) than someone trying to serve his own people's interests. The long term is another matter. The replacement of Mossadegh, the Iranian nationalist, in 1953 with a puppet, the shah, may have given temporary economic benefit to the United States, but certainly it helped produce a long-term disaster.

The United States invaded Nicaragua thirteen times in the twentieth century before turning the murderous Contras loose on them in the

1980s, when the Nicaraguans finally voted for socialism. The country remains the second-poorest in the Americas, second only to Haiti, another country that has enjoyed frequent US invasions (including a twenty-year occupation). The Brazilian adventure was typical. A US-supported military coup in 1965 overthrew the democratically elected and mildly socialist government, instituting a reign of terror and laying the groundwork for similar events in Argentina and Chile, with combined mortality running into the hundreds of thousands. The US ambassador to Brazil at the time put the matter succinctly, in the best tradition of false historical narratives: The coup was "the most decisive victory for freedom in the mid twentieth century." The "democratic forces" now in power would "create a greatly improved climate for private investment." Thus is a false historical narrative maintained and embroidered. We start with the notion that it is our right—nay, our duty—to intervene in the internal affairs of our neighbors because we thereby create freedom, democracy, and (most important) improved investment opportunities for ourselves that we then imagine benefit the Brazilians apace. In fact, it is only now, after the military dictatorships have long withered away, that under a fully democratic (and mildly socialist) government, Brazil is making rapid economic strides in the world, much more so than is the United States.

Much more recently, George W. Bush said the United States was going to war with Iraq. Congress said they wanted evidence that Iraq was a threat. The CIA provided the evidence. Congress voted to go to war. My guess is that most Americans now remember the sequence as: The CIA provided evidence that Iraq was a threat. Based on this evidence, Bush and Congress decided to go to war. If so, a false historical narrative was born, another aggressive war turned into a defensive one.

One cost of US attachment to international intervention and war is the growth of the military-industrial complex famously warned against by President Dwight Eisenhower fifty years ago—or military-industrial-congressional complex, as he first called it. Its appetite seems insatiable; the United States alone now spends almost as much on warfare ("defense") as the rest of the world put together. Many of the chief US export industries are military as well: fighter jets, helicopters, rifles, bullets. We arm the world at every level, from criminal gangs in our own hemisphere to entire states throughout the world. The collapse of the Soviet system gave only a temporary respite from these forces, and the United States is now spending relatively more than ever. At the same time, an enormous and very expensive intelligence system is being created.

Note that the Soviets provided a counterweight to rapacious capitalism. With their collapse, the past twenty years have seen intense American wars, an accelerated shift of wealth to the already wealthy (a trend that began a few years earlier), and gross financial thievery by the wealthy and their agents leading to near economic collapse.

US HISTORY TEXTBOOKS

A useful part of understanding false historical narratives is seeing what efforts are made to instill them in schools, and we shall try to do this for each of our examples. In the United States, high schools were first required to teach US history around 1900 as part of a nationwide, flag-waving frenzy. Although by logic, one might easily imagine that the function of teaching one's own history would be to learn and prepare oneself for the future, the nationalistic origin reveals the deeper force that operates in country after country—toward building a positive, patriotic story, one that encourages group cohesion, self-congratulation, and superiority vis-à-vis others, a self-serving false historical narrative available to rationalize every action.

What we have now in the United States is instructive. Several huge books compete for a very large market. The average weight of each book exceeds six pounds and contains more than one thousand pages. This is partly due to pressure to mention every state and president, every event big and small, thus precluding any study of history's larger patterns and events. To help the teacher get students to read these bloated books, multiple free teaching aids are offered, crisscrossed with organization. One book has 840 "main ideas within the main text," 310 "skill builders," and 466 "critical thinking" questions. No system of human thought is known to produce coherent patterns with so many variables. Students have been described as memorizing material for each chapter, only to forget it to free up neurological space for the next chapter.

In short, US history is sliced and diced right out of existence. Main themes and topics are easily lost. One book offered little more than a paragraph on all of slavery. Conflict of any kind, or even suspense, tends to be removed. The story is one in which every problem has been solved or is about to be. The present is almost never used to illuminate the past, and we learn nothing from the past that would help us with the future and very few lessons of any kind. Thus, the study of US history has become an exercise in rote memory and self-glorification, with almost no relevant learning. Not surprisingly, students routinely describe history as the most boring subject in school, easily beating English and chemistry, yet interest in history in other contexts, including general books, museums, and films, remains high.

When I was an undergraduate major in US history at Harvard in the early 1960s, the names of the texts gave away the game: *The Genius of American Democracy.* You did not need to read the book; the content was right there in the title. The chief problem in American historiography was: Why are we the greatest nation that was ever conceived and the greatest people who ever strode the face of the earth? Competing answers had to do with the value of a receding frontier (a benign metaphor for territorial expansion), of having upper-class Englishmen design the society, of building a country on perpetual immigration, and so on. The key is what was assumed in advance, and of course high school history texts reflect this as well: *Triumph of the American Nation*, *Land of Promise*, *The Great Republic.* Meta-message—you have a proud heritage, certainly nothing to be ashamed of, look at what the United States has accomplished and just imagine what it will soon do. Be a good citizen; be all you can be.

LARGER VIEW OF US HISTORY

The pervious sections are not meant to be a representative history of the United States. US history has many virtues, among which is the fact that the US population is reconstituted every generation through a roughly

10 percent admixture by external immigration from throughout the world. Although in its history rules of immigration have favored some groups over others, all have had some opportunity. And with illegal immigration, such opportunities are sometimes greatly enhanced. From a biological standpoint, the resulting outbreeding (insofar as it takes place, as it inevitably must) will tend to be genetically beneficial. The US population is perpetually heterogeneous, about to be infused with 10 percent more genes from around the world. This continual level of in-migration, outbreeding, and cultural diversity is unusual for most countries.

One other feature of US history is highly unusual and largely positive. Its most costly war to itself—700,000 dead out of a population of about 18 million—was the Civil War, a most ironic war in that one side wished to free slaves to whom they were less related than were the slaves' owners. The owners cared primarily about maintaining these people as their property (rather than, in some cases, their children), so they fought to maintain this right even though this sometimes harmed their own flesh and blood. In short, the Civil War was fought in great part as a moral crusade to end something that was seen as a moral evil. Loss of life was mostly suffered by European Americans and roughly equally on both sides, those fighting for justice and those against it. The later history of African Americans was in some ways more dreadful than under slavery, since not counting as property they could be hanged or "lynched" by the thousands as a form of social control. Nevertheless, the subpopulation had become strong enough by the middle of the twentieth century to begin a political and social movement that led to eventual legal liberation, and with this yoke lifted, the intrinsic benefits of strong outbreeding associated with strong selection has produced a vibrant and powerful subgroup. African Americans are the melting-pot population par excellence in the United States, genetically roughly 25 percent European in origin, 70 percent African, and the remainder Amerindian and Chinese. At the same time, social policies such as the war on drugs amount to a war on lower-class African Americans, greatly increasing incarceration rates, with destructive effects on their communities. So the racist attack continues, but in the long run it can only strengthen the biological power of its target.

THE REWRITING OF JAPANESE HISTORY

In the past ten years, Japan has shown a very interesting retrogressive approach to its own past, in which critical events, sometimes formerly acknowledged, are now denied despite massive evidence to the contrary. With each opposing revelation, the denials are then tailored downward but always with the intent of minimizing official complicity in the historical crimes being assessed.

It is well documented that the Japanese government, mostly via the army, ran a vast, forced system of sexual slavery throughout conquered sections of Asia during World War II in which local women—Chinese, Koreans, Filipinos, Indonesians, and others—were forced, often at the point of a bayonet, to serve the sexual needs of the invading Japanese soldiers (often more than fifty men per day). They were given the euphemism "comfort women." The matter was well researched immediately after WWII based partly on interrogation of Japanese prisoners in connection with possible war crimes. Dutch investigators described the forcible seizure of Indonesian women, who were beaten, stripped naked, and then forced to sexually service large numbers of Japanese soldiers every day. Their sufferings were vividly described by some of the women themselves who had long hidden the true facts in shame but spoke out in the early 1990s, when the Japanese government initially refused to acknowledge the crime, much less make any amends. And of course this is one of the benefits of denial: the lack of any need for restitution.

Initially, the Japanese government was forced to accept these conclusions as part of a peace treaty signed with the Allied powers in 1951. It was thus more difficult later to deny them, but conservatives (nationalists) do deny the tribunals' conclusions, calling them "victor's justice." They assert that the role of teaching history is not to dwell on the dark and "masochistic" side but to teach history, however false, in which Japanese can feel pride. This is exactly what a false historical narrative is supposed to do: replace a potentially negative personal self-image with a positive one—or, more accurately, a negative image of one's ancestors with a positive one. Of course, with simple (largely erroneous) genetic assumptions, the two images are the same.

In 1993, the Japanese government finally acknowledged that it had managed the "comfort stations" but still refused to pay compensation. Even this meager step forward was contradicted by a recent prime minister who denied that the military had forcefully recruited the women, saying instead that "employment" had been arranged by "brokers." A prominent Japanese historian summarized the state of affairs in 2007 by saying that the system obviously was one of sexual slavery but "the movement to openly deny this has grown stronger in the government and elsewhere." This is only one of several examples of false Japanese historical narratives, including crimes such as the Rape of Nanjing and mistreatment of prisoners of war—not to mention the slaughter of more than twenty million Chinese in the 1930s and 1940s, a fact that has disappeared from world memory except in China and Korea.

Of course, there is an irony here, since the teaching of false history is merely a new source of shame, a new dark history, so there is no redemption but a deepening moral problem. By contrast, the Germans long ago confessed their crimes, with numerous benefits in improved relations with neighbors and others. They can be faulted only for being overly solicitous toward Israel, but this is at least an understandable reaction to their own past crimes. Note again the role of the honest and often courageous historian who tells the truth. In all the cases of false historical narratives, we know they are false because of the work of historians in the societies themselves, often a small minority and often risking their jobs and sometimes their lives.

The Japanese controversies highlight a larger problem in the teaching of history: to what degree is its function, especially in the young, to foster feelings of patriotism (self- and within-group love), and to what extent is its function to provide an objective view of the past, warts and all? Periodically the issue will burst forth in the UK press, for example, with some arguing that patriotism requires that Oliver Cromwell be taught as an exemplar of English manhood at its most manly while others say it would be better to emphasize that he was a warmongering, genocidal murderer who perpetrated huge atrocities on the Irish in the name of God and empire.

Or take an interesting case from within Japan. Okinawa, the southernmost island, was the last annexed into Japan itself (in the late nineteenth century), and there is a long history of derogating the Okinawans within Japan. Even the huge US Army base located (and unwelcome) there is a gift from the larger country. One little problem that recently arose concerns how to teach the end of World War II to Japanese children. The land invasion was aimed at Okinawa first and one-fourth of the civilian population was killed. Japanese imperial troops treated the locals brutally. They were indifferent to their safety, used civilians as shields, and finally, in March 1945, urged them to commit mass suicide before US forces started landing on the main islands. This was said to benefit the Okinawans because they would thereby escape the horrors and humiliations the Americans had in store for them: rape, torture, and murder. The benefit to the imperial Japanese (besides continuing to decimate a Japanese sub-breed regarded as inferior by their overlords) allegedly was to prevent Okinawans from actively helping the advancing Allied forces. This was both a hostile projection and a guilt-ridden one. If Okinawans had not for so long been mistreated, would their loyalties be so easily questioned? Some Okinawans fell in line and committed suicide, some even bashing sisters, brothers, and mothers to death. Others politely declined.

In the most recent twist, the Japanese legislature got reinvolved in 2007 and passed a law that promotes the teaching of patriotism in schools. Shortly thereafter, new textbook guidelines were announced that required the deletion of all references to the role of the Japanese Imperial Army in inducing mass suicides of Okinawans. Demonstrations ensued in Okinawa against this revisionism, which denied the cause of a particularly painful injustice at the hands of their overlords. More than 100,000 people massed in September 2007, the largest rally in Okinawa since its reversion to Japan from the United States in 1972. Two key pieces of evidence were that (1) mass suicides took place only where Japanese army units were stationed, and (2) grenades, which were precious weapons against the invaders, were given instead to the Okinawans to encourage group suicide. Textbook companies then petitioned the government to reverse the regulation, a change soon granted. This is a general (and welcome) feature of the three major "textbook controversies" in Japan. Nationalistic and right-wing forces arguing for a reversion to false historical narratives are often overcome by other forces in the society. Not so in Turkey.

TURKEY'S HOLOCAUST DENIAL

What about Turkey? What is this country's problem admitting to a historical crime now nearly one hundred years old? I refer, of course, to the mass extermination of nearly the entire Armenian subpopulation. Some of the ancestors of the present inhabitants indubitably launched a brutal campaign of genocide against their relatively successful and middle-class ethnic subgroup, the (Christian) Armenians. About 1.5 million were put to death in the space of a year and a half. In other words, 100,000 Armenians were being murdered every month. This decision was taken at the highest level of the Turkish government, and a key figure was later assassinated for his role. Yet to tell the truth about this monstrous crime now is to risk assassination on the streets or incarceration for "insulting Turkishness." It is explicitly against the law (article 305 of the Turkish penal code) to ask for "recognition of the Armenian genocide." As in Japan, official school curricula also ordered teachers (in early 2004) to denounce to their children "the unfounded allegations" of the Armenians, that is, to openly attack the truth. With this kind of historical amnesia and enforced falsehoods, it is perhaps not surprising that the great majority of Turkish people seem offended at the very notion of an Armenian genocide.

As we have seen, the younger the child, the stronger the force to teach lies. It may be fine for university students to learn that one's country was founded on genocide and that slavery was a horribly degrading arrangement, but surely we should spare our children such negative self-images. Elementary-school students across Turkey were recently forced to watch a film in which Armenians are portrayed as having stabbed their own country in the back during World War I, massacring thousands of (non-

Armenian) Turks, cooking their babies alive, and using civilians as firewood. This, of course, is the crudest kind of propaganda, reminiscent of the ancient claim that Jews killed Christian babies to use their blood to bake matzo, yet this new Turkish film is an official product of their Ministry of Education, ordered to be shown to all children.

Genocide is presumably never pretty, but just so we know what this one looked like, consider the following. The Turkish army might appear in a town and demand all Armenians to the center. Grown men would be removed at once and killed elsewhere. Babies' and children's skulls would be cracked on the pavement in front of their shrieking mothers. Attractive young women might be removed for later rape and reproductive use while the others were either killed or set on long marches without food, water, or protection from the elements. Sometimes the Turks would ask for all the children so they could care for them, but this care consisted of piling the children on top of one another and setting them afire. With the ruthless efficiency that genocide often brings, eighty people might be tied together at the neck, one shot dead, and all pushed over a gorge into a river below, the dead one sure to drag down all the others. Detailed accounts were dispatched at the time by diplomats and others. Survivors, even into the 1990s, had vivid memories of the atrocities they witnessed. Remember: three thousand a day were dying. The Turks even devised primitive gas chambers, in which large numbers of Armenians were herded into huge, low-hanging caves and then fires were lit at the entrances to suck the oxygen out of the caves and from their occupants.

What is perhaps more extraordinary than Turkish genocide denial is how governments around the world under pressure from Turkey fail to call genocide, genocide. It becomes instead a wartime "tragedy," so that on the world stage the country of Turkey can maintain this falsehood. Turkish spokespeople often talk of Armenians starving to death during warfare. They do not mention that the Armenians were driven from their homes and properties and forced into long death marches without food or water. Naturally, if one does not die of dehydration, one dies of starvation. Successive US presidents (including Barack Obama) have promised to use the word "genocide" on the official commemoration date in April only to turn coward when the date arrives. Eight former secretaries of state argued against using the dreaded word in a proposed US congressional statement (which was not passed).

Not only does Turkey threaten consequences for any such honesty, but it also follows through, as when it canceled more than $7 billion worth of military contracts with the French when their Senate passed a law in 2000 acknowledging the Armenian genocide. In 2010, it threatened to expel 200,000 Armenians said to be living in the country illegally in return for low wages. Thus, Turkey offers a false historical narrative to its own people and then insists that everyone else fall in line. Here it has been more successful than with the Japanese, whose rewriting of history is met by immediate hostility from its near neighbors Korea and China.

Even Israel and some Jewish Americans have joined in Armenian holocaust denial, the more surprising because the Jewish holocaust and the Armenian one share many features in common, including the eradication of a commercially successful group of different ethnic/religious persuasion (Armenians/Christians slaughtered by Turkish Muslims, and Jews by European Christians). Hitler consciously patterned his behavior after the Turkish example, including perhaps his gas chambers. "Who," he is alleged to have said, "still remembers the Armenians?" before launching his all-out assault on Jewish people. Fortunately, people still remember his victims, but Israel joins in denying the Armenian genocide partly because of pressure from Turkey (a close ally) and in part because the Armenian genocide is imagined to detract from the uniqueness of *the* holocaust. But there is nothing unique about the German holocaust of Jews per se, as events in the same century in the Congo, Turkey, Cambodia, Rwanda, and Sudan have shown. The notion of *the* holocaust has spurred the growth of an industry designed to extract long-ago costs of this event, which flow not to the camp survivors but to their distant cousins, usually nowhere near the camps, while serving to justify Israel's frequent attacks on its Arab neighbors.

Yet the Turkish genocide must in principle have had large indirect benefits for the remaining population—and this I believe is the key. The immediate loss of these skilled people puts you at a disadvantage in competing with neighboring groups, but their destruction allows you to occupy niches formerly denied to you because Armenians already occupied these places. The first is temporary, while the second is for keeps. That is, for a while you may do worse against other groups, but you will soon develop to fulfill the functions of those you have destroyed. Failing counter-genocide, your new position is all but secure. A whole series of non-Armenian Turks are benefiting every day from the absence of their former compadres, and this must make admitting to the genocide especially threatening—to the legitimacy of their own positions in society. Almost everyone must have moved up after the removal of the local Armenians, while a large Armenian country now sits next door.

A LAND WITHOUT PEOPLE FOR A PEOPLE WITHOUT LAND

A key original Zionist falsehood was the slogan popularized in the 1880s that the Jewish people needed to settle in Palestine because it was "a land without people for a people without land." Alas, there were plenty of people in Palestine. Even by 1920, after a wave of Jewish immigration, there were about 80,000 Jews in Palestine and more than 700,000 Arabs. Most of these Jewish people seemed content to live with their Arab neighbors the way they had for generations, but the Zionists had other ideas—a simple colonial project to occupy ("reclaim") places important in their religion. The Zionist project seems to have been set from the beginning, to entice enough Jewish people to Palestine with support from the colonial power of England and from Jewish people worldwide until they had enough power to seize Israel, which, when they did, involved expelling large numbers of Arabs, destroying or confiscating their property, and refusing them any right to return or compensation of any sort so as to produce a (now) 80 percent homogeneous Jewish state in lands some of their forebears had occupied a few thousand years ago. The Zionists were nothing if not consistent. Maps they drew up in the 1920s of their future state reveal later Israeli behavior remarkably well—they show Israel as including the West Bank and Gaza, as well as southern Lebanon, which

Israel did indeed occupy from 1982 until 2000, before Hezbollah finally drove it back out of Lebanon.

The notion of a people without a land occupying a land without people has been reinforced repeatedly since then. Of particular note was a book published in 1984 and widely applauded in the United States, where it was reprinted seven times in its first year alone because, among other things, "it could also affect the history of the future," which of course is precisely the purpose of such narratives. The brand-new and far-reaching claim of this book (*From Time Immemorial*) was that there had been a massive—but hitherto undetected—illegal immigration of about 300,000 Arabs into Palestine during the British mandate (1920–1947), attracted by the flowering economy produced by the industrious and intelligent Jewish immigrants. This explained away roughly half the Arab population.

The book argued that the inherent superiority of the Jewish immigrants attracted Arabs in search of economic opportunities, who then illegally occupied space that with any justice should have gone to new Jewish immigrants. In addition—absent a history of Palestinians occupying their own land—there is no current refugee problem or problem of compensation. The Arabs should simply return to where they came from in the first place. Is it any wonder that Zionists in the United States fell over themselves in praise of this pathbreaking and remarkable book, and still do? But in Israel the praise has been somewhat muted, since many know that the author cooked her demographic facts thoroughly to generate her novel results. The book is, in fact, a hoax. All the available evidence shows that a natural increase of about 2.5 percent every year, augmented by minor immigration (about 7 percent in total, primarily legal), explained the Arab population increase. In other words, most Arabs living in Palestine when Israel was formed had been there for their entire lives, as had their ancestors—if not from "time immemorial," then at least for several centuries.

The denial of Palestinian history is also built into Israel's school curriculum. As an Israeli historian has pointed out, the land that would become Israel has no history from the destruction of the Second Temple until the onset of Zionist settlement. It is only a religious image surviving from biblical times, the subject of Zionist yearning but (with the exception of the occasional arrival of Crusaders) it has no occupants. The Palestinians first appear during Zionist colonization in the early twentieth century, but then only as external obstacles to the Zionist project. Even the most recent textbooks (which delete some of the overtly racist content of earlier ones) do not have a single map of the land during Zionist colonization that includes all the human settlements, showing only the Jewish ones (and occasional mixed Arab/Jewish ones). There are no Palestinian towns or villages, no people with their own desires, aims, and conflicts. Instead, the Palestinians appear first because of opposition to their de-employment in the late 1920s, but the fate of the banned laborers receives no attention in retrospect (as it did not at the time). Palestinians then reappear only because of their later opposition to Zionist projects, opposition that is portrayed in racist terms.

THE FOUNDING OF THE STATE OF ISRAEL

Once the United Nations agreed to set aside a section of Palestine to form a Jewish state in 1947, the Zionists launched an ostensibly defensive war against surrounding Arab armies. The conflict ended up expanding the size of Israel from the UN-mandated 56 percent of Palestine to 78 percent. Israel's subsequent history is one of expansion outward and relentless attacks on the hapless, displaced Palestinians and on the nearby Lebanese—all to seize additional land and water and to terrorize the Arabs into submission. The policy continues to this day, with such regular events as the five-week bombing of Lebanon during the summer of 2006 (killing at least 1,300, while Israel lost about 160, mostly soldiers) and the slaughter of another 1,300 Arabs in Gaza in late 2008–early 2009 (Israel losing only 11). By the way, a 100:1 kill ratio is considered a successful war but a 10:1 ratio is a failure, and a 3:1 ratio drives Israel out of seized territory (southern Lebanon in the late 1990s). The United States is also willing to tolerate similarly gross disparities in mortality, with US deaths limited to fighting personnel. When America loses three thousand civilians in one day, the entire world trembles—

each dead 9/11 victim has been redeemed now by almost one hundred victims elsewhere.

This is not of course the story Israel told—to its own citizens or to the world at large. In its version, a brave set of souls set about reclaiming their natural birthright, that is, all the land, part of which their distant ancestors may once have occupied. They had a book that the same ancestors were said to have written that gave them the land in perpetuity from the God they worshipped. If this absurd rule were applied generally, it would require the wholesale resettlement of the world's peoples, with re-resettlement required by extending the time horizon backward. European Americans would be forced to return to their "homeland" in Europe so America could be returned to its rightful owners, the Amerindians, from which it had surely, and very recently, been stolen through wholesale slaughter and lies. But the Jewish Zionist dream resonated with aspects of what can be called Christian Zionism, especially in the United States. This, combined with horror at the recent genocide of six million European Jews, permitted the rule of "right of return" to lands through which one could claim an ancient connection, to be enforced in this particular case. But the reality is that a racialist (and then racist) country was shoehorned into the Middle East, so that Jewish people (half and quarter also) from around the world can immediately claim citizenship to this land but none of those who were so recently expelled could do so. This ethnic definition of Israel could only create pressure for expansion.

One consistent feature of the mythology is that the Zionists have always reached out in peace to their Arab neighbors, wanting only their fair share of the land, but these overtures always have been met by hostility and rejection. "The Palestinians never miss an opportunity to miss an opportunity," in the memorable phrase of Abba Eban. The first happened in 1928, when the British offered an assembly of Arabs and Jews. The Zionists accepted the proposal; the Arabs rejected it and grew restive. Yes, indeed, but the assembly was to be divided 45 percent to the Arabs and 45 percent to the Jews with the controlling 10 percent going to the British, who were already on record as pro-Jewish and were busily promoting Jewish migration (to the Holy Land). But the Arabs were roughly ten times as numerous as the Jews, so they would have been disenfranchising themselves by 90 percent as well as giving up all control. This is typical: a grossly unfair offer is made to the Palestinians, who reject it, and their rejection is described as a rejection of peace.

Again, as the story goes, Israel made a far-reaching compromise in accepting the UN Partition Resolution in November 1947, thereby recognizing Palestinians' right to their own state—all in the hopes of achieving peace with the Palestinians. But the Palestinians totally rejected partition and decided to launch a war on the new Jewish state, forcing it into a defensive war lest it be killed off before it could even begin. Actually, the war was launched in violation of the UN mandate by the Zionists as part of a tactic in a larger strategy of expansion and dispossession. The main aim was to increase the size of Israel, while preventing the formation of a Palestinian state. The latter was helped by a secret agreement with Abdullah of Transjordan, whose annexation of the territory meant for a Palestinian state was part of his own dreams of territorial expansion.

Although the war of 1947–1948 is often presented in retrospect as Israel barely escaping a new holocaust, the fact is that the Zionists were better armed, organized, prepared, and motivated than the surrounding Arab armies, and everyone knew it. The Arabs almost never launched an attack on Israel itself and did not try to intervene in the ongoing ethnic cleansing, even when they were observing it directly from a safe distance. Their function was to protect their own borders against Zionist encroachment. Israeli policy since then hardly seems to have changed. The country pretends in public to be interested in peace and a fair settlement, but these appear to be delaying tactics to camouflage the exact opposite: complete dispossession of the Palestinians and continual seizure of everything of value, especially land and water.

VOLUNTARY FLIGHT OR ETHNIC CLEANSING?

The original myth asserted that the flight of Arabs before and after the birth of the state of Israel was in response to calls from the surrounding

Arab armies and occurred despite strenuous efforts by Jewish leaders to persuade them to stay. This is complete nonsense. An invading army far from home has a serious supply problem, and a receptive local population is exactly what it needs. More to the point, directly after World War II, the Zionists appear to have adopted a secret plan for the ethnic cleansing of Palestine, by force of arms, terror, encirclement, starvation, and murder. At no time did they beg the Palestinians to stay behind. When the time came, the expulsion was followed by the deliberate destruction of the deserted villages to prevent the return of the displaced people to their homes and their land. There was an immediate economic incentive for the latter. The cost of settling fresh Jewish arrivals was nearly five times as high if they were settled on land that had not been recently "cleansed."

Then the Israelis set up the special Minority Unit to prevent the return of the Palestinians, even those merely trying to retrieve their possessions or harvest crops they had planted and their trees bearing fruit. Designated "infiltrators," they were shot on sight or "successfully shot," as it was put in official reports. This euphemism is echoed in current Israeli policy of assassinating Palestinian leaders or (much more often) low-level "militants" by specially trained assassination squads, said only to be engaged in "targeted killings" (a policy and terminology recently adopted by the United States). Actually, the policy of assassinating Palestinian leaders started long before Israel was founded, being a Zionist ploy used from the late 1970s onward, a kind of mal-genetics, in which the top end of a society is regularly purged to weaken the group generally.

Many of the larger cities were deprived of their Arab populations ("de-Arabized") very directly. Haifa was a particular horror, right after the massacre of the entire village of Deir Yassin. The orders to the Jewish troops were simple and direct: "Kill any Arabs you encounter: torch all inflammable objects and force doors open with explosives." On the April 22, 1948, the Arabs were streaming to the market and harbor under orders to evacuate. To make sure everyone got the idea and moved in the right direction, the Zionists then stationed three-inch mortars on the mountain slopes overlooking the market and port. The idea was to force the Arabs into the sea. When the shelling of the market began, the crowds did indeed rush in a panic toward the port, trampling one another to death and, in a desperate attempt to survive, attempting to commandeer all boats, many of which soon were swamped and sank. It is more than some passing irony that the Israelis often claim that the goal of the Arabs is to drive them into the sea, when historically the movement has been entirely in the opposite direction.

It is understandable why a people so recently and completely traumatized could believe that the end of their own safety justified any means. But what about today's people? Do they really wish to repeat these crimes? The challenge now is to talk about all these events honestly. At no time up until the present has Israel allowed any consideration of a "right to return" or to receive monetary compensation of any sort. Indeed, this has been explicitly ruled out while continuing to assert Israel's divine right, in principle, to all of Palestine. Jewish people have energetically sought and received compensation for property stolen by Nazis (or, say, their Swiss bankers) some sixty years later, but they fail to acknowledge any contradiction between this policy and the one they have taken when the shoe is on the other foot. They have the right to return to truly ancient land and the right to compensation for gross theft immediately prior to 1946, but Palestinians have no right to return to land stolen from them in 1948, land that they and their ancestors occupied for centuries. Nor do they have any right to compensation. To buttress both arguments, the Palestinians had to be stripped historically of ownership of their own land. This, as we have seen, was achieved both by denying their history and creating false versions of it.

ARAB DECEIT AND SELF-DECEPTION

The principles we are describing are universal. Surely if Jews in Palestine practiced deceit and self-deception, then so did the Arabs, and surely if Zionists do it today, anti-Zionists do it as well. There are, however, two important variables that one overlooks at one's peril: relative power and relative justice. If there is a growing difference in power, with the powerful more prone to self-deception, their unjust behavior requiring cover-up and rationalization, then there will be a positive association between power, injustice inflicted, and degree of self-deception. If you are the victim of injustice, simply telling the truth about it may be your best move. *Nakba*, or disaster, is the Arab word for what happened in 1948, and it is no accident that some right-wing Israeli politicians echo their worst Turkish counterparts in now insisting that the use of the word itself be made illegal.

Nevertheless, I see several strains of Arab self-deception. Certainly the Palestinians were slow to realize the danger they faced, they were slow to organize in response to it, and, perhaps worst of all, they often put their faith in neighboring Arab countries, whose leaders were too corrupt to act positively, instead often posturing and promising while secretly sabotaging them. King Abdullah of Transjordan was an early example, Hosni Mubarak of Egypt a more recent one. This kind of public posturing continues to this day, with Arab leaders, for example, begging the United States in private to attack Iran while maintaining a public illusion of impartiality.

Under Mubarak, though the Israelis denied vital supplies to those living in Gaza, so did the Egyptians, who shared a border with Gaza but showed scant concern for the welfare of their Arab brethren. Indeed, Egypt long ago sold out to the United States, whose large annual subsidies serve to build up "crony capitalism," which tends to enrich a favored few at a cost to the larger nation. In addition, Egypt's plutocrats faced the Muslim Brotherhood, a much more serious and principled set of people than themselves, and these people resembled Hamas too much for the comfort of the then-rulers of Egypt. Better to starve their Arab "brothers" in Gaza. The same pattern is true of much of the Arab world, interests of the people sold out to an elite often based on explicitly anti-Islamist views, in tandem with US and European interests. Sometimes conflict with Israel is also used to rationalize the suppression of one's own citizens. A forty-nine-year state of emergency in Syria has included every kind of arrest, torture, and murder based on the theory that they are at war with Israel. Yes, Israel still has Syria's Golan Heights (and is busy "settling" it), and yes, Syria has been very good at arming and supporting the only successful anti-Israel force, Hezbollah of Lebanon. But what does this have to do with suppressing their own people? Why a state of emergency to arm Hezbollah? And what is being done about the Golan Heights?

Perhaps the greatest Palestinian daydream has been the belief that Israel might actually live up to agreements made. In 1994 at Oslo, Palestinians made major concessions in exchange for Israeli promises, which were not kept, and the Palestinians had no leverage to prevent this. Israel continued to settle the occupied territories, install Israeli-only roads, mark off sections of the West Bank as Israeli security zones, and so on, all while pretending not to.

CHRISTIAN ZIONISM

The elephant in the room of Israeli behavior is the United States. No way would Israel act as it does to its neighbors if it did not have the active, massive support of the world's great superpower. If you have a dispute with your neighbor, and you have a large, ferocious dog behind you, while your neighbor stands alone, you may be tempted to overstate your case. In that sense, Israel has repeatedly acted in a much more aggressive fashion because the United States gave tacit or full support, while underwriting the Israeli military to the tune of more than $1 billion a year. Where does this support come from?

It is usually not appreciated that long before there was the Jewish Zionism of the 1880s, there was something called Christian Zionism. It was alive and well in the United States in 1810 and has been a powerful force ever since. Its roots in Europe go back well into the sixteenth century. The movement has transmuted into various forms, but underlying it is the Bible and a shared story of expansion and ethnic cleansing glorified as God's will. As the American writer Herman Melville enthused, "We Americans are the peculiar, chosen people—the Israel of our times; we bear the ark of the liberties of the world."

This was a neat trick, stealing the mantle of the chosen people away from the chosen people, and here's how it worked. The self-proclaimed chosen people were, indeed, the chosen people because they were chosen to give rise to Jesus Christ the Savior (God incarnate). But when Jewish people then rejected Jesus, they became the unchosen people, while those who embraced Jesus became the new chosen people. The new chosen people had an ambivalent relationship with the old chosen ones. On the one hand, the usual out-group derogation and racism ("Christ killers!") was practiced. On the other hand, the two shared a book. Jewish people had not just given rise to Jesus, whom they rejected, but also to the Old Testament, which the Christians wanted. Common elements of history only deepened this connection—genocide of surrounding peoples celebrated in the Bible, new land being settled, racial superiority, a shared creed based on God's own word.

In 1891, four hundred people signed a petition and delivered it to the US president, Benjamin Harrison, calling on him to induce the Ottoman Empire to turn Palestine over to the Jews. Most signing the document were not Jewish but included the country's elite in all realms, the chief justice of the Supreme Court, the Speaker of the House, key chairs of committees, a future president, the mayors of major cities, owners and editors of major newspapers, major industrialists, and top Christian clergy of every stripe. This was no plot hatched by a Jewish subgroup. This was US Christianity rising to its full moral heights and anointing Israel the chosen land for the unchosen people. And this ambiguity continued. One advantage for Christians of having Jews return to Israel is that there would be fewer of them nearby.

Harry Truman worked tirelessly after World War II against both his own State Department and Great Britain (the colonial power that had created the mess in the first place) to establish the country of Israel. He was a biblical literalist and a Christian Zionist. He also noted that there were almost no Arab voters (nor rich ones) in the United States. The Old Testament said Jews belonged in Israel. That and being appalled at the Jewish holocaust and the postwar treatment of European Jews was enough to get him on board, contrary to the UN's plan. Israel circumvented the latter by at once declaring itself a state, going to war, and then using ethnic cleansing to achieve a largely homogeneous state, itself more than 50 percent larger than the one the UN envisioned.

In what must count as one of the more bizarre recent scenes from US Christian Zionism, Defense Secretary Donald Rumsfeld in April and May 2003 fed President Bush a daily set of intelligence briefings on the Iraq war whose covers juxtaposed dramatic wartime scenes—a US tank rumbling through the desert—with exhortations from the Old Testament: "Therefore put on the full armor of God, so that when the day of evil comes, you may be able to stand your ground, and after you have done everything, to stand." Or along with a picture of Saddam Hussein striking a dictatorial pose: "It is God's will that by doing good you should silence the ignorant talk of foolish men." Some in the Pentagon were conscious that if these covers were to become public, Muslims (among others) might well interpret this as evidence of yet another Crusade against them backed by biblical prophecy. Rumsfeld appeared intent on manipulating Bush, who was known to frequently quote the Bible (unlike Rumsfeld).

FIRST LINE OF DEFENSE: CRY "ANTI-SEMITE"

The naive reader should be aware that in criticizing Israel for its racist and/or unjust policies toward Arabs, you at once risk being called an anti-Semite, that is, someone who has a (racial) bias against Jewish people (or, if Jewish, a "self-hating Jew"). The term has now been so degraded by its frequent use in defense of injustice that its actual meaning is inverted—it is now usually a racist term used by those who support racist policies against those who do not. Or, better put, "an anti-Semite" used to mean someone who hates Jews; now it means anyone Jews hate—a simple case of denial and projection.

It takes more than showing that a person speaks out against Jewish-perpetrated injustice to show that he or she is anti-Jewish. Perhaps he or she is merely anti-injustice. But the anti-anti-Semites have an answer for this. Why are you picking on *us*? Are there not worse people in the world? According to this view, you must rank the world's injustices from biggest to smallest, then criticize everybody above Israel before you are permitted to criticize Israel itself. When you have finally reached Israel, though, a new rule is imposed: balance. If you concentrate only on Israel's manifest injustices—let us say its regular attacks on its northern neighbor, Lebanon (1976, 1982, 1996, and 2006) or its remorseless theft of Palestinian land, water, indeed life itself, all based on terror and subjugation—you are being unbalanced. For every Israeli transgression, you must show a parallel Palestinian one to demonstrate lack of bias. But this is of course impossible (given reality). The best you can come up with are suicide attacks and some poorly guided missiles that claim fewer than one-thirtieth of those being killed by the Israelis during the same time period. So much for balance. Finally, should you come up with an argument that is strong in logic and content, you are said to make "tendentious" statements against Israel. This is a possible case of a malphemism treadmill (see Chapter 8).

Many first-class minds in mathematics, the sciences, and many other intellectual pursuits are Jewish (or partly Jewish). But this intellectuality can have a downside. Greater intellectual talent may be associated with more deception and self-deception (see Chapters 2 and 4). Where Israeli misbehavior is concerned, this has the unfortunate effect that an enormous amount of blather in defense of the indefensible pours out from every corner. This ranges from the truly rabid and racist—with full bells and whistles—to much more subtle arguments in which small, key errors are well concealed. UN Resolution 242 calls for Israeli withdrawal from lands occupied in the 1967 war—but not "the" lands. Even though "the" appears in the French version of the resolution and there is no mistaking the UN's intent, this missing article is used to assert that the UN deliberately called for Israeli withdrawal from some but not all of its occupied land. And because the UN never specified which land should be relinquished, any withdrawal would satisfy the UN—a few square meters if put to the test. Or take another piece of sophistry. Israel declares that it is necessary for its neighbors to acknowledge Israel's "right to exist" before diplomatic relations can be sought, but nowhere else in the world is this a prerequisite. You recognize that a government exists and you set up diplomatic relations—nowhere do you assert that the government has a right to exist. In addition, Israel is unusual in failing to define its own borders, so recognizing its right to exist may have hidden implications regarding future ownership of land. To take but one example, Israel has taken care to build about 85 percent of its security wall outside of Israel, creating new borders and a larger country by fiat.

Thus, on the subject of Israel, a vast wave of biased argumentation washes over people who have not had (or taken) the chance to study the matter carefully. The key is a fundamental inversion of reality: The Palestinians are not displaced people, driven from their homes and their land and persecuted ever since. They (and Arabs more generally) are terrorists—virulent anti-Semites—against whom all is permitted. What looks like Israeli terrorism and relentless theft of land and water is really just a proactive campaign to prevent another holocaust (apparently by inciting the very feelings that would invite one).

The truth about Israel's theft of Arab land and water since 1967 via "settlements" was well put by a pair of Israeli historians:

> Deception, shame, concealment, denial, and repression have characterized the state's behavior with respect to the flow of funds to the settlements. It can be said that this has been an act of duplicity in which all of the Israeli governments since 1967 have been partner. This massive self-deception still awaits the research that will reveal its full magnitude.

As is so often true, what can be said in Israel is usually more honest and detailed about Israel than what can be said on the same subject in the United States.

WHY FALSE HISTORICAL NARRATIVES?

False historical narratives are important because every country has one, they are often fiercely defended (and regularly upgraded), and they provide a strong underlying system of logic (easily biased) for interpreting social and historical trends and truth. In short, they are available to justify all action—contemplated, under way, or accomplished. Deception is often involved in their construction. That is, people consciously lie to create them, but once created, false historical narratives act as self-deceptions at the group level. Most people are unconscious of the deception that went into constructing the narrative they take to be true.

A true historical narrative might force us to make reparations for past crimes and to confront more directly their continuing effects. A false one permits us to continue a policy of denial, counterattack, and expansion at the expense of others. Why do we continue to attack our Arab neighbors? Well, because they have long harbored racist animosity toward our Bible-ordained project. Why are we attacking Iraq? Because it is part of our divine mission, our "American exceptionalism" that requires us to interfere and sacrifice for the good of the world.

Inevitably, false historical narratives will have their deepest connection with religion: Where did we come from and with what aim? To that subject we shall return, but first we consider self-deception and war, to which false historical narratives make their own contribution.

CHAPTER 11

Self-Deception and War

It has been said that truth is the first casualty of war. Actually, truth is often dead long before war begins. Processes of self-deception make an unusually large contribution to warfare—especially in the decision to launch aggressive ones. This is as depressing as it is important: one of our most critical behaviors, often with huge, widespread costs, appears to be strongly ruled by forces of self-deception. There is, indeed, a subdiscipline of military studies devoted to the study of military incompetence, and this does not usually refer to computational error. It refers to biased and self-deluded mental processes. Think Custer's Last Stand.

Faulty decisions are said to arise from four main causes: being overconfident, underestimating the other side, ignoring one's own intelligence reports, and wasting manpower. All are connected to self-deception. Overconfidence and underestimation of others go hand in hand, and once self-deception is entrained, the conscious mind does not wish to hear contrary evidence—even when provided by its own agents, whose express purpose is to provide such information. Indeed, the old rule was to shoot the messenger. Likewise, self-deception will make it more likely that manpower is underestimated (vide US invasion of Iraq in 2003) or employed along illusory lines of attack. In the military it is said that "amateurs talk about strategy, professionals talk about logistics." False logistics easily feeds overconfidence and vice versa.

For faulty logistics (and other acts of self-deception), Napoleon's invasion of Russia provides the classic example. In an extreme act of overconfidence, he grossly underestimated the enemy, the harsh conditions of the Russian winter, and, most critically, the problem of supply. When he reached Moscow, he was more than a thousand miles from home and his men and horses required 850 carts a day for their care alone, never mind the additional carts needed to transfer weapons, medicine, the injured, and so on. There was no way such a feat could be sustained, so the French were forced to live off the land, but of course the Russians did their best to make this difficult. Stuck without resources far from home, with no ability to seize Moscow (or clear advantage in doing so) and the Russian winter closing in, Napoleon was forced to withdraw. He marched in with 450,000 men and returned with 6,000. Even worse, he lost 175,000 horses. The men could be replaced; the horses could not. After another disastrous foray by Napoleon into Russia a year later, the Russian army stood outside of Paris. It was the problem of supply that broke the back of the overconfident warmonger. Napoleon had been very successful before his disastrous Russian adventure. This is a deep feature of self-deception: success entrains confidence but also overconfidence. How many of us have taken success *one* step too far? (Bill Clinton and his women?)

Here we will gain an overview of the evolution of warfare in humans and the growing role self-deception plays. Besides such classics as World War I, we will concentrate especially on recent wars where the facts are well known: the US war on Iraq in 2003 and the US-supported Israeli assault on Gaza in 2008. These are not to suggest that the war in the Congo is not more hideous than all other ongoing wars put together, and probably more deserving of analysis in terms of deceit and self-deception, but the relevant information for this is far more meager than for the recent US and Israeli wars.

CHIMPANZEE RAIDING → HUMAN WARFARE

Chimpanzees reveal a likely route to human warfare. Chimps raid other groups, or, more precisely, usually three or more chimpanzee males working together will watch a neighboring group until they spot a chance to make a lightning strike on an isolated male (or occasionally more), who is attacked and killed. The marauders quickly return to the relative safety of their own territory. If over a period of time enough males are killed, the killers may take over their neighbors' territory, along with some of the surviving females, but with even a single rival dead, the killers can expect to gain a little more territory and, thereby, food. At the Gombe Preserve in Tanzania in the 1970s, one group of chimps appeared to pick off and kill isolated males in a neighboring group until, after four years, all seven were gone. In another area of Tanzania, five adult males in their prime disappeared under similar circumstances, and after ten years the entire group was gone, with most of the females (and territory) absorbed by the larger (murderous) group. Attacks appear to be carefully planned, that is, launched when there is a clear likelihood of success—an isolated male is quickly overwhelmed by a superior force acting in silence.

In both chimpanzees and our own lineage, primitive warfare—or raiding—was a male territorial strategy based on the coordinated murder of neighboring males. The benefits were increased access to resources, including, in some cases, adult females—in either case, a net increase in reproductive rate. Deception by attackers was based primarily on hiding and surprise, with traps on the other side unlikely. Recently, remarkable evidence has surfaced of ten to twenty males engaging in regular warfare against a neighboring group. About every two weeks, males are drawn by some unknown signal to walk very quietly, single-file, into a neighboring territory to attack a vulnerable male. Infants are often killed, as in animal infanticides more generally, the better to bring the mothers into reproductive readiness. Likewise, an adult female is sometimes killed, but the overwhelming targets are other males.

This pattern of intergroup male raids leading to murder and later territorial expansion probably lasted in our lineage for several million years, undoubtedly increasing steadily in subtlety and design. Detailed studies of surviving hunter-gatherers suggest that intergroup war was widespread and dangerous. The best data from both archaeological sites and current hunter-gatherers suggest an astonishing 14 percent of human mortality every generation due to war (a percentage that thankfully has declined steadily since then). Killers were almost always men, as usually were the victims. Circumstances varied from massacres of vulnerable strangers encountered by chance to deliberate forays in search of victims in distant groups. The key was usually overwhelming advantage in surprise, numbers, or technology. Sometimes surprise consisted of inviting people to a peace banquet and then slaughtering them. Evidence from slash-and-burn agriculturalists (such as the Yanomamo of South America or the Dugun Dani of New Guinea) suggests that raiding often resulted in killing but that battles were rare, largely ceremonial, and ended badly only when one displaying group discovered to its dismay that it was greatly outnumbered, after which it might well be massacred. Participation in warfare was voluntary, but since attackers were rarely killed, it was, for them, not very dangerous.

The emergence of battles—conflict between massed warriors on each side—is much more recent, almost certainly connected to the large increases in the size of human societies about ten thousand years ago associated with the introduction of agriculture and animal husbandry. With these battles involving large numbers of soldiers, several new elements came into play. Relevant information was apt to be much more scarce, the outcome harder to predict, the opportunities for fooling the opponent greater, all of which are more congenial to self-deception. Overconfidence emerges as a key variable, a factor that by itself can create wholesale slaughter, especially when practiced on all sides (witness World War I).

Perhaps worst of all—from an evolutionary perspective—there is now lower negative biological feedback on those making bad decisions. You decide, hundreds die—but do you also die or even suffer? If you choose to attack an apparently isolated male chimp in a neighboring group and you miscalculate, you may lose your life. That is, natural selection acts directly back on any self-deception that helped produce the mistake. The same was probably often true of primitive warfare. Of course, it is sometimes true of those initiating large-scale wars: not only may your own country be invaded and your relatives slaughtered or suppressed, but you too may be killed—vide Adolf Hitler, whose thousand-year *Reich* ended with his own pathetic death by suicide in a concrete bunker only six years after he launched his disastrous wars. Still, in terms of natural selection, this was one, or a few men, who launched wars with aggregate costs of probably more than sixty million people killed.

Even minimal evolutionary feedback to leaders is not necessarily the case. The war on Vietnam was a disastrous miscalculation, violating a fundamental US Army doctrine: no land war in Asia. It cost more than fifty thousand US lives and well over a million in Vietnam, and another million in Cambodia and Laos, while bringing on the Khmer Rouge in Cambodia with an additional million or so slaughtered. It also left behind an ecological disaster that to this day is producing, among other effects, horribly mutated and deformed children. It produced no known "strategic" benefit. But those who designed and propagated the war in the United States suffered no such adverse effects. Neither JFK's advisers—"the best and the brightest"—nor LBJ and his, nor Nixon and Kissinger suffered, so far as we know, any adverse consequences to their inclusive fitness. In other words, there may well have been stronger selection against warlike stupidity and self-deception in chimpanzees than in ourselves where the decision-makers are far removed from the biological consequences of their decisions. Herbert Spencer summarized the general effect: "The ultimate effect of shielding men from the effects of their folly is to fill the world with fools."

The switch to large-scale battles and warfare that may stretch for weeks and even years has several important consequences for self-deception. Predicting the future is far more difficult than in a single cross-border chimpanzee raid, and there is opportunity for bluff on a large scale. One may bargain in bad faith. It may also be necessary to convince the home population or onlookers that the war is worth fighting or supporting—in any case, not opposing. These generate a whole host of new opportunities for deceit and self-deception. Recent wars, such as the 2003 US war on Iraq, for example, are more of this kind: not fooling your opponent but your own citizens and, if possible, the larger world.

SELF-DECEPTION ENCOURAGES WARFARE

Evolutionary logic suggests that self-deception is expected to be especially likely (as well as costly) in interactions with members of other groups. In interactions *with* group members, self-deception is inhibited by two forces. Partial overlap in self-interest gives greater weight to others' opinions, and within-group feedback provides a partial corrective to personal self-deception. In interactions *between* groups, everyday processes of self-enhancement are uninhibited by negative feedback from others or by concern for their welfare, while derogation of the outsider's moral worth, physical strength, and bravery is likewise unchecked by direct feedback or shared self-interest. These factors result in systematic faulty mechanisms of assessment, in turn making aggression more likely and contests more costly (without any average gain). Processes of group self-deception only make matters worse. Within each group, individuals are often mis-oriented in the same direction, easily reinforcing one another, while absence of contrary views is taken as confirming evidence (even silence being misinterpreted as support).

When you and an opponent who are fairly equally matched face off in an escalating fight, each has to decide how long to persist—given that if one is going to lose anyway, it is better to lose early and thereby lower the costs. There seem to be several reasons why in an even match, a positive illusion that exaggerates your own competitive abilities and chance of prevailing may reduce your chance of losing (along with the cost of battle). The positive illusion increases self-confidence and, therefore, apparent competitive ability and motivation. It decreases signals indicating fear and other emotions that would undermine the effectiveness of any threats. It therefore increases the chance that the opponent will view you as unbeatable and will give in outright (or be so scared that he fights poorly). The positive illusion may actually make you more effective mentally, because it reduces cognitive load by making you focus on positive strategies that may work instead of the full range (although there is, of course, a risk in inattention to the downside). In short, positive illusions may be important in a fight, because we partly commit more resources to it. On the other hand, we will suffer less ability to read our opponent and fail to respond appropriately to negative information.

Sports would provide a useful parallel, but there has been precious little useful study of self-deception in sports. It would be interesting to have data from sports. Are more fearful individuals worse at sports, since to be good at competition it helps to think you are going to win, which is easier the less fearful you are about losing? The only evidence I know of comes from swimming. Individuals who are more likely in a choice situation to concentrate on negative rather than neutral stimuli do worse, while those who concentrate on positive over neutral do not do any better.

It is a striking fact that almost every category of self-deception we have described in this book is conducive to aggressive wars. Modern war is conducted against an out-group by powerful people who have an exaggerated opinion of themselves and their degree of morality, are overconfident, often have an illusion of control, enjoy taking risks, and are almost always male. Let us briefly review these biases.

The general bias to consider oneself superior to others is obviously congenial to waging war, where these positive traits include strength, endurance, fighting ability, and so on. Both sexes display this bias. Derogating others is especially dangerous if it both incites your aggression and prevents you from seeing the power and tenacity of the resistance your aggression is likely to engender. Overestimating your own morality is a critical bias since it naturally leads you to overemphasize the strength of your own position and to underemphasize that of your opponents. After all, when you invade your neighbor's country, there is already a prima facie case in favor of the neighbor and an expectation of a "home field" advantage (see page 255).

All of the above feed into overconfidence, one of our deeper and deadlier delusions where fatal aggression is concerned. Men seem especially prone to overconfidence, as we have seen already in financial trading, where they trade too often and lose more money compared to women (see Chapter 8). It is likely that a much longer history of being evaluated for degree of confidence, both in male/male interactions and in courtship of females, has led to greater degrees of overconfidence supported by deeper structures of self-deception.

A related variable is thrill seeking. A tendency toward thrill seeking can be measured by choice of risky driving, risky sports, drugs, and gambling. Measured this way, men are much more prone to thrill seeking than are women. Of course, wars can be very thrilling, at least at the beginning. It is also easy to imagine that an illusion of control gives greater impetus toward war. If you think you can control events favorably after initiating an (often surprise) attack on your neighbor, you are more likely to do so.

War is waged by the powerful. They decide and typically send others to die. Being put in a position of power—made to feel powerful—reduces one's orientation to the viewpoint of others, their welfare, and their emotions (see Chapter 1). So warlike decisions will be helped along by the biases that power induces. Except for some civil wars, typically wars are fought by an in-group against an out-group. (Even in civil wars, in-groups and out-groups can be, and are, quickly formed.) As we have seen, few distinctions are as powerful in our psychological lives as that between in-group and out-group, with the latter easily inviting derogation, dehumanization, and overt attack, with the aim of elimination or subjugation. This must have begun long before human warfare, in intergroup violence in our chimpanzee (and more distant) past. But warfare presumably intensified the negative consequences and connotations of being an out-group member.

An additional sex difference is highly pertinent. There are several lines of evidence to suggest that men are likely to be less compassionate toward others than are women. They are less likely to read emotions correctly from facial expressions (the sex difference persists even when the time given per expression is only one-fifth of a second). They are less likely to remember emotional information and relate it to the emotional reaction of others. And there is evidence that men are much less likely than women to show compassion toward others who are perceived to have acted unfairly against them. For example, women treated either fairly or unfairly by a partner in an artificial economic game show similar evidence of compassion toward the two when either is being given an electric shock. By contrast, men show no neurophysiological evidence of compassion toward the unfair person when that individual is subjected to electric shocks. Indeed, they show pleasure in inflicting pain. This predicts a male bias toward self-deception whenever it can be based on perceived unfairness. Moralistic outrage in men is expected to be especially heartless and easy to manipulate—toward war, for example.

False historical narratives also contribute. An honest narrative might force people to make reparations for past crimes and to confront more directly their continuing effects. A false one permits a continual policy of denial, counterattack, and expansion at others' expense.

DEROGATION OF OTHERS → FATAL OVERCONFIDENCE

The derogation of the abilities—never mind moral value—of others can have immediate, dangerous consequences when contemplating war, especially if the other people are assumed to lack fighting ability or motivation. Contrary to conventional wisdom, for example, it has been shown that planning for the US war on Vietnam in the 1960s was rational and calibrated on almost every point except one—the United States underestimated the discipline of its opponents and their willingness to absorb punishment.

This mistake is especially striking, since there is a near-universal rule in animal behavior (including that of humans) of a "home field" advantage: the lizard wins in his or her own territory against an opponent he or she would lose to in the opponent's territory. Motivation is stronger to protect what you have than to seize what you don't. Why would one ever assume that the locals cared less about your invasion of their land than you do? But nationalist and/or racist conceptions tempt us to exactly this position. Home-field advantage is true of sports teams in front of their own fans—where they show both a boost in testosterone not enjoyed when playing on the road, as well as a greater chance of winning. Anything that tends to increase fan effect (such as domed stadiums) tends to increase home-field advantage. But much of the effect may be due to the referees who show an increasing bias toward the home team with the same factors. Incidentally, in baseball there is something called "home-field clutch." In the World Series, the home team wins more often during the first six games, but if the playoff goes to a final, seventh game, win or lose, home teams lose more often than outside teams. Apparently the pressure is just too much.

Regarding self-deception and war, it is easy to forget that a human holocaust such as World War I, which lasted a full four years with twenty million dead in the fighting alone, was launched in a festive, holiday spirit, partly grounded in nationalistic and racist views of the opponents. All sides were convinced that the war would be short, they would win, and victory would bring benefits, none of which (excepting victory for a few) turned out to be true. There was dancing in the streets throughout Europe and men rushed to recruiting offices, lest the war end before they could enjoy the fun. In August 1914, hundreds of thousands of people in Paris and Berlin celebrated the outbreak of the war. Only three months later, 300,000 French and Germans were dead, with a further 600,000 injured—and the war still had four years to run. When the first (enthusiastic) British infantry arrived in France in October, they started losing five hundred men a day until three weeks later, scarcely an eighth of the original soldiers remained. Thus does fantasy collide with reality.

Each expected to best the other one. The Germans believed the French were not prepared for a fight, while the French expected a quick victory, and a British officer predicted that Germany would be "easy prey" for the British and the French. Austria and Russia both expected to beat each other. Russian officers believed they would reach Berlin within two months. Turks got caught up in the frenzy and imagined that after victory in the Caucasus, they might very well march through Afghanistan into

India. There were exceptions, of course—midlevel German military officers were in no way convinced of quick or even ultimate victory, but no one paid them any mind.

A classic case of overconfidence was based in considerable part on deeply held racist attitudes by the British toward Turkish people in general and their army in particular. The notion that any and all British troops must be superior to their Turkish counterparts was widespread throughout all levels of British society, and the commander at the disastrous Suvla Bay invasion believed that British soldiers must win every time because they were superior to Turks in such well-defined traits as "ideals" and "joy in battle." The British soldier was worth several dozen Turks, he declared, though the actual statistics of the battle suggest the reverse, one Turk worth about ten British when fighting on Turkish soil.

THE 2003 US WAR ON IRAQ

From the outset, the US war on Iraq in 2003 was drenched in deceit and self-deception. Using the false pretext of 9/11, it was a war of choice and aggression apparently designed for control of oil and related economic assets, as well as to build a regional power base and to support its joined-at-the-hip ally, Israel. It was of course sold under patently false pretenses. If the world survives, this war will surely be taught as a textbook case of a colossal military blunder entrained by deceit and self-deception.

One nice feature is how much of the internal deliberations are already known to us so that the underlying processes can be studied in detail. Although as usual, overconfidence was a key factor, another was the one we have seen so vividly with NASA (see Chapter 9). When you are selling a lousy product under false pretenses, you do not wish to hear about the downside. This was not a war in which the adversary needed to be fooled or in which capturing the capital and routing the enemy could be in any kind of doubt. So there was little or no self-deception to deceive the enemy on this point—all of the self-deception was directed toward internal and international consumption and had to do with the aftermath of this action and its beneficent effects, for which no rational planning was seen as either necessary or desirable, turning a blunder into a catastrophe.

As has well been said, if Iraq's major exports were avocados and tomatoes, the United States would have been nowhere near the country. Of course, this had to be denied, but two small facts alone symbolize the truth. When looting broke out in Baghdad within days of US arrival, the United States did nothing to defend the treasures of this great civilization, libraries and museums of immense value (despite repeated pleas from the relevant Iraqis), but they did station guards in front of the oil ministry (of trivial importance even to the oil industry). Likewise, before all hell broke loose in Iraq, the United States announced that any country that did not participate in the invasion would be frozen out of bids for oil reconstruction and redevelopment projects. Nothing was said about the avocado industry.

The war's rationale was based on two unlikely falsehoods: that Saddam Hussein had weapons of mass destruction (WMDs), including nuclear warheads, and that he had somehow been involved in perpetrating 9/11. Evidence, however feeble, was organized to support both claims that were then widely trumpeted as true. Was the administration lying or simply mistaken? A very nice linguistic analysis suggests lying. When making statements about the (in fact missing) WMDs or Iraq's (in fact nonexistent) connection to bin Laden, the statements showed the classic signs of deception we saw in Chapter 1 (compared to statements on neutral subjects made by the same people). That is, the first-person pronoun ("I," "we") was sharply reduced, the better to reduce personal responsibility. Exclusive words ("although it was raining") were also reduced, the better to avoid complexity, cognitive load, and the need to remember. Negative words were increased, perhaps due to denial or even unconscious guilt. The only variable that ran in an unexpected way was action words. These were reduced, perhaps because planned action was then being denied. It is notable that the linguistic features predicted lying better in real life than in the lab—just as expected, since the consequences of lying in the lab are usually trivial compared to those of lying on the international stage.

These falsehoods and the underlying aggressive logic of the war also entrained a series of self-deceptions with very unfortunate consequences. Chief among these was the denial of the enormity of the undertaking and the need for careful, adequate advance preparation, including the use of far more troops on the ground. An outline of the key events goes as follows.

The war was decided on very quickly, with a minimum of deliberation. Although "regime change" in Iraq was declared official US policy in 1998, no planning for an invasion seems to have occurred prior to 9/11. Iraq, in turn, had nothing to do with this event—indeed Saddam Hussein was more anti–bin Laden than was the United States. Nevertheless, the US government immediately turned its attention to Iraq. Bush asked counter-terrorism expert Richard Clarke and others to gather any evidence of Iraqi complicity, and the next day (September 12) Secretary of Defense Rumsfeld made a revealing remark. "Afghanistan," he said, "has no targets worth bombing, while Iraq has many." Presumed translation: Afghanistan has no resources worth coveting, while Iraq has them in abundance. Within weeks, lower-level Pentagon generals knew that the United States intended to attack Iraq and that an ambitious five-year plan had been drawn up for the successive attack of a series of countries *after* Iraq: Syria, Lebanon, Libya, Somalia, Sudan, and finally (the prize) Iran. Under one popular neoconservative version, the inspirational example of Iraqi freedom would lead to a domino effect, similar to the fall of the Berlin Wall, causing many of the above countries to embrace democracy and the United States without the need for invasion. The immediate stimulus for all of this cogitation was an attack on the United States emanating from Afghanistan.

Here was an imperial fantasy to fit the full grandeur of American exceptionalism and manifest destiny. In a unipolar world, the United States is now an empire, the greatest ever, and an empire creates its own reality, something people in the "reality-based" world do not appreciate. The United States will seize on 9/11 as a pretext to launch a series of interconnected, aggressive wars with lightning speed and increasing beneficial effects on themselves and others, including those being invaded. Here self-deception is directed at the nation and the world. The more we are convinced by our fantasy of imperial action, the more easily we can unite ourselves and others in pursuing this fantasy. But fantasies, as we have seen, are intrinsically dangerous—we do not wish to hear information, however important and true, that would disturb the fantasy.

Six weeks after the attack from Afghanistan, in mid-November 2001 a secret, formal order was given by President Bush to the secretary of defense to initiate detailed planning for the invasion of Iraq. Within months, resources and personnel were being moved from Afghanistan to just outside Iraq. The United States decided to go to war very quickly and then maintained a long (and unconvincing) public posture that war was the very last resort. As we know, this decision had a catastrophic effect on the future of Afghanistan—the second US abandonment in twenty years, with effects reverberating to this day. Psychologists long ago showed that when we are deliberating a decision—such as whom to marry or what job to take—we are willing to consider contrary evidence and to evaluate alternatives rationally, that is, with reference to benefits and costs. But once we have decided—to marry Susie, or take that job in Beirut—we no longer wish to hear about the choices not made or the possible downside to the decision we have made. We are now in the instrumental phase; we are carrying out our decision. By deciding within hours to attack Iraq, on entirely fictional grounds, with hardly any follow-up appraisal, the period of rational decision making was truncated to nothing at all. Once the decision was made, not only were alternatives no longer embraced, but there was no appetite to hear any evidence regarding the potential costs—the downside to this whole enterprise. Quite the contrary, there was active avoidance of any such evidence, while seeking out the flimsiest kind of evidence to hype the upside.

In the process, the United States easily imagined that the invasion was seen as good for the Iraqi people. The United States will naturally act in the interest of the Iraqis, who will appreciate this fact, so the United States will be seen as liberators and not occupiers. This was backed up by such reliable informants as "Curveball" and the notorious con artist Ahmed Chalabi. Here the US thinkers were swallowed by their own self-deception—denying that this was an aggressive war to grab control of very valuable resources, they embraced the countervailing fallacy that they were doing this for everyone else's benefit—ridding the world of the threat of nuclear war, weakening terrorists worldwide, and (especially when these rationalizations became implausible) freeing the Iraqi people of their longtime oppressor and our former dear ally, Saddam Hussein. American exceptionalism out there pulling off another exception.

The decision had to be sold to the larger country. This consisted of emphasizing that the goal was important and legitimate—preventing an imminent terrorist attack—as well as safe and inexpensive. There would be no casualties (or very few), asserted Bush to White House visitors days before the attack, and Paul Wolfowitz of the Pentagon assured Congress that the war would be inexpensive, perhaps on the order of a few billion dollars, that Iraqi oil would pay for the reconstruction, and that a modest number of soldiers could easily do the job. The war has cost more than four thousand US lives alone and its direct and indirect costs to the United States, including lifetime care for more than twenty thousand grievously wounded soldiers, exceed $2 trillion and counting. So much for a safe, inexpensive excursion into Iraq. As for WMDs, before the war UNSCOM (the UN organization tasked with investigating nuclear activity) had virtually proved their absence, and the occupation soon confirmed this. As Hans Blick, the chief UN inspector, later put it, "Could there be 100 percent certainty about the existence of weapons of mass destruction but zero-percent knowledge of their location?"

A striking illustration of the power of denial in public opinion occurred in 2003–2004, some six months after searches for WMDs in Iraq turned up nothing. A strong split appeared in the US population over knowledge of this new—and apparently incontrovertible—evidence that there were no such things, WMDs being apparently nothing more than a US/UK fantasy used to justify invading another country. Democrats found it relatively easy to believe they had been lied to, and most knew about the new evidence. What was striking was that more than half the Republicans (the war party) had either not heard of the new evidence or had dismissed it out of hand and believed that WMDs had been found.

With this strength of confirmation bias, you need merely state the lie of your particular group to get almost everybody aboard. And counterevidence ends up being cited as evidence.

Some people have argued that you can't infer self-deception, that US spokespeople could simply be bald-faced liars, and that is indeed a general problem when trying to interpret official behavior, but I do not think this is plausible here. Some exaggeration surely the actors were aware of, but did any of them imagine the scale of the disaster they were producing? Did Wolfowitz appear before Congress conscious that the war might easily cost more than $1 trillion, with no known economic benefit (or indeed any other kind) while killing more than a hundred thousand Iraqis and displacing from their country an additional four million? Seems doubtful. Of course, it could simply be a mistake—computational error—or a symptom of underlying mental illness or whatnot. By definition, one can't prove self-deception through examining people's behavior alone, but I find the notion of simple, unbiased "error" naive on its face. Wolfowitz and others at the top practiced one of the most elementary forms of self-deception: they made sure they were not exposed to information that conflicted with their optimistic fantasies.

It is certain that among the decision-makers, little effort was made to learn relevant information about ruling Iraq once the invasion succeeded. No national intelligence estimate was made on the conditions to be expected during and after the war, yet such estimates are routinely produced for a large range of less important (and certain) contingencies (such as invading Bolivia). The CIA began war-game exercises in May 2002, to plan for what might happen after the fall of Baghdad, and people from the Defense Department attended the first of these sessions, but when their superiors found out, they were ordered not to attend again. The key is that postwar planning was seen as an obstacle to war itself. Paul Pillar, who was the national intelligence officer in the CIA for both the Near and Far East, points out that no one had any appetite for such assessments and gives two general reasons:

> Number one was just extreme hubris and self-confidence. If you truly believe in the power of free economics and free politics, and

> their attractiveness to all populations of the world, and their ability to sweep away all manner of ills, then you tend not to worry about these things so much. The other major reason is that, given the difficulty of mustering public support for something as extreme as an offensive war, any serious discussion inside the government about the messy consequences, the things that could go wrong, would complicate even further the selling of the war.

These are the two great drivers of self-deception: overconfidence and active avoidance of any knowledge of the potential downside to one's decisions. The contrast with World War II is instructive. Before the United States even entered that war, teams at the Army War College were studying what went right and wrong when Germany was occupied after the previous world war. Within months of the attack on Pearl Harbor, an entire School of Military Government was created at the University of Virginia whose mission was to plan for the occupation of both Japan and Germany. But of course this was much closer to a just war and was designed and thought through without the need to sell the war under false pretenses likely to induce self-deception. Injustice always requires justification and special pleading, justice less so.

CREATING KNOWLEDGE AND THEN WALLING IT OFF

As in the *Challenger* and *Columbia* disasters, where top management did not want to know about safety problems, during the Iraq war, no one wanted to hear about the problems on day two in Baghdad. Such knowledge interferes with the sales job. In the case of NASA, the safety unit was degraded to a caricature of what such a unit should look like. In the case of planning for Iraq, a truly bizarre partitioning took place. A set of working groups was duly created, then walled off from the decision-makers and the rest of the government to render the working groups impotent. Each involved knowledgeable people from throughout the US government, the army, the CIA, State Department, USAID, and so on. The State Department's Future of Iraq project began a month after 9/11 and was publicly announced in March 2002, eventually comprising seventeen working groups and producing fourteen volumes of detailed findings. The project was headed by a man who was skeptical of the wisdom of the war but certain that planning for its aftermath was essential.

Given the chaos that ensued, it is worth noting that in the summary report, stress was placed on (1) the need to get the electrical grid up and running and with it the water system, (2) the need to plan carefully for the change in the Iraqi army by removing its leaders without losing the ordinary soldiers, and (3) the need to plan for civil disorder, including the emergence of common criminals bent on rape, murder, and looting. All of these turned out to be exactly on the mark but were overlooked entirely. In September 2002, USAID also began what would soon become the Iraq Working Group to study the problems of postwar occupation. Drawing heavily on expertise from numerous nongovernmental organizations, it too highlighted some of the obvious problems forces occupying Iraq would face.

These working groups were walled off from the rest of the operation, in particular from those actually making the decisions; information flow was halted in both directions. Individuals close to power who happened to attend meetings of either study group were chastised, and higher-ups were ordered not to attend any of these meetings. Why bother with a detailed analysis that is deliberately overlooked? This seems extraordinary. Perhaps those in charge learned that the news was pessimistic and did not wish to hear the details, or perhaps they set up this whole charade in the first place to look as if they were taking seriously problems that they were not.

Curiously enough, identical forces were at work in the UK. Prime Minister Tony Blair insisted on limiting key war decisions to a very small (like-minded) group within his cabinet because of the danger of information leaking to the press. Precisely the problem one would expect with a questionable or unethical decision: let's avoid group introspection and do it on our own. In the effort to limit information leaving the small group, information is diminished within the group.

In any case, in for the dime, in for the dollar. When Jay Garner, the first person put in charge of "administering" the new Iraq, asked that the head of the Future of Iraq study group be assigned to him as an aide, the request was denied at the highest level of the administration. This is an extraordinary self-inflicted wound. It is one thing to exclude him in advance, while you are still hyping the adventure to self and others, but quite another to spitefully deny yourself his knowledge when you desperately need it—first blinding, then beheading yourself. Self-deception seems like a drug: once you start, it's hard to stop.

Having sold the public on the war through two lies, the US government then wasted at least six months' time and resources in Iraq looking for the nonexistent weapons of mass destruction and torturing prisoners to try to get them to give up the nonexistent link between Saddam Hussein and Osama bin Laden. The intelligence agencies and interrogators were told to do "whatever it takes" to get the information from their detainees, and when they still came back empty-handed, they were told to push harder. Of course, the one sure property of torture is that it creates in the victim a desire to say whatever the torturer wishes to hear. Many Iraqis described the situation as surreal. The first question posed was, "Where is Osama bin Laden?" and the answer was, "How should I know? I am in Iraq." From this point the relationship went downhill. It seems absurd that such a farce could be entrained by an initial simple lie; months after the invasion, people were being tortured to generate evidence in support of the lie, even though the lie had done its job.

CAN WARS BE WON THROUGH BOMBING?

War is a rapidly changing phenomenon, driven in part by continual technological development, which is surprisingly susceptible to change in the wrong direction. For example, the invention of armored knights on armored horses led to hundreds of years of investment in Europe in a manifestly foolish strategy: the knights were easily overcome by soldiers on foot.

World War I was (roughly speaking) the last major war in which troops were sacrificed to protect civilians, and the majority of those dying directly in the war were soldiers (more than eighteen million versus five million civilians, not counting the influenza pandemic). World War II reversed this pattern, which has remained reversed ever since. Careful estimates from World War II suggest that at least fifteen million troops and more than forty-five million civilians died.

The key to this change was massive aerial bombing of civilian populations. Although the Allies began by emphasizing tactical bombing—of military targets and key industrial areas—by the end of the war, they had perfected large-scale bombing of cities with or without any military value. Among the cities were Hamburg, Cologne, and Dresden in Germany, and almost all the major cities of Japan, including Tokyo, in which more than 100,000 people were incinerated in a single night's conflagration, the deliberate firebombing of a largely wooden city in order to create a massive inferno with temperatures reaching 1,800 degrees and winds more than fifty miles an hour. In all, more than sixty Japanese cities were destroyed by intensive bombing, leaving almost nothing of the cities on the list that bore the names Hiroshima and Nagasaki.

An enduring fallacy of surprising strength is that wars can be won through airpower, bombing from the relative safety of the skies, decoupling the killer and the killed. Among other virtues, it is claimed that bombing can turn a population against their leaders, whose activities are said to have induced the bombing. Proven wrong in World War II (with the single exception of the nuclear attacks on Japan) and repeatedly since then, nobody seems able to drive a stake through the heart of this fallacy. As recently as 2006, both Israel and the United States imagined that a devastating bombing campaign against all of Lebanon would turn the country against Hezbollah as the presumptive cause of the bombing. In fact, the campaign had been conceived about a year earlier and extensively wargamed by Israel and the United States in the six months leading up to the onslaught. As usual, the bombing had the opposite effect: the country rallied behind Hezbollah, which enjoyed its highest levels of general support ever during the bombing itself.

For the United States, the Lebanon war was apparently meant as a prelude to a larger bombing attack on Iran based on the same absurd logic—in this case, that the Iranian people would rise up against their leaders for having somehow provoked the United States into mass attacks on their people. The notion that the Iranian people might regard the United States instead as a mass murderer seems scarcely to have reached the consciousness of these decision-makers. And the mistake continues. In Afghanistan, US murder of civilians from the sky may be driving the population into the arms of their murderer's enemies. Or as one headline (May 17, 2009) put it: DEATH FROM ABOVE, OUTRAGE DOWN BELOW. In 2009, the United States acknowledged that killing fourteen alleged al-Qaeda members cost seven hundred civilians their lives in Afghanistan for a 98 percent cost on the innocent.

One of the most spectacular failures of mass aerial bombing occurred in the US war on Vietnam, with truly horrific consequences, quite apart from the hundreds of thousands of civilians directly slaughtered. In Cambodia alone, an entirely innocent bystander in the US assault on Vietnam, the United States dropped more than 2.75 million tons of bombs between 1966 and 1973, during 250,000 missions on more than 100,000 sites. That is to say, more tonnage was dropped there than by all the Allies on Germany and Japan in all of World War II, including the two atomic bombs. Put differently, an average of almost 1,000 tons of bombs were dropped every day on Cambodia for about 2,900 straight days, during 100 separate attacks each day. This was on a small rural country that had not attacked or threatened a soul.

More than 10 percent of these sites were completely indiscriminate, in the sense of never having been targeted or described. The stated purpose of this horror? To deny the Vietnamese troops safe haven in the Cambodian forests. Survivors describe scenes out of hell, massive bombs looking like lightning bolts and producing terrifying explosions that ripped trees and people apart, leaving deep craters and victims walking around dazed and haunted by the recent horror from above—some struck speechless for days, only to be revisited by fresh attacks. The full scale of the assault emerged to general knowledge only when President Clinton (in a gesture of reconciliation) released in 2000 data on bombs dropped, in order to help the Vietnamese and Cambodians find unexploded ones still waiting to claim lives and limbs.

And what was the effect on Cambodia's social and political life? A small radical fringe group in the countryside—the Khmer Rouge—grew in five short years of US carpet-bombing from a poorly organized force of 5,000 to a raging and well-organized army of 250,000. When it seized control of the capital, the group turned on its own population, fulfilling, in effect, the genocidal words of Henry Kissinger when he transmitted President Richard Nixon's orders to carpet-bomb Cambodia: "Hit anything that is moving with anything that is flying." Subsequently, more than a million Cambodians were slaughtered by the Khmer Rouge in less than two years.

War these days has all sorts of unintended and unvisualized consequences, such as the entire destruction of Cambodian society, its evisceration from the inside. The horrors being inflicted are usually deliberately hidden from view of the society inflicting them. No one is describing the horrors and, in any case, a mentality of "we are at war" rules at home, discouraging interest in such "side effects." Efforts are made to hide the truth from the larger world and from later historical memory, the better to maintain a false positive collective view of one's past. The United States did not slaughter Asian people wantonly in the 1960s and 1970s. Instead it fought international communism with some collateral damage.

BOMBING TO ERADICATE HISTORY AND TO REINFORCE IT

Bombing can be used to alter history and to impose it. The importance of eradicating history is nicely illustrated by two bombing runs during Israel's assault on Lebanon in 2006. In the first, the Khiam prison near the Israeli border was obliterated. The prison had no military value, no rockets were being fired from its midst, and there weren't even any civilians inside to terrorize. The attack was not an accident; it was a systematic bombing raid designed to obliterate the prison, as indeed it did.

The prison was a museum, the actual prison in which southern Lebanese men and women not cooperating with the Israeli occupation were housed in the 1990s. I visited it myself six months before the war. Except for two guides, it was empty. Signs on the wall said that women were allowed to bathe for fifteen minutes once every two weeks but men only once a month. And there was a torture chamber, somewhat resembling the old electrocution chambers in the United States. Noam Chomsky was photographed sitting in such a chair. And that is precisely the point. The living memory of the prison invited Israeli destruction, the better to destroy the past and build future, false narratives.

The second bombing run destroyed another museum, built on the site of the Qana massacre of 1996. This massacre took place during one of Israel's periodic assaults on its northern neighbor Lebanon. As shown conclusively later, successively calibrated shells fired by the Israelis finally landed on a UN refuge, their intended target, overhead drones providing the key information. Within the refuge were huddled women and children trying to survive the general bombing while their men remained at large. Of those sheltered by the UN, 106 were incinerated, mostly women and children. Many of the victims were then buried there and a museum erected with photos and exhibits of that dreadful day: ghastly burned corpses, shredded bodies, blood everywhere—the pitiful remnants of human beings cowering under the wings of the UN for false protection. (They would have been better off widely dispersed in the open. Many more sitting ducks would have survived.) Yet why does Israel now bother with the museum, with rebombing dead people? Precisely to kill the victims for good, so their memories do not come back to haunt or to indict. Rewriting history through rebombing.

An extraordinary—if very ugly—bombing also happened in 2006, with very different intended effects on group memory. The Israelis perpetrated a second Qana atrocity—with similar loss of young life. From their drones in the sky, the Israelis had been carefully tracking the residents of Qana, who were forced to join together in ever-tighter clusters as homes were successively destroyed. Children played outside in the daytime, easily visible to the drone. Then one night Israel attacked the final house where twenty-seven were huddled, claiming a fresh set of seventeen dead children. This was deliberately grinding salt into an old wound. So here we have Israel acting oppositely regarding the history of 1996: to the outside world it wants to destroy direct physical evidence of the atrocity, but to the local Arabs it wants the dreadful memory of 1996 to be reinforced, as if to say, "We will murder you as piteously as we choose, when and where we choose. We will then do everything to deny and eradicate in the minds of outsiders what we have just done to you."

CARNAGE IN GAZA

In what seemed to me, a mere bystander, to be a stupefying act of violence, Israel attacked the people of Gaza at eleven o'clock on a Saturday morning, December 27, 2008, when the streets were bustling with people and a graduation ceremony for police cadets was under way. Within the space of twenty minutes, almost two hundred people lost their lives, including many young men who were about to become police officers. In one sense, it was beautifully planned, two days after Christmas, the last day of Hanukkah, a few days before the new year. Who would have guessed that Israel would choose this time to launch a savage attack long in the planning? Israel's biggest-selling newspaper called this "a stroke of genius" since "the element of surprise increased the number of people who were killed." "Genius" seems a curious word to describe what is only a well-timed act of malevolence. Another commentator echoed earlier US self-praise in Iraq: "We left them in shock and awe." Or as another put it, "Israel can and must mete out a severe punishment to Hamas, one that sears its consciousness (yes, sears its consciousness) and causes it to hesitate before it fires again."

American editorialists and commentators rushed to Israel's defense. As gathered from a single day's commentary (January 4, 2009), the following talking heads on US television justify Israel's brutal assault: three governors, one senator, and one newspaper columnist. "The missile firing into Israel, I think, brought the proper response from the Israelis." "All Americans know what we would be doing if missiles were landing . . . from Vancouver, Canada, into Seattle." "It is inconceivable to me that if missiles were coming out of Cuba into south Florida that we wouldn't respond." "Israel has no choice but to take military action." "Israelis are doing the only thing they can possibly do to defend their population."

It would be much more accurate to ask: If Vancouver were imposing an economic stranglehold on Seattle, and regularly assassinating and kidnapping its city leaders while murdering civilians at will, would we consider a barrage of primitive rockets from Seattle—incapable of landing with any precision—to be an appropriate response? And if, in alleged response to such crude rockets, Vancouver bombed and invaded Seattle, with horrific effects on civilians, women and children alone numbering in the hundreds, what would we then do? And which objective observer would put credence to any claim that the attack was Vancouver's *only* option in self-defense?

If you are inverting reality, it helps to have many others inverting it, too, making it more plausible by repetition alone and by excluding alternative voices. To me, the near-unanimity of the political class and most of the news commentators in the United States in supporting Israeli terror on a mass scale is astonishing—evidence of the stranglehold that a good false historical narrative can exert on an entire group. It is, after all, the unconditional support of Israel by the world's military superpower that permits Israel to act in ways it would never dare if it had to relate to its neighbors on a level playing field. Here the false historical narrative is so powerful—the hate-filled terrorist Palestinians, anti-Semitic to the core, need regular punishment—that hideous attacks on one's neighbor are both moral and praiseworthy.

The savagery continued for seventeen days and included numerous well-known Israeli devices, including attacks on UN positions, on ambulances, on mosques, on infrastructure, on civilians ordered to leave buildings, on young men, women, and children. That numerous assaults on civilians were deliberate is evident on inspection even if routinely denied. A war that began with lies ended with them. Indeed, it was not actually a war, more like a massacre of any and all. "Shooting fish in a barrel," as one observer aptly put it.

Certainly if it was meant as a surgical strike on an alleged terrorist group (Hamas), someone forgot to tell the soldiers. When leaving homes they had vandalized after murdering the occupants and their neighbors, the soldiers left behind graffiti more honest than any Israeli government spokesperson: "Arabs need 2 die," "Die you all," "Make war not peace," "1 down, 999,999 to go," and written over the image of a gravestone, "Arabs 1948–2009." Hamas, like Hezbollah, is a fictitious enemy, or at least a demonized one. The whole point is to frighten—indeed terrorize—the enemy into submission.

Someone also forgot to tell soldiers afterward not to tell the truth about the onslaught. As we know from Israeli testimony alone, massive firepower was used to cover advances, there was systematic demolition of housing, indeed constant destruction through massive firepower regardless of whether buildings were known to be occupied, the use of Palestinians as human shields, the use of deadly white phosphorus, rules of engagement in which "any movement must entail gunfire. . . . No one is supposed to be there. If you see any signs of movement at all, you shoot." These rules referred not to combatants but to everyone.

Military rabbis were busy whipping the troops up to a pitiless state because "the Palestinians are like the Philistines of old, newcomers who do not belong in the land, aliens planted on the soil which should clearly return to us." Not too different from the hordes of clerics who proclaimed passionately on each side of America's Civil War over slavery, or the clerics who have risen up to support numerous other wars. In this case, ancient genocidal logic is called upon to justify the current form. Of course, the Israeli military kept up the usual pretense. Its actions were governed by "uncompromising ethical values," which is certainly true, but are they just values or purely Israel-benefiting ones? Far from wantonly slaughtering innocent people, Israel made "an enormous effort to focus its fire only against the terrorists whilst doing the utmost to avoid harming uninvolved civilians." But quite the opposite was true.

The attack was sold under patently false pretenses. Hamas and Israel had agreed to a six-month cease-fire, during which rocket attacks on Israel (by primitive devices lacking guidance signals and killing all of seventeen Israelis in the previous seven years) were reduced by 97 percent, and the remaining few may well have been beyond Hamas's control. It was Israel that broke the truce, first by failing (as required) to open access to Gaza when in fact it slowly tightened its stranglehold. Israeli forces then entered Gaza contra agreement on November 4 and murdered six so-called Hamas militants; the next day, Israel cut imports to Gaza to one-tenth of their former trickle. Hamas responded by not renewing this one-sided "truce," and Israel in turn unleashed its long-planned assault, killing about 1,300 Palestinians in three weeks, roughly the number it had killed in the previous five years. The vast majority were civilians, and more than half were women and children. There can be no doubt that the operation's chief function was to terrorize, once again, its Arab neighbors. Only thirteen Israelis died, for a kill ratio of 100:1.

The brutal opening assault on police cadets held a special irony, because one of Hamas's signal achievements since gaining power through democratic elections was its ability to sharply reduce common street crime, including murder, robbery, and rape. But when the Zionists are in full blood, the Arabs are not allowed even police guns to protect themselves. They are seen instead as irremediable terrorists intent on inflicting suffering on their innocent prison wardens, the Israelis.

What is striking about the Gaza attack is how radically differently it is viewed by most of the world—the account given above is a fair summary—as opposed to the version believed by Israel and the United States. Why this incredible blindness in the United States to the moral outrages being perpetrated by its client state of Israel? The key factor, I believe, is the false historical narrative that links Christianity, Judaism, and American exceptionalism—that is, Christian Zionism (see Chapter 10). Fortunately, the excesses of the Gaza operation seem to have caused enough general revulsion that a prominent Jewish-American critic of Israeli behavior can ruefully title his new book on the subject *This Time We Went Too Far.*

Likewise, there are welcome signs in the US Jewish community alone of a change in the narrative. A revulsion at the old "anything Israel says is fine and anyone who says otherwise is an anti-Semite" style of arguing is seen in such organizations as J Street in Washington, D.C., Jews-4-Peace in Los Angeles, and Justice in Palestine, popular at US universities, to name but a few. There are parallel, though unpopular, organizations in Israel.

SELF-DECEPTION AND THE HISTORY OF WAR

In one sense, Israel's attack on Gaza was a stupefying act of violence. In another, it was merely the latest example of heartless intergroup murder and warfare that stretches back perhaps some five million years when, chimpanzee-like, we started to regularly kill neighboring group members. That is, by the standards of behavior within a group, the attack was stupefying, heartless, and cruel, but by the standards of behavior between groups, it was routine. We have all done it: Christian, Jew, Muslim, Hindu, animist, and atheist. In Sudan, people are raped and killed by the thousands, in the Congo, by the millions. Every ethnic group, it seems, in every corner of the globe indulges this ancient habit.

Of course, we have made enormous progress since chimpanzee days—in technological sophistication, for example, and in the scale of the adventure, but also in the use of language, both prior to the attack (permitting planning and coordination) and afterward (permitting rationalization). The latter is especially important in the face of onlookers, members of the rest of the species who may witness the attack or learn about it, yet another novel feature of recent warfare, depending also on language.

Chimpanzees appear not to face this problem of verbal self-deception or to have any verbal component to warfare. Indeed, it is hard to see any communication system, though one surely must exist. Nobody has been able to detect a signal by which male chimps organize for war. So far as we can tell, when they are about to go to war, they do not suddenly start looking at one another or otherwise communicate before setting out on territorial patrols, a common prelude to attacking neighbors. They do so either spontaneously or in response to a sound or smell from a neighboring territory. When on a border patrol, males are quiet—they stop, sniff, listen as a group, and are at all times alert. Sometimes they make deep incursions into neighboring areas, the signal or means of coordination being completely unknown to us. Nobody has seen anything resembling group discussion afterward—certainly no need to justify the behavior in the face of external observation—so that the verbal element seems entirely missing from chimpanzee warfare start to finish and the basis for the cooperative synchrony is completely unknown.

By contrast, consider the contrary set of arguments unleashed after Israel's attack on Gaza. Either, at one extreme, the attack was a fully justified assault on a terrorist entity and its critics were Hitler's newest anti-Jews or, at the other, it was an Israeli terrorist attack on the human remnants of the ethnic cleansing on which it is itself based. So language, which permits the past to be expressed, communicated, and remembered, both vividly and in detail, adds immense opportunities to dress up the past or deny it, to one and all, present and future. Perhaps no aspect of language acts as a more powerful force for war than religion, the topic to which we turn next.

CHAPTER 12

Religion and Self-Deception

A book could be written on this subject—no, a twelve-volume treatise. Religion is a deep and complex subject, and so are its interactions with deceit and self-deception. Religions range from animists to monotheists to nontheists to atheists and then from Christian to Hindu to Buddhist to Muslim to Jew, with many subspecies. Here I can only hope to sketch out some of the major biological forces favoring religion and some of the important ways in which religion may encourage deceit and self-deception. This is a very tentative chapter, one that is heavily biased toward my own limited knowledge, namely that of the monotheistic religions of the West, chiefly Christianity, and not toward polytheism or the great Eastern religions. Nevertheless, I hope to show how religion and self-deception can interact in important ways and invite others to make the more important advances.

Some people think of religion itself as complete self-deception, all of it nonsense on its face, counterfactual, and in the extreme having nothing but negative side effects. In this view, the entire enterprise is self-deluded at the outset, so religion should be studied as a well-developed system of self-deception, as it certainly is—but is it only that? This may in fact be true, but these people have no theory for how this malady could have spread so far—to every culture and almost every human being in every culture—by self-deception alone.

What some have is a metaphor. Religion is a viral meme; that is, it is not an actual virus, which can easily bring a population to its knees, but rather it is merely a thought system that happens to propagate as if it were a virus, to the detriment of those with the belief system. Despite its negative effects, it apparently generates insufficient selection pressure to suppress the spread of this non-coevolving nonorganism. This is not a very impressive foundation for an evolutionary theory of religion, and it easily invites undue optimism regarding the life span of current religions. For example, from one such proponent of the meme-centered view: "I expect to live to see the evaporation of the powerful mystique of religion. I think that in about twenty-five years almost all religions will have evolved into very different phenomena, so much so that in most quarters religion will no longer command the awe it does today." I think it more likely (though not the most likely) that twenty-five years from now, evolutionary biologists and philosophers will be in hiding from the then-dominant religious groups. Fifty years from now, no one stubbing his or her toe will say, "Charles Darwin, Charles H. Darwin, this hurts!" but they will still be saying, "Jesus H. Christ, this fucking hurts."

At the same time, many, many people believe religion is the received truth from the Almighty—or, more to the point, that *their* religion is. Some have a book—a Torah, a Bible, a Koran—all of whose words are true, often literally so. Their view has as little backing as the viral-meme story and appears at first to be nothing more than a deep form of self-justification. If their own religion is God's own truth, then competing religions are often seen as anti-truth, or the work of the devil, the ultimate target. So we begin with two very extreme views of religion, with the truth probably somewhere in between, but where exactly and why?

First we need to separate the truth value of religious statements from the possible benefits of believing in them, and likewise separate partaking in religious ceremonies from the truth value attributed to them. Then we need to analyze beliefs and behavior in a more fine-grained way so that we can evaluate the meaning and function of particular beliefs. In my own view, there is often an internal struggle within religions between general truth and personal or group falsehood. That is, the essence of religion is neither self-deception nor deep truth, but a mixture of the two, with self-deception often overwhelming truth.

Religions tend to increase within-religion cooperation at the cost of lowered cooperation with outsiders. Often this involves a false historical narrative and shared group self-deception: "We are the chosen people or the original people from creation or those whose beliefs [e.g., in the divinity of Jesus] cause God to favor us [or whatever]." In short, religions often act as templates for in-group/out-group biases. Insofar as they encourage in-group cooperation, many benefits may accrue, but insofar as they encourage in-group cooperation in aggressive attacks on out-groups, they both inflict harm on others as a price of their cooperation and inflict harm on self when they fail (which, in warfare, is roughly half the time).

At the same time, certain features of religion provide a recipe for self-deception, removing nearly all restraints from rational thought. The universal system of truth espoused by a religion usually gives special status to the believer. Various phantasmagorical things are easily imagined, and "faith" is permitted to supersede reason.

Religion has a complex relationship with health and disease. On the one hand, health may be a major selective factor favoring religious behavior and beliefs. Not only do religions often preach healthy behavior, but there also is evidence that religious belief and association improve individual survival, immune function, and health. Even music, so common in religion and courtship, has positive immune effects. Medicine was originally embedded within religion, and both provide strong placebo benefits to at least part of the population.

A completely unexpected association between disease and religion emerges when we study the entire globe for degree of religious diversity (number of religions per unit area) as a function of parasite load (roughly, degree of human loss due to parasites). Here we find many more religions (and languages) per square inch when parasites are high. Since splitting of religions is also naturally associated with ethnocentrism and ethnic differentiation, parasites are a factor expected to degrade general religious truth value over time and thus to be positively associated with in-group deceit and self-deception. This may be especially true of the polytheistic religions, but with monotheism came additional forces of self-deception associated with global conquest and a single, dominant spirit.

Finally, we consider the role of prayer and meditation, specific teachings against self-deception, and the contrast between the social and internal sides of religious devotion.

COOPERATION WITHIN THE GROUP

By logic, religion ought to increase altruistic and cooperative behavior among group members—of obvious potential benefit—but it may do so along with reduced such behavior toward nonmembers and, worse still, outright aggression and murder. That is, an increased degree of hostility toward neighboring groups can heighten the within-group bias (and vice versa). This is the double-edged sword of religion, inside and outside: a religion urges its own members to treat each neighbor as they would treat themselves, yet also to slaughter every nonbeliever and outsider, as is ordered in the good book, for group after group, down to every last man, woman, and child. At the extremes, some religions advocate in-group love and out-group genocidal hatred.

In some religions, people imagine that God is watching and evaluating their every action. Reputational concerns are expected to have obvious effects on human cooperative tendencies. One study shows that even a pair of eyelike objects on a small part of a computer screen can unconsciously increase cooperative behavior in an anonymous economic game. An awareness of observing, judging god(s) may have similar effects. Indeed, providing a "God prime" hidden in a game of sentence creation increases cooperative tendencies to about the same degree that primes of secular retribution do (police, courts, etc.). Insofar as fear of God's judgment entrains more moral behavior on our part toward others, it can be seen either as a device that costs us some occasional selfish behavior but protects us from the greater cost of such behavior being detected by others and of being aggressed against, or as a form of imposed self-deception by others, in effect, scaring us into greater group orientation.

A tendency to detect agency in nature likely supplies the cognitive template supporting belief in supernatural agents transcending the usual limitations of nature. Since only in some religions do these gods watch, monitor, and respond to human behavior, it would be most interesting to know which religions do so and why. Is this, in part, a means of increasing in-group cooperation?

Although those Christians who frequently pray and attend religious services reliably report more altruistic behavior—such as charity donations and volunteer work—it is uncertain how much this applies only within the religious group or even whether it applies at all. This is because various measures of religiosity repeatedly have been shown to correlate with higher false opinions of self, suggesting an obvious self-deceptive effect of religion: you think better of yourself than you otherwise would. In Islam, it is mandatory to give to the poor, but there must be variability in doing so, and it would be most interesting to know what such variability correlates with.

One interesting fact on the effect of religion on cooperation emerges from comparing small religious organizations—"sects"—with small nonreligious communes. There is a striking tendency for the religious to outlast the secular (at least in the United States). In each year, the religious sect is four times as likely to survive into the next year as the secular. So religion provides some kind of social glue that makes organizations based on them more likely to endure than those based on nonreligious themes. Living in a cohesive and mutually supporting organization would be expected to have immune benefits as well, since one is less isolated and more likely, in a crisis, to be able to draw on the resources of others. As we have noted, the placebo effect is based partly on its expected association with caring acts by others.

Another interesting difference between the two kinds of communes is that the more costly the requirements imposed on group members in a commune (regarding food, tobacco, clothing, hairstyle, sex, communication with outsiders, fasts, and mutual criticism), the longer the survival of a religious commune, though there is no association between cost and survival in the nonreligious. This raises two questions: Why should cost be positively associated with commune survival, and why should this hold only for religious ones? According to cognitive dissonance theory, greater cost needs to be rationalized, leading to greater self-deception, in this case in the direction of group identity and solidarity. Why do religions provide more fertile ground for this process than secular communes? Perhaps because religions provide a much more comprehensive logic for justifying beliefs and actions. In religious communes, men's participation in group prayer predicts their degree of sociality in an experimental economic game.

RELIGION: A RECIPE FOR SELF-DECEPTION

Whether religion is entirely devoted to self-deception from its very foundation to its every last branch seems unlikely, but the fact that this is even a theoretical possibility suggests the degree to which religion has been infected by forces of self-deception. Even a casual glance at most religions suggests that there is far more nonsense than revealed truth. Some of the key features of Western religions (and some Eastern ones) are the following.

A Unified, Privileged View of the Universe for Your Own Group

Most religions propose this view. Either you are the founding people and all others degenerate dogs, or else yours are the "chosen people" either by ethnicity (Jewish) or by attachment to this or that prophet (Jesus, Muhammad). Of course, any general system of thought that places you at the center is useful to you in interactions with others. In defense of religion's inadequacies, it should be remembered that for many thousands of years, there was nothing else other than religion. Certainly no organized science, no Newton or Darwin, but still this alone can't justify the strong egocentric biases of religion.

There May Be a Series of Interconnected Phantasmagorical Things

For example, there may be an afterlife; a giant spirit who controls all but is amenable to human persuasion on the most trivial matters; a prophet capable of performing miracles, whether parting the seas, raising the dead, or feeding the masses; a prophet who is born without a human father, only God himself, and who stays dead for only three days; and so on. Once you have signed on to a few of these notions, there are hardly any boundaries left, and very small details can turn out to be critical features of dogma.

The supreme spirit (or God) is typically given a masculine name that on biological grounds seems most dubious. Besides imparting an image of God as a fearsome tyrant, there is no such thing as an all-male species in nature. Not a single one. Only females can reproduce by themselves, females preceded males in evolution, and to this day they are still the critical sex as far as biological work is concerned. God should be interpreted as mostly female, and I will do so throughout. A male God has many unfortunate features, including the heartlessness and aggression associated with men and their divorce from reproduction, producing a series of horrors—pedophilia in all-male "celibate" castes; hostility toward women's interests, especially efforts to control their reproduction and sexuality (banning sexual activity, abortion, in vitro fertilization, etc.); honor killings; indeed every kind of anti-female horror including mass rape during warfare in the name of God.

The Deification of a Prophet

The deification of Jesus is unlike the treatment of prophets in either Islam or Judaism. His birth by unheard-of means, miracles ascribed, and of course, his very brief death, so that now he is one-third of the show: Father, Son, and Holy Ghost. The basic story was put together after his death in the years that Christianity was a small, persecuted sect. To believe in his divinity became the key test, one that automatically shrank and exalted the group. The bigger you make Jesus, the smaller you make God. Not only are other gods no longer real but also God herself has lost a good part of her powers to a (dead) human being.

It is also ironic that the more you deify the prophet, the less attention you pay to his actual teachings, since the key distinction then becomes whether you believe in his divinity, not whether you believe in any of his teachings. "I *believe*, Jesus, I *believe* in you as the Lord, my personal savior."

Yes, but do you believe that the meek shall inherit the earth, that blessed are the peacemakers, that you should treat *all* others as you wish to be treated yourself, and so on? I doubt it. Deification of Jesus also makes more likely patently absurd beliefs, such as intercessory prayer, since Jesus now joins God as someone you can beg favors from (and the Catholics add yet another layer, Jesus's mother, the Virgin Mary), no matter how many laws of nature need to be violated in the process. Among prophets, Jesus was an extreme case—hung on a cross until he gave up his life—but do not imagine that the earlier prophets in the same tradition were welcomed (whether Isaiah, Jeremiah, Ezekiel, or whoever). During their time, they were often persecuted and rebuked, only later restored in memory to prophetic status.

Sometimes a Book Is Treated as Received Wisdom Direct from God

This allows plenty of room for interpretation. Sometimes every word is literally true, even if this results in numerous contradictions within the book itself, never mind the larger world. Other times, metaphor is permitted, and indeed encouraged, giving plenty of latitude for how this divinely generated document is interpreted. The key is that you—or your group—control the document and the interpretation. If God literally created the world in seven days about six thousand years ago, then all of astronomy, geology, and evolutionary biology must be nonsense. Did God really give "the land of Israel" in perpetuity to a people who wrote a book a few thousand years ago saying he did?

Faith Supersedes Reason

Sometimes anti-logic is directly pushed, as in the notion that "by faith ye shall know them"—indeed, an attachment to reason may be evidence of sacrilege. The degree to which we believe something now becomes a determinant of its truth value. Once again, this joins a long line of features that tends to remove all rational boundaries from religious thought, permitting any and every deceptive ploy and self-deceptive concept.

We Are Right

And here comes the critical, all-encompassing self-deception: we are the measure of what is good, we represent the best, we have the true religion, and as believers we are superior to those around us. (We have been "saved"; they have not.) Our religion is one of love and concern for the world, our God a just God, so our actions can't be evil when they are done in God's name.

Given the ease with which religion slides toward self-deception, what are the larger forces that might propel a religion toward more or less self-deception? One important factor is the degree to which the religion is associated with the powerful in a society. Another important force has to do with religious fragmentation. Because religions almost always preach within-religion mating, fragmentation is expected to lead to intergroup conflict over minor religious distinctions. I will argue that parasite load—average pressure on a society every generation from coevolving parasites—may be an important force fragmenting religions and thus encouraging parochial self-deception. The evidence for an association with parasite load is strong, but the evidence for a connection to self-deception not nearly so strong. First, let us turn to the positive association between religion and health.

RELIGION AND HEALTH

Religious behavior and practice appear to be positively correlated with health, a well-established fact with dozens of careful studies in support, on both sick people and well. Longitudinal studies suggest that variables such as degree of attendance at religious service are positively associated with survival years into the future.

Part of this effect may result from the tendency of religions to establish rules related to health: avoid tobacco and alcohol, pork, top predators such as sharks and lions (which tend to concentrate toxins as they move up the food chain), and generally risky or unwise behavior, such as gambling. One long-term study of US Christians showed that degree of religious attendance in 1965 predicted a change to more positive health behaviors thirty years later.

Under Islam, some behavior is prohibited, some encouraged, and some required. The forbidden (*haram*) tend to relate directly to health:

- Gambling
- Alcohol
- Eating pigs or dogs
- Eating dead meat
- Eating meat of animals not slaughtered the Islamic way (cutting throat at aorta and bleeding animal)
- Eating predatory fish
- Eating shellfish
- Usury (charging interest on money)
- Saying *oiff* to parents (an expression of impatience or annoyance), or yelling at them
- Suicide

All of the prohibitions regarding eating probably reduce parasite acquisition. Predatory fish are like sharks and lions in other religions—top predators that may be forbidden because they strongly concentrate toxins. Bleeding presumably reduces exposure to blood parasites. Only avoiding usury and saying *oiff* may not be directly related to personal health.

It is perhaps interesting to note that of the requirements in Islam (*wajeb*), three have positive connections to health (among other effects):

- Daily prayer (five times per day)
- Cleanliness (must be clean to pray: use only running water or sand)
- Fasting
- Alms to poor
- Pilgrimage to Mecca (if possible)
- Testifying ("there is only one God and Muhammad is his prophet")

The latter three are clearly social: two showing off, and one helping a group member, all with unknown possible immune effects.

But the relation between religion and health goes deeper than health-related behavior. Some effects may come from the benefits of positive belief itself—for example, on immune function—as well as benefits that flow from being a member of a mutually supporting group, including musically supported activities that raise group consciousness, a very common feature of religion. As we have seen (Chapter 6), music has positive immune effects, while noise has negative ones. The exalting, positive music of so many religions is probably on the high end for positive immune effects (in contrast to, say, jazz or rap). Even confessing sins to God and disclosing trauma may have beneficial immune effects. The private confessional in the Catholic Church facilitates this, as do numerous public rituals of confession common to Amerindian religions. It seems likely that private, verbal confession in prayer has similar immune benefits, an example of a personal benefit to private religious behavior because it mimics a social interaction.

Whatever the precise causes, the links between religion and health seem strong enough on their own to select directly for religious behavior and belief. As biologists, we need not view religion phobically, as some negative, nonliving force of unknown nature that has us in its viruslike grip. We might remember that before the advent of modern science, almost all medicine was practiced within religion, often by special castes, medicine men and women, faith healers, and so on. Some medicinal benefits were certainly real, for example, consuming plants for their real chemical effects, a behavior that reaches deep into our monkey past (although the causal connection was usually unknown to the actors), and some may merely be the blessed placebo effect, itself probably the dominant benefit throughout two thousand years of Western medical "science." Belief kills and belief cures.

One benefit of religion is that it does provide a framework for understanding and acting within our world, a framework we might expect to provide some psychological and mental benefits. Recent work in neurophysiology suggests one such benefit. Scientists concentrated on the anterior cingulated cortex (ACC), a region involved in many processes, including self-regulation and the experience of anxiety. EEG neural activity in the ACC was recorded while people were taking the Stroop test (name the color in which words are written, though the words denote a different color). The stronger people's religious zeal (as measured by a scale) or the more they professed a belief in God, the less their ACC fired in response to errors and the fewer errors they made. It was as if religion was providing them a buffer against error. There must be many such possible effects.

PARASITES AND RELIGIOUS DIVERSITY

Religions have repeatedly split into subreligions that are sometimes at one another's throats. Religions occasionally join together, but this occurs much more rarely than splitting. There is thus a bias in the propagation of religions over time, with a tendency for major faiths to split into subgroups, which may split further, typically emphasizing relatively minor doctrinal differences on which to disagree: from universal truth widely shared to smaller within-breeding units at war with each other on the basis of intellectually false distinctions. If this is an important feature of splitting—corruption of the religion's generality and logic—then we need to understand its origins.

Recent work suggests that parasites and, in particular, parasite load may drive religions to split. These splits, in turn, entrain changes in doctrine to justify them, and thus tend to degrade the universal truth value of religion with parochial arguments whose true meaning is usually hidden. Parasite load is meant as an aggregate measure of the number of parasites and their degree of damage on a local population. Ideally, parasite load would be measured as something like the degree of overall mortality (or loss in reproduction) due to disease, but it is usually measured as a simple count of the major diseases present and the relative strength of their negative effects.

The argument goes as follows: Where parasite load is low, an in-group and out-group member may be almost equivalent where risk of transmitting a new infection is concerned, namely, low. But where parasite load is high, an asymmetry emerges. An in-group member will in general have been exposed to the same set of parasites as the other members and will carry some of the same genes that give at least partial resistance to many of these parasites. But an out-group member will be subject to selection from a slightly different set of parasites and will carry a subset to which it may be partly resistant but in-group members are not. From the standpoint of each group, the other is a threat—you may transmit your parasites to one another far faster than the genes that would protect against them. Hence, individuals in both groups may be selected to avoid one another. In short, other things equal, high parasite load is expected to increase ethnocentrism, within-group love, and hostility toward strangers. By this argument, degree of self-deception across religions and cultures is expected to correlate positively with parasite load.

What is the evidence? Two broad factors are of interest: religious and linguistic diversity. That is, how many languages and religions coexist per unit area? With high parasite load, we expect many of each, since splitting into smaller groups facilitates language formation. Regarding the evidence, there can be little doubt. Across the entire globe, religious and linguistic diversity map directly on parasite load, as does ethnic diversity—the higher the parasite pressure, the more religions, languages, and ethnic groups per unit area. The exact overlap between religion and language has not been described, but these results have been corrected for numerous possible confounding variables, and the associations remain strong and unambiguous. For language, the correlations are significant for all five of the great continents.

Canada and Brazil are roughly the same size, yet Canada has 15 religions and Brazil, 159. Canada is located in the far north, where parasite load is low; Brazil is in the American tropics, high in parasite load. Likewise, Norway, in the far north, has thirteen religions, while Cote d'Ivoire is the same size but is located in the parasite-rich African tropics and has seventy-six religions. Of course, if there is a bias toward interactions based on shared language and religion, this ought usually to intensify within-group mating, with resulting ethnic differentiation (and hostility).

It is certainly striking how often out-groups are characterized as if they were flea-ridden and scabrous, if not syphilitic.

Whether this argument applies to major splits in religion is unknown. Did Shia and Sunni really split in response to parasites? And Roman and Greek Catholics? The peeling off of various Protestant sects from Roman Catholicism was associated with the publication of the Bible in modern languages, as well as with a great European outward surge of warfare, plunder, and colonialism. Where is the parasite connection, if any? Did the newly fragmented groups interact less frequently? In short, the general trend seems clear, but particular major cases may have little or nothing to do with this rule, at least given our current understanding.

One subject that requires analysis is the degree to which the formation of cities and more widespread trade conspired with monotheism to create a world less fractured along parasitic lines. We know that the appearance of agriculture and the subsequent explosion in both population numbers and rate of adaptive evolution preceded the invention and spread of monotheism, but we know little about the interaction with parasite pressure. In general, higher density increases parasite pressure, resulting in such horrors as the Black Plague, which wiped out one-third of Europe in the Middle Ages, or the influenza incubated in the trenches of World War I that consumed twenty million lives before it was done. On the other hand, we are completely ignorant of the subtler dimensions of this subject. A series of other variables have been shown to covary with parasite load, so these will very likely covary with religious features as well. High-parasite-load societies appear to be more xenophobic, more in-group oriented and homogeneous, more suppressive of women, less permissive of casual sex—in short, a suite of characteristics that can at least by logic be linked to parasite defense. So far as I know, no one has studied the interaction of these variables with religion, yet surely we would expect many connections: the more numerous religions there are in parasite-rich areas, the more the religions are expected to be xenophobic, harsh on women, conformist, and so on.

In this situation, underlying correlations are expected to bubble up from the unconscious, requiring post-hoc justification. Presumably, no one is saying, "Look, worm density has increased alarmingly in ourselves in this area for the past ten years. Perhaps it would be wise for us to be more focused on in-group interactions, including mating. Let's up our racism level." Instead, as I imagine it, religion provides substitute logics with similar effects—let's emphasize minor doctrinal differences: "We scratch our asses with our right hands, they with their left [note the parasite implications], so let's avoid the nasty left-scratchers entirely."

WHY THE BIAS AGAINST WOMEN?

There is one important problem hidden in the above account: the assumption that in-group mating will be as strongly selected for as in-group favoritism. This is counter to expectation. We know that sexual reproduction—and the recombination it promotes—is strongly associated with evolutionary protection from coevolving parasites. Thus, parasite load may generate impulses toward in-group favoritism while at the same time heightening interest in sex with an out-group member.

Consider greater sexual promiscuity, or diversity of mating partners, well known to be higher in both birds and humans in the tropics, and presumed to represent an adaptive response to parasite load by increasing genetic quality of offspring. So why should this kind of sex be *more* prohibited in parasite-rich regions? Is it precisely because in these situations women would benefit more from such activity (improved genetic quality of their offspring) and thus provoke greater male countermoves, the kind of behavior we described so vividly in Chapter 5: mutilation, beating, terror, and murder? Certainly religions are overwhelmingly patriarchal in logic and structure, with numerous resulting effects.

One such effect is the bizarre recent claim from the Holy Roman Catholic Church that male celibacy does not contribute to priestly pedophilia, but homosexuality does. Certainly the latter should bias molestation toward male children, but what could be more conducive to sex with children than a complete prohibition on sex between adults? And what is more conducive to abusing boys than an all-male priesthood that presumably attracts men who like men? What continually haunts me when I think about such matters is the function of all this nonsense. Who benefits from an all-male priesthood? As for the priests' being nonreproductive, at least this guards in principle against narrow kin interests. There are very few genetic dynasties in the Catholic Church (contrast North Korea, Syria, Egypt, Jordan, India, Haiti, and the United States), so the Church is likely to be corrupt but not nepotistically so.

But why all-male? This makes celibacy easier, but all-male priesthoods coexist with priestly reproduction in Islam, Judaism, and many Protestant sects, among others. And why the association with distortions against women's interests? Is female reproduction to be subordinated to male interests for group benefit or male benefit—and at what cost to females? The Catholic Church outlaws all control by a woman over her own reproduction short of abstinence from sex at the very moment that she is most eager for it. She is not allowed to prevent conception if copulation occurs, and she is not allowed to terminate a pregnancy, however induced (rape and incest included). This appears to be a simple strategy for maximizing group reproduction, or at least male group interest. Female interests appear to count for little.

POWER CORRUPTS

As we have seen, power corrupts: the powerful are less attentive to others, see the world less from their standpoint, and feel less empathy for them. The converse is that the powerless are more apt to see things from the other person's standpoint, to be committed to the principle of fairness, and to identify with people like themselves. The religious effects are that humility, fairness, forgiveness, and neighborly love are more apt to be virtues preached among the powerless. It is no accident that in both Christianity and Islam, this dynamic has been played out. The Christian gospels were all written while the church was a small, underground, persecuted sect. Islam's more peaceful injunctions came when it was an oppressed minority, its more assertive when it reemerged with military power.

It has been said that when after three centuries Constantinople elevated Christianity to the state religion, it went in one century from being the persecuted church to the persecuting church. This is a recurring theme in monotheistic religions: with state power comes a new source of bias. They change from emphasizing the universal principles of brotherhood that would especially benefit the oppressed and those needing alliances with other groups to emphasizing principles of dominance and imperialism—the lesser orders should remain so and unbelievers and outsiders may be attacked more or less at will. Racism is a valuable handmaiden. If the others are biologically inferior, is it then not God's will that they should be supplanted by their superiors? How else is evolution supposed to work?

Islam provides a nice example of these forces, because we know the order in which the Sura of the Koran were written, its verses—that is, the actual words of the prophet—recorded while he lived. (In contrast, all of Jesus's teachings were written long after he died.) Just like Jesus, Muhammad began as a marginal prophet of the marginalized, but unlike Jesus, he ended up as the head of a reinvading army of true believers. Muhammad began his ministry in Mecca, where he formed a small sect, often persecuted and vulnerable, so he preached an ideology of peace, respect for other groups, humility, and universal brotherhood. He then moved to Medina, where he initially faced the same situation and talked the same talk, but he then came to power in Medina and was able to head an invading army back to Mecca, which he promptly took over. During all this time, his Sura became more self-assertive and less tolerant the more powerful he became, sometimes urging attack on the infidels by the faithful. Similarly, in the Jewish tradition, it is King Josiah who is said to have both consolidated monotheism and been the first bloodthirsty advocate of it.

Consider a much more recent example from the Catholic Church. Pope Paul XXIII and Vatican II inspired in the Latin American Church a new "liberation theology" in the 1980s closer to the humble, persecuted church (prior to Constantin), the time when Jesus's teachings were actually written down. This liberation theology explicitly favored "the preferential option for the poor" and urged their organization into self-supporting communes. This entire movement was crushed by the US military, explicitly so, and of course by the Catholic Church itself, always eager to bend theology to local power. Assassinations were the preferred means of enforcing orthodoxy, especially in El Salvador, whether of nuns traveling innocently on the road or of the Archbishop (Romero) while saying Mass, or of a courageous Jesuit priest who cried out in prophesy, "Very soon the Bible and the gospel will not be allowed in our country. We'll get the covers and nothing more." If Jesus were to reappear, he would be arrested as subversive, said the priest a few weeks before his own assassination. Thus is religion degraded by very regressive forces.

RELIGIONS IMPOSE MATING SYSTEMS

Religions tend to impose their own mating systems, and these in turn affect degrees of relatedness within and between religions. Religions typically ask (or require) of their adherents that they marry within the religion or subreligion: Catholic with Catholic, Protestant with Protestant, Shia with Shia, Jewish with Jewish, and so on.

The pressure to breed within the group leads to a degree of inbreeding, that is, nonrandom mating with those to whom one is (at least marginally) more closely related. We are not here talking about close inbreeding—parent/offspring, brother/sister—but typically more distant, first cousin to second cousin and beyond. But repeated generation after generation, inbreeding inflates degrees of relatedness between group members (that is, genetic similarity because of common ancestry). At the same time, it creates a chasm in relatedness to other groups—one is *less* related than one otherwise would be.

Two important kinds of migration are important here. People may outbreed—that is, marry outside their group—and people may convert, or join another group.

When a man (for example) outbreeds and his children are raised outside his original group, his out-migration is experienced as a "selective death" to his original group. Whatever genetic traits he has are lost to that group, including his outbreeding tendencies. To put a fine point on it, if he is on average less ethnocentric, less self-loving, and less narrow in outlook than members of his original group, his out-migration lowers the frequency of these traits in that group as surely as if he had died young.

On the other hand, his arrival in the new group has the opposite effect. It is experienced as a selective birth, as if someone had been born (at full reproductive age) with the same traits we just described. Returning to the composition of the original group, the key question is, how much in-migration occurs and under what conditions? If a man marrying a woman has his children accepted as being of her faith, then the same kind of traits lost through out-migration will tend to return through in-migration. But are the two processes equally strong? If there are more men leaving than arriving, then this group will become more inbred. I have not had the chance to pursue this subject in greater detail, but if I were interested in the genetics of religious diversity, I would pay attention to biases in between-group transfers, by sex and by magnitude.

As for genetics, inbreeding has well-known effects. Products of inbreeding show less internal variability than do products of outbreeding. This genetic similarity can have two detrimental effects. On the one hand, relatively rare negative traits that require two copies of the same gene for expression (for example, sickle-cell anemia, Tay-Sachs disease) become more common. On the other, greater genetic variability has well-known benefits in defending against rapidly coevolving diseases, so that outbreeding becomes a genetic defense.

The second form of in-migration is simple conversion (initially unconnected to marriage), and religions differ in their rules regarding this. Thus, Christianity has usually been a proselytizing religion, continually seeking converts, wherever and however, as has Islam. Sunni and Shia Muslims may stretch from Senegal to Sudan to Lebanon to Pakistan to India to Indonesia, with similar opportunities for interbreeding all along the continuum within each group, but limited exchange between the two. With some notable exceptions, Judaism has not been a proselytizing religion, although Jews have been subject to forced conversion (for example, in Spain in the sixteenth century).

RELIGION PREACHES AGAINST SELF-DECEPTION

Many religions have teachings that are either explicitly or implicitly against self-deception. It is often argued that self-deception interferes with one's ability to know not only oneself and others but also God herself. For one thing, there is a presumptive case for the utility and validity of general principles. What is true here should be true there. What applies to you should apply to me. The very universality argues against the usual biases of deceit and self-deception. If you are told to treat others as you wish to be treated, then you have a rule, which, if actually followed, would counter much of your unconscious self-deceptive tendencies in favor of self over others. Similar general rules could reduce self-deception further. Of course, as we have seen, the generality of these "general" principles is easily undercut by forces of fragmentation, in-group and out-group formation, and the rule of the powerful.

Religions also preach explicitly against self-deception. Consider Jesus's famous teachings about not judging others (Matthew 7:1–5):

> Judge not that ye be not judged. For with what judgment you judge, you shall be judged: and with what measure you mete, it shall be measured to you again. And why do you behold the mote that is in your brother's eye, but consider not the beam that is in your own eye? Or how will thou say to your brother, let me pull out the mote out of your eye; and behold, a beam is in your own eye? You hypocrite, first cast out the beam out of your own eye; and then you shall see clearly to cast out the mote out of your brother's eye.

I translate this directly into the language of self-deception. Beware of self-righteousness, because it easily invites self-deception. You may be projecting onto others your own faults. And beware lest you come to be judged by the same criteria you are enforcing on them. Why do you see the minor fault in your neighbor but fail to see the major one in yourself? Instead of denying your own fault and projecting it onto others, admit your fault, the better to see whether any fault lies elsewhere. Otherwise, you are a hypocrite, criticizing the wrong person, in the wrong order.

Another argument against the speed—and injustice—with which we judge others comes from the case where Jesus is presented with a woman about to be stoned to death for committing adultery. His reaction? "Let he who is without sin cast the first stone." And it is said that everyone left the room in reverse order of age, the oldest—who had the most sins—the first. In both of these cases, it is internal contradictions that drive the argument, precisely the reason that universally valid principles tend naturally to argue against self-deception.

Other teachings are less explicitly opposed to self-deception but have similar implications just the same. Here is one that is opposed to the in-group/out-group bias. In the parable of the Good Samaritan (really the good Arab or Palestinian), Jew after Jew passes by the badly injured fellow Jew. It is an outsider, an Arab, a Samaritan, who responds to the sufferer's needs by binding his wounds, giving him water and food and finding him safe lodging. Who is the admirable person here, the heartless in-group member or the otherwise hated out-group one? Or what about Nicodemus, the man who came by night? It is precisely his willingness to meet Jesus at night, out of sight of others, that made him a hypocrite, one who eventually voted to condemn Jesus but then made sure to help bind the body for burial.

Another example is the structure of the Lord's Prayer, which has interesting features where self-deception is concerned. First, it is short. Then it is divided into only three parts, the first an assertion of humility: "hallowed be thy name" and "thy will be done." When landing at an airport, I often pray that "thy will be done" and add the hope that this does not include flipping the plane upside down on arrival but, if so, *thy* will be done. In other words, let us accept a larger plan than our own and not seek to change the plan through personal begging. Let us humble our own self-interest to the larger plan. In any case, if the plane is going to flip, the plane is going to flip; the only thing we can pray for is to be calm upon arrival.

The second part of the prayer has an interesting feature—you are allowed to beg for only two things on your behalf, and one of these is contingent. You can ask for your daily handout, what every creature needs: its daily bread. And then you may ask that your own sins be forgiven but *only insofar* as you forgive those of others. This is critical: no blanket amnesty. You must give to get; you must forgive to be forgiven. This binds you to a psychological commitment—one that ought to reduce self-deception on the spot.

Then comes the final part, where you ask not to be led into temptation—really an injunction against allowing yourself to be tempted—and to be protected from all evil (self-induced included). No intercessory prayer here. No "and may the president continue to make wise decisions and may God bless America," so commonly heard in US churches (or, more absurdly, in President George W. Bush's words, "may God continue to bless America," as if an obligation had developed). Indeed, the ability on Sunday of so many Christian preachers to forget the only teachings by Jesus on prayer is astonishing, were it not for the power of deceit and self-deception.

Sometimes the teaching against self-deception is only a metaphor. In the twenty-seventh psalm (vs 8), David says, "When Thou said, seek ye my face, my heart said unto Thee, Thy face, Lord, will I seek." It is hard to imagine looking God straight in the face and lying—to God or to oneself.

For a parallel in Islam, there is an important distinction in Sufi thinking between the jihad (struggle) against the outside world, called the small jihad, and the jihad against oneself, called the greater jihad. The small jihad is relatively simple: one struggles in group activities against an out-group in order to convert them. In the extreme case, either they are killed or you are, at which point you ascend to heaven. No great problem. But the struggle against oneself is far more difficult, and to reach God's light, one must succeed in controlling one's own body. This is a personal struggle that requires controlling your bodily desires (for money, pleasure, satisfaction) in order to purify your soul. These desires occlude self-knowledge, in our system of logic, by encouraging self-deception. In the Sufi system, you must enslave your desires or they will enslave you. And finally, controlling the self is also a useful tool for controlling the outside world. The Greek sage Thales once put the general matter succinctly. "Oh master," he was asked, "what is the most difficult thing to do?" "To know thyself," he replied. "And the easiest?" "To give advice to others." Various Eastern religions also sometimes urge rather extreme systems of physical self-denial to free the individual from its egocentric center.

INTERCESSORY PRAYER—DOES IT WORK?

A bizarre belief widespread in many Christian circles is that of the power of intercessory prayer. That is, many people seem to believe that a group of people in a room, scrunching up their foreheads in intense concentration on behalf of someone miles away about to undergo surgery, can have a positive effect on the outcome. Were this to be true, the laws of physics would have to be violated on a daily, even minutely basis, by a deity who chooses to alter reality in response to the pleas of petitioners according to some unknown criterion—a most unlikely structure to the real world. The matter has been put to a test a number of times but often with poorly controlled studies and small sample sizes, precisely the conditions expected to produce a conflicting array of positive and negative findings, feeding the illusion that something may actually be going on.

Then came a multimillion-dollar study, carefully organized with six hospitals in which groups prayed for given patients from the day before they entered surgery until two weeks later, while another group of patients received no such prayer. Meanwhile, some of those being prayed for were told that they were being prayed for and others were not. Patients were followed for a month after surgery. The results were unambiguous: no effect whatsoever of intercessory prayer on the outcome, no hint of a benefit. So our first question is answered: it has no direct effect.

But does it have a placebo effect? Does belief in the efficacy of intercessory prayer by the victim give any kind of efficacious benefit? Quite the contrary. Those told they were prayed for had *more* postoperative complications of every sort than did those who did not know they were being prayed for. One hypothesis is that when told people are praying for you, you interpret your situation as being more dire than it really is, with associated stress. The patients are not being offered anything more than a useless prayer: no talk of cleaning the apartment or keeping their dog alive, no investment in their future, nothing—just the claim of people in intense wishful thinking on their behalf.

Note that the truly devout have no problem with these new scientific results—God responds to these experiments by simply withholding the usual benefits of intercessory prayer the better to keep scientists (and unbelievers more generally) in the dark. Did not Jesus say, "I will reveal unto babes what I will keep hidden from the wise"?

RELIGION AND SUPPORT FOR SUICIDE ATTACKS

There has been an exponential increase in suicide attacks worldwide, at least as measured over the past twenty years. This is a device by which a member of one group sacrifices his or her life to inflict damage (death and otherwise) to numerous or highly important members of another group. There is no question that this behavior could in principle be an effective political (and reproductive) strategy with return benefits to the martyr's much larger kin group, but there is also no doubt that such behavior easily induces massive return spite. In any case, suicide bombing can serve as a sensitive measure of the degree of willingness to commit violence against an out-group at great personal cost.

It is of some interest to know the role of religion in all this—pro, con, or otherwise. Recent work has provided a most interesting answer. When measured one way, religious activity makes the participation in (and support of) suicide bombings more likely. When measured another way, religion has no effect. What is the difference? Religion has an external, social aspect and an internal, contemplative one. Across a variety of suicidal conditions (Palestinian surveys, a hostile prime for Israeli settlers), religious attendance (the social aspect) is positively correlated with support for suicide bombings, but prayer (the contemplative) is not. This holds for study after study. In a summary of six religions in as many countries, regular attendance at religious services predicted both out-group hostility and in some cases willingness to be a suicide martyr—

but prayer did not. The Sufi outer jihad is run by social interactions, the greater inner jihad by achieving independence through prayer. This is, I think, the double face of religion—outward, hostile, and egocentric; internal, contemplative, and anti-egoistic.

RELIGION → SELF-RIGHTEOUSNESS → WARFARE

Religions tend to contribute to war in several ways. They encourage an in-group mentality, backed up by a breeding system that increases within-group relatedness (while decreasing between-group relatedness) and they readily provide the shared self-deceptions on which to base group action. But there is one final gift of many religions: self-righteousness. Murder is not only not prohibited (as it is within the group), but it is also sometimes required. It is your moral duty to kill the infidel, the unbeliever, the other. You are doing the Lord's work—not just your own or that of your group. You are fulfilling more than your manifest destiny—you are the Lord's executioner. You are helping natural selection along its ordained path. The Bible, as it turns out, warns against this path: "Vengeance is mine," saith the Lord.

CHAPTER 13

Self-Deception and the Structure of the Social Sciences

There is structure to our knowledge. Take science, for example, with its various subdisciplines of mathematics, physics, chemistry, biology, psychology, and others. Or history and philosophy and philology. Or literature, biography, and poetry. How do processes of self-deception affect the structure of knowledge? We have already addressed history; here I wish to focus on social biology and the social sciences, economics, cultural anthropology, and psychology. If we believe, as we have seen over and over, that self-deception deforms human cognitive function among individuals, airline pilots, governmental agencies, war planners, and so on, how can we not imagine that our very systems of knowledge are not likewise systematically deformed?

Of course, I can pretend no overview of this immense subject—all of knowledge itself—but several points strike me as important. First, we expect knowledge to be more deformed, the more deformation is advantageous to those in control. If you are trying to land a missile more accurately or transmit knowledge more quickly, you will be drawn to science itself, which is based on a series of increasingly sophisticated and remorseless anti-deceit and anti-self-deception mechanisms. It seems likely that the enormous success of science in part reflects this feature. Second, it seems manifest that the greater the social content of a discipline, especially human, the greater will be the biases due to self-deception and the greater the retardation of the field compared with less social disciplines. It may be that the intrinsic complexity of social phenomena impedes rapid scientific progress, but modern physics is very complex, and its findings were unearthed by procedures relatively unimpeded by self-deception. The study of history seems to be a conflict between a few honest historians trying to gain a true picture of the past and the greater number, who are primarily interested in promoting an uplifting view of the group past—in short, a false historical narrative.

Another possibility regarding the development of social disciplines is that a prior moral stance regarding a subject may influence the development of theory and knowledge in that subject—so that, in a sense, justice may precede truth (and false justice, untruth). Let us begin with this topic.

PRECEDENCE OF JUSTICE OVER TRUTH?

The usual assumption within academia is that we will derive a theory of justice from our larger theory of the truth. But what if our prior stance regarding justice impedes our search for the truth? For example, an unconscious bias toward an unjust stance will invite cognitive biases in favor of this stance. The "truth" that one produces on the justice of a situation will have been distorted by the prior commitment to an unjust position. In short, injustice invites self-deception, unconsciousness, and inability to perceive reality, while justice has the opposite effect. This can be a very pervasive effect in life. That is, we can construct social theory—at the microlevel, marriage, family, job; at the macrolevel, society, war, etc.—and think we are pursuing the truth objectively, but we may only be fleshing out our biases. This suggests that an early attachment to fairness or justice may be a lifelong aid in discerning the truth regarding social reality. Of course, if your attachment is to pseudo-justice, one may have exactly the opposite effects. It is possible to use an alleged attachment to justice defensively—for example, to prohibit outside knowledge from entering your discipline—which may lead you far from truth, as we shall see for cultural anthropology. Behavior may cause belief, as I have been arguing, but that still leaves open the question of what causes the behavior in the first place, that is, the just or unjust stance.

SUCCESS OF SCIENCE IS BASED ON ANTI-SELF-DECEPTION DEVICES

The success of science appears in great part to be due to a series of built-in devices that guard against deceit and self-deception at every turn. First, everything is supposed to be explicit. Famous mathematical proofs (Godel's theorem) begin with a set of all the symbols used and what they mean. By contrast, in the social sciences, entire subdisciplines may flourish in the interstices of poorly defined words. *Scientific* work is supposed to be described explicitly in detail, with terms and methods defined to permit the work to be repeated exactly in its entirety by anyone else. This is the key guard against untruth: repeating work to see whether the same results emerge. Think of the number of tantalizing hoaxes that are dismissed because they can't pass this first hurdle—for example, achieving atomic energy via cold fusion. Of course, full-time hoaxes, such as psychoanalysis, preclude experimental tests at the outset (in favor of such bedrock data as clinical lore). The requirement for exact description permitting exact repetition applies not just to experimental work but also to any way of gathering data that reveals patterns of interest.

Experiments are conducted under controlled conditions—that is, with certain key variables held constant and/or varied in a logical and systematic manner. The results are then subjected to a statistical apparatus that has grown very sophisticated in the past one hundred years. Very complex sets of data can now be rigorously searched for information regarded as statistically significant. By convention, data that can be generated by chance more than 5 percent of the time are rejected as unreliable. For important results, such as medical findings, we prefer an error rate of 1 percent or less. Finally, meta-analyses can be performed on large numbers of related studies to see what statistically valid generalizations can be made across the full range of evidence. Every single one of these advances tends to minimize the opportunities for deceit and self-deception. They also permit us to rank information by degree of reliability (statistical significance) and effect size (weak or strong).

The acid test of science is its ability to predict the future, in particular, hitherto unknown facts. Yes, light really is bent by gravity (per Einstein); in an eclipse of the sun, the apparent position of stars in the nearby background was altered by the sun's gravity. The same principle operates on much more humble work. That ants produce a 1:3 ratio of investment in the sexes (unlike almost all other animals) was first predicted on kinship grounds alone (the female ants producing the ratio are three times as related to their sisters as to their brothers, unlike almost all other species) and has been confirmed by detailed evidence from dozens of scientific studies. Of course, scientists will pretend that "predictions" lack any foreknowledge, when in fact they are "post-dictions." This is the beauty of Einstein's prediction compared to that concerning ants: How on earth could Einstein have had advance information about the apparent positions of stars during a solar eclipse more than ten years into the future? By contrast, one can easily bone up on ant sex ratios before launching one's prediction.

There is one more key requirement to true science. Science asks that, whenever possible, knowledge be built on preexisting knowledge. Key assumptions may already be contradicted (or supported) by preexisting knowledge, and where no such knowledge exists, science suggests the value of producing it. Errors in the foundation—of both buildings and disciplines—are the most costly. Yet there is surprising resistance in many quarters of social science to adopt—much less embrace—this feature of real science.

The structure of the natural sciences is as follows. Physics rests on mathematics, chemistry on physics, biology on chemistry, and, in principle, the social sciences on biology. At least the final step is one devoutly to be wished and soon hopefully achieved. Yet discipline after discipline—from economics to cultural anthropology—continues to resist growing connections to the underlying science of biology, with devastating effects. Instead of employing only assumptions that meet the test of underlying knowledge, one is free to base one's logic on whatever comes to mind and to pursue this policy full time, in complete ignorance of its futility.

By contrast, mathematics gave physics rigor, physics gave chemistry an exact atomic model, and chemistry gave biology an exact molecular model. And biology? You would think it would have much to give—most important, an explicit, well-tested theory of self-interest, but also a vastly expanded set of evidence, including a detailed understanding of many underlying variables (immunological, endocrinologic, genetic) that would otherwise remain obscure.

THE MORE SOCIAL THE DISCIPLINE, THE MORE RETARDED ITS DEVELOPMENT

In physics, we imagine precious little self-deception. What difference does it make for everyday life whether the gravitational effect of the mu meson is positive or negative? None at all. So the field is expected to advance relatively unimpeded by forces of deceit and self-deception—with one exception. Physicists will overemphasize the importance and value of their work to others. They will talk of producing "a theory of everything" and make other grand claims, but their social utility, in my opinion, is primarily connected to warfare. Their major function has been to build bigger bombs, delivered more accurately to farther distances, and this has probably been their main function reaching back into prehistory. When I read of nine billion euros spent on a supercollider in which tiny particles are accelerated to incredible speeds and then run into one another, I think "bombs." This factor may lead to more resources being directed toward physics and to some subareas than is objectively sensible, but it is unlikely to have much effect on constructing theory.

In my opinion, a key to the development of the very solid and sophisticated science of physics is the complete absence from its subject matter of social interactions or social content of any sort. More generally, I imagine that the greater the social content of a discipline, the more slowly it will develop, because it faces, in part, greater forces of deceit and self-deception that impede progress. Thus, psychology, sociology, anthropology, and economics have direct implications for our view of ourselves and of others, so one might expect their very structure to be easily deformed by self-deception. The same can be said for some branches of biology, especially social theory and (separately) human genetics. Many of these illusions have in common that function is interpreted at a higher level than is warranted (for example, society instead of individual).

SELF-DECEPTION IN BIOLOGY

For roughly a century, biologists had the social world analyzed almost upside down. They argued that natural selection favored what was good for the group or the species, when in fact it favors what is good for the individual (measured in survival and reproduction), as Darwin well knew. More precisely, natural selection works on the genes within an individual to promote their own survival and reproduction, which is usually equivalent to what is beneficial for the individual propagating the genes. Yet almost from the moment Darwin's theory was published, scientists in the discipline reverted to the older view of benefit as serving a higher function (species, ecosystem, and so on), only now they cited Darwin as support for their belief. In turn, the false theory was just the kind of social theory you would expect people to adopt in a group-living species whose members are concerned with increasing one another's group orientation. This theory also can be used to justify individual behavior by claiming that such behavior serves group benefit (for example, murder justified as population regulation) and can be used to create the ideal of a conflict-free world.

For example, take the classic case of male infanticide, first studied in depth in the langur monkeys of India, and now known for more than one hundred species. Male murder of dependent offspring (fathered by a previous male) was rationalized as a population-control mechanism that kept the species from eating itself out of house and home. Male murder thus served the interests of all. Of course, it did no such thing. Since a nursing infant inhibits its mother's ovulation, murder of the infant brought the bereaved mother into reproductive readiness quicker, which aided the male's reproduction but at a cost to the dead infant and its mother. In some populations, as many as 10 percent of all young are murdered by adult males—each murder gaining on average only two months of maternal time for the new male. These deaths are unrelated to population density (as would be expected if they served a population-regulation function), but they are correlated with the frequency with which males take over new groups. What this work shows is that an enormous social cost can be levied every generation by natural selection on males, even though there is only a modest male gain (two months of female labor) compared with the female loss (twelve months of maternal care).

It was famously argued that male aggression is intrinsically good for the species, since it is always better for the species if the stronger of two males takes control of a favored female. But this is precisely what is not known. Whether an aggressively successful male has genes at other loci that are beneficial to his progeny is an open question that must be answered in each separate case (especially by the choosing female). Perhaps the success of aggressive males spreads genes only for aggressiveness, which are otherwise useless for the species (or a female's daughters). In any case, male elephant seals fighting for access to females clumped together on breeding islands typically kill about 10 percent of the young every year (fathered by other males) by trampling them to death during fights. In what sense is male aggression good for the species? Are they eliminating inferior genes underfoot?

Close relationships are also easily imagined to be conflict-free. Thus, mother/offspring coevolution is allegedly favored—each party evolving to help the other. As we saw in Chapter 4, nothing like this is actually true of real families. Even in the formation of the placenta, the mother does not help the invading fetal tissue—she puts up chemical and physical obstacles (the better to avoid later excess investment). Likewise, in the 1960s, bird watchers liked to imagine that the families they loved to observe were free of conflict, but this was soon proven wrong when rates of extra-pair paternity exceeding 20 percent were regularly reported.

Thus, for years evolutionary biologists have used a form of argumentation that helped cement in the social sciences and elsewhere the notion that evolution favored what was good for the family, the group, the culture, the species, and perhaps even the ecosystem, while minimizing the reality of conflict within any of these entities. Anthropologists soon rationalized warfare itself as favored by evolution because it too was such a nifty population-regulation device. Note that the error is virtually irrelevant for nonsocial traits. The human locking kneecap allows us to stand erect without wasting energy in tensed legs. It evolved because it benefited the individual with the new kneecap, but if you said it evolved to benefit the species, you would not misinterpret the kneecap. Not so for social traits. Here, as we have seen, we can exactly invert the meaning of a trait by failing to see how it is favored among individuals, even though it may be more costly to others. Instead we imagine that everyone benefits. This often amounts to reaffirming Pangloss's theorem—that everything is for the best in the best of all possible worlds.

Likewise, altruism toward others presents no great problem for species-advantage thinking, because as long as benefit is greater than cost, there is a net benefit for the species. Of course, at the individual level, altruism is a problem to explain and requires special conditions, such as kinship or reciprocal relations, with internal conflict in both cases. The latter generates a sense of fairness to evaluate nonreciprocal relations, an adaptation unnecessary under a group-selected view.

IS ECONOMICS A SCIENCE?

The short answer is no. Economics acts like a science and quacks like one—it has developed an impressive mathematical apparatus and awards itself a Nobel Prize each year—but it is not yet a science. It fails to ground itself in underlying knowledge (in this case, biology). This is curious on its face, because models of economic activity must inevitably be based on some notion of what an individual organism is up to. What are we trying to maximize? Here economists play a shell game. People are expected to attempt to maximize their "utility." And what is utility? Well, anything people wish to maximize. In some situations, you will try to maximize money acquired, in others food, and in yet others sex over food and money. So we need "preference functions" to tell us when one kind of utility takes precedence over another. These must be empirically determined, since economics by itself can provide no theory for how the organism is expected to rank these variables. But determining all of the preference functions by measurement in all the relevant situations is hopeless from the outset, even for a single organism, much less a group.

As it turns out, biology now has a well-developed theory of exactly what utility is (even if it misrepresented the truth for some one hundred years) based on Darwin's concept of reproductive success. If you are talking about utility (that is, benefit) to a living creature, then it is useful to know that this ultimately refers to the individual's inclusive fitness, that is, the number of its surviving offspring plus effects (positive and negative) on the reproductive success of relatives, each devalued by its relatedness to them. In many situations, the added precision of this definition (compared to reproductive success alone) makes no difference, but by resolutely acting as if they can produce a science out of whole cloth, that is, independent of noneconomic scientific knowledge, economists miss out on a whole series of linkages that may be critical. They often implicitly assume, as we noted in the first chapter, that market forces will naturally constrain the cost of deception in social and economic systems, but such a belief fails to correspond with what we know from daily life, much less biology more generally. Yet such is the detachment of this "science" from reality that these contradictions arouse notice only when the entire world is hurtling into an economic depression based on corporate greed wedded to false economic theory.

The mistake is partly related to the fact that "utility" has ambiguity built into it. It can refer to utility of your actions to you or to others, including the rest of your group. Economists easily imagine that the two kinds of utility are well aligned. They often argue that individuals acting for personal utility (undefined) will tend to benefit the group (provide general utility). Thus they tend to be blind to the possibility that unrestrained pursuit of personal utility can have disastrous effects on group benefit. This is a well-known fallacy in biology, with hundreds of examples. Nowhere do we assume in advance that the two kinds of utility are positively aligned. This must be shown separately for any given case.

One recent effort by economics to link up with allied disciplines is called behavioral economics, a link with psychology that is most welcome. But as usual, economists resolutely refuse to make the final link to evolutionary theory, even when going through the motions. That is, even those economists who propose evolutionary explanations of economic behavior often do so with unusual, counterlogical assumptions. For example, a common recent mistake (published in all the best journals) is to assume that our behavior evolved specifically to fit artificial economic games.

To imagine how bizarre this is, consider the ultimatum game described in Chapter 2. People often reject unfair offers of a split of money by anonymous others (for example, 80 percent to the proposer and 20 percent to the recipient) even though they thereby lose money. Thus, the game measures our sense of fairness: How much are we willing to suffer in order to punish someone acting unfairly toward us? But a group of economists (with some anthropologists thrown in for added rigor) has made the extraordinary argument that people are acting as if they had evolved to fit this unusual lab situation. Put differently, that we reject unfair offers at a cost to ourselves in order to punish the perpetrator in a completely anonymous exchange means to them that the bias evolved to fit exactly this situation—one-time exchanges with no possible return benefit for the actor, or relatedness, only a group benefit. Once again, group trumps individual. But this is as logical as arguing that our terror watching a horror film evolved to fit movie showings. Biologists have brought living creatures into the laboratory for centuries to study their traits, but no one I know of has shortcut the study of the *function* of the trait by imagining that the trait evolved to fit the laboratory.

A recent Nobel winner in economics wondered how it was possible for his well-developed science to fail completely to predict the catastrophic economic events that started in 2008. One part, of course, is that economic events are intrinsically complex, involving many factors, and the final result, the aggregate of the behavior of an enormous number of people, though not quite as complex as the weather, is almost as difficult to predict. As for the cause the economist located, it was infatuation with beautiful mathematics at the cost of attention to reality. Surely this is part of the problem, but nowhere does he suggest that the first piece of reality they should pay attention to—and this has been obvious for some thirty years now—is biology, in particular evolutionary theory. If only thirty years ago economists had built a theory of economic utility on a theory of biological self-interest—forget the beautiful math and pay attention to the relevant math—we might have been spared some of the extravagances of economic thought regarding, for example, built-in anti-deception mechanisms kicking in to protect us from the harmful effects of unrestrained economic egotism by those already at the top.

Finally, when a science is a pretend science rather than the real thing, it also falls into sloppy and biased systems for evaluating the truth. Consider the following, a common occurrence during the past fifteen years. The World Bank advises developing countries to open their markets to foreign goods, let the markets rule, and slash the welfare state. When the program is implemented and fails, the diagnosis is simple: "Our advice was good but you failed to follow it closely enough." There is little risk of being falsified with this kind of procedure.

CULTURAL ANTHROPOLOGY

Cultural anthropology made a tragic left turn in the mid-1970s from which it has yet to recover (at least in the United States). Before then, the field was called social anthropology and included all forms of human social behavior, especially as displayed by different cultures and peoples. The field was meant to partner with physical anthropology, the study of the body, including fossils and artifacts from the past. But suddenly in the early 1970s, strong social theory emerged from biology and a variety of subjects were addressed seriously for the first time: kinship theory, including parent/offspring relations, relative parental investment, and the evolution of sex differences, the sex ratio, reciprocal altruism and a sense of fairness, and so on. Social anthropologists had a choice: accept the new work, master it, and rewrite their own discipline along the new lines, or reject the new work and protect their own expertise (such as it was). As has been noted, "Faced with the choice between changing one's mind and proving that there is no need to do so, almost everyone gets busy on the proof." This is perhaps especially true in academia.

Consider your dilemma as a social anthropologist. You have invested twenty years of your life in mastering social anthropology. Along the way, you have completely neglected biology. Now comes the choice: acknowledge biology (painful), invest three years in catching up (nearly unimaginable), then compete with people twenty years younger than you and better trained (impossible)—or instead ride the old horse for all she is worth, whipping social anthropology until she bleeds? Even in physics, it was famously said that the field advanced one funeral at a time—only death could get people to change their minds. But notice the intermediate path not taken. They could have said, "I will not retool myself; it is too late. But I will make sure my students learn something useful about the new work in biology (they can even teach me) while I continue to do my work." Complete rejection is redolent of self-deception. Outright denial is the easiest immediate path but entrains mounting costs, now onto the third generation, making it ever harder to resist each new wave of denial.

Certainly the social anthropologists rose to the challenge, even renaming their field "cultural anthropology" to more explicitly rule out the relevance of biology in advance. Now we were no longer social organisms but cultural ones. The justification, in turn, was moral. Out of biological thinking flowed biological determinism (the notion that genetics influences daily life), whose downstream effects included fascism, racism, sexism, heterosexism, and other odious "isms." To mention natural selection was to imply the existence and perhaps even utility of genes, which was prohibited on the moral grounds just given. Thus an entire new area of social theory would be ruled out based on the alleged pernicious influences of its assumptions, which were, in fact, widely accepted as true (genes exist, they affect social traits, natural selection alters their relative frequencies, and this produces meaningful patterns). Once you remove biology from human social life, what do you have? Words. Not even language, which of course is deeply biological, but words alone that then wield magical powers, capable of biasing your every thought, science itself reduced to one of many arbitrary systems of thought.

And what has been the upshot of this? Thirty-five wasted years and counting. Years wasted in not synthesizing social and physical anthropology. Strong people welcome new ideas and make them their own. Weak people run from new ideas, or so it seems, and then are driven into bizarre mind states, such as believing that words have the power to dominate reality, that social constructs such as gender are much stronger than the 300 million years of genetic evolution that went into producing the two sexes—whose facts in any case they remain resolutely ignorant of, the better to develop a thoroughly word-based approach to the subject.

In many ways, cultural anthropology is now all about self-deception—other people's. Science itself is a social construct, one among many equally valid ways of viewing the world: the properties of viruses may also be social constructs, the penis may, in some meaningful sense, be the square root of –1, and so on. As a result, most US anthropology departments consist of two completely separate sections, in which, as one biological colleague put it, "they think we're Nazis and we think they are idiots"—hardly a platform for synthesis and mutual growth.

PSYCHOLOGY

In the 1960s, psychologists often explicitly disavowed the importance of biology. At Harvard, to get a PhD in psychology, you were required to take one semester of physics. This was to give you an idea of what an exact science looked like. No biology was required. Like economists, psychologists were going to create their field out of itself: learning theory, social psychology, psychoanalysis—essentially competing guesses about what was important in human development, none with any foundation. Psychoanalysis was a long-running fraud, as we shall see below, and learning theory made far-reaching and implausible claims about the ability of reinforcement to mold all behavior adaptively. It was soon shown on logical grounds alone that reinforcement could not produce language, or even just associations of actions and their effects when the latter were delayed more than a few moments.

On the positive side, psychology has always concentrated on the individual and was thus congenial to an approach based on individual advantage. Recently a school of evolutionary psychology has developed, while psychology has been increasingly integrated with other areas of biology, sensory physiology long ago but now neurophysiology and immunology. So psychology is rapidly becoming the branch of evolutionary biology it always wished to be.

Social psychology somewhat lags the rest of psychology, another example perhaps of the retarding effects of deceit and self-deception on disciplines with more social content. It has generated artificial methodologies meant to shortcut work and achieve quick results, the curse of psychology for more than a century: wishing to say more than available knowledge permits. A key such method was that of self-reports, or questionnaire-answering behavior—what people say about themselves. In retrospect, it seems unwise to have tried to build a science of human behavior on people's verbal responses to questions. For one thing, forces of deceit and self-deception—or call them issues of self-presentation and self-perception, if you prefer—loom large. We often do not tell the truth about ourselves to others and we often do not know the truth in the first place. In using these measures, exactly how were they screening out deception, never mind self-deception, to arrive at the truth? And how is this possible in the absence of an explicit theory of deceit and self-deception? Building a science on this foundation led to numerous significant correlations between ill-defined variables that are poorly measured, but little or no cumulative growth over time. Instruments (that is, questionnaires) were said to be well-validated, predictive, and internally consistent, that is, people answer the same way a month apart, the measures correlate with some other measures, and all questions point in the same direction (or are reverse scored). Not a very impressive nod toward methodology, but fortunately this era is coming to a close, with new methodologies that access unconcious biases directly.

PSYCHOANALYSIS: SELF-DECEPTION IN THE STUDY OF SELF-DECEPTION

Freud claimed to have developed a detailed science of self-deception and human development: psychoanalysis. But one measure of a field is whether it grows and prospers or wilts and withers, and psychoanalysis has not prospered. As it turned out, the empirical foundation for developments in the field was something called clinical lore, essentially what psychiatrists told one another over drinks after a day's work. That is, when you asked a psychiatrist what his (as he almost always was) basis was for believing that a key part of the female psyche was "penis envy" or that the route to understanding males lay in something called castration anxiety, you were told that the basis was shared experiences, assumptions, and assertions among psychoanalysts about what went on during psychotherapy—something inaccessible to you, unverifiable, and, as a system, providing no hope for improvement. Indeed, the failure to state or develop methodologies capable of producing useful information is almost the definition of nonscience, and in this regard, psychoanalysis has been spectacularly successful. When is the last time you heard of a large, double-blind study of penis envy or castration anxiety?

Freud's theory consisted of two parts: self-deception and psychosocial development. The theory of self-deception had many creative concepts—denial, projection, reaction formation, ego defense mechanism, and so on, but these were wedded to a larger system that made no sense at all, the id (instinctual forces heavily based on alleged critical transitions in early life—anal, oral, and oedipal), the ego (roughly, the conscious mind), and the superego (the conscience, or something like that, formed by interaction with parents and significant others).

His theory of psychosocial development was corrupt, in the sense that it was built on weak and suspect assumptions that had little or no factual support. The argument was heavily centered on sexual attraction within the nuclear family—and its suppression—but there is good reason to doubt that this should be a major offspring concern. Almost all species of animals are selected to avoid close inbreeding, which has real genetic costs, and they have evolved mechanisms—for example, early exposure to parents and siblings causes sexual disinterest—that minimize inbreeding. This is especially true from the offspring's viewpoint. That is, fathers may gain in relatedness by forcing sex on a daughter (and therefore a child) sufficient to offset the genetic cost, but the daughter is unlikely to benefit sufficiently in relatedness to offset her cost. The son could in principle benefit from impregnating his mother, but selection would be weak at best, the one ending her reproduction while the other is beginning his, and there are other very good reasons for showing deference to one's mother (especially for a male's maternal genes).

Thus, with Freud's claim that sexual tendencies in the family arose from the unconscious needs of the child, he was committing a classic case of denial and projection—denying the inappropriate sexual advances toward young women by their male relatives (as his women patients were describing to him) and imagining instead that the women were lusting after precisely these couplings.

He was also obtuse to harsh parental treatment as a cause of offspring malfunction. Once again, his tendency was to blame the victim. One of Freud's celebrated analyses was that of "Wolf Man," psychotic since adulthood with sensations of being tormented physically, bound and restricted, and unable to control his fears. Freud conjured this whole syndrome as resulting from the child's failure to mature properly, getting stuck in some early stage of development, but he never considered the father's possible role in this—indeed speaks warmly of him as a highly regarded educator with numerous books—even though he was a sadistic educator and parent. He advocated tying children into bed at night and using a series of other torture devices, all in the name of good posture. Alas, he applied his theories to his own children. One boy committed suicide; the other survived to become Freud's "Wolf Man."

The degree to which Freud's habit of cocaine abuse during his early years helped fuel his grandiosity is impossible to know, but he certainly easily believed in other phantasmagorical things, for example, that the number twenty-nine played a recurring and decisive role in human life, or that thought could be transported instantaneously across wide distances without the use of electrical devices, and so on. What is truly extraordinary is that he was able to build a cult that took over whole sections of psychiatry and psychology, and provided employment for generations of like-minded people, charging high fees, four times a week, to misinterpret the lives of those they were talking to.

Freud's own attitude toward empirical verification was nicely summarized when he responded to someone asking if after thirty years of theorizing, perhaps it was time for some experimental testing. Though allowing that experiments could do no harm, Freud said:

> The wealth of dependable observations on which these assertions rest, make them independent of experimental verification.

This is an unusual assertion, since it suggests that counterevidence cannot count as actual evidence. Put differently, the worlds of experimental truth and psychoanalytic truth are independent, as indeed they are. Contrast the position of the famous physicist Richard Feynman:

> It doesn't matter how beautiful the guess is, or how smart the guesser is, or how famous the guesser is; if the experiment disagrees with the guess, the guess is wrong. That's all there is to it.

SELF-DECEPTION DEFORMS DISCIPLINES

We have seen numerous ways in which self-deception may deform the structure of intellectual disciplines. This seems obvious in both evolutionary biology and the social sciences, where increasing relevance to human social behavior is matched by decreasing rates of progress, in part because such fields induce more self-deception in their practitioners. One common bias is that life naturally evolves to subserve function at higher levels. Not genes but individuals, not individuals but groups, not groups but species, not species but ecosystems, and, with a little extra energy, not ecosystems but the entire universe. Certainly religion seems to promote this pattern, always tempted to see a larger pattern than is warranted. Science provides some hope, since it has a built-in series of mechanisms that guard against deceit and self-deception, but it too is vulnerable to the construction of pseudo-sciences (Freud), not to mention outright fraud. Over the long haul, however, falsehood has no chance, which is why over time science tends to outstrip competing enterprises.

CHAPTER 14

Fighting Self-Deception in Our Own Lives

There are two divisions in my life regarding self-deception: the personal, affecting how I relate to those around me, and the general, which refers to my scientific work and the problem of interpreting society more generally. One is much more intimate and is bound up with the biology of those relationships most important to me. The second is an enterprise that affects the thinking of many more people, but these are usually much more distantly connected to me.

Regarding one's personal life, the problem with learning from living is that living is like riding a train while facing backward. That is, we see reality only after it has passed us by. Neurophysiologists have shown that this is literally true (Chapter 3). We see (consciously) incoming information, as well as our internal intention to act, well after the fact. It seems as if it is difficult to learn after the fact what to predict ahead of the fact; thus, our ability to see the future, even that of our own behavior, is often very limited. I believe I have learned a lot about my self-deceptions but not in ways that prevent me from repeating them—often exactly. Take one common problem I have involving both conflict and self-deception: Someone does me harm, and I imagine a spiteful response, a nasty letter or some other gesture of contempt. Then the submerged side of me says, "But, Robert, you have been in this situation 614 times already and you have talked yourself into the spiteful action, yet in every case shortly afterward you regret your action. This is no different. Do not do it." And then the dominant part of my personality comes roaring back. "No, *this* time is different. This time I will feel satisfied and happy." And there goes number 615. One form of this error is nicely captured in an ancient Chinese expression: "When planning revenge, build two graves, not one."

By contrast, I do imagine—although this may be complete self-deception—that a life dedicated to the pursuit of truth, especially via science and logic, has honed my mind through the years, so that I practice relatively little self-deception in the work domain of my life. In fact, I have become somewhat more critical and exacting, requiring higher levels of significance and better methodologies before committing to evidence. Of course, my logical mind is weaker now but I believe I rarely bend logic to suit personal need. For most scientists, this bending results from competition with fellow scientists for recognition, and here the well-known "tender ego syndrome" of academics leads many of them to downgrade the work of those competing for similar niches in their discipline or outshining them more generally. It has always seemed absurd to me to let such petty personal concerns get in the way of understanding the truth, when that is the entire alleged purpose of your work, yet the tendency toward self-aggrandizement and diminution of others' achievements seems as strong here as elsewhere in life.

On the other hand, I have noticed that the standards regarding my own arguments I am willing to push forward have dropped. I care less about appearing the fool, so I am willing to live with a higher ratio of foolish thought to true insight in my statements. I believe this is a function of age. Get a reputation for being foolish when you are young and people will have very long memories. Being foolish in old age may merely lead people to say, "Well, of course he did get dotty toward the end." On the other hand, old age can comfortably coexist with some wisdom, most of your relatives being much younger and therefore more equally related on both sides of your genome, with important effects in the deeper future to which you may wish to attend.

TO FIGHT ONE'S OWN SELF-DECEPTION OR NOT?

Before we begin, we may well ask whether we should bother fighting in the first place. Self-deception has been favored by natural selection, the better to deceive others and ourselves, so why should we fight such tendencies in ourselves? They are advancing our own evolutionary interests. Surely it must be useful to adjust our self-deception strategically—toward situations where it is most likely to be effective—but oppose it in general. Why? Does this not violate our attachment to evolutionary self-interest?

My own answer is simple and personal. I could not care less. Self-deception, by serving deception, only encourages it, and more deception is not something I favor. I do not believe in building one's life, one's relationships, or one's society on lies. The moral status of deceit with self-deception seems even lower than that of simple deception alone, since simple deception fools only one organism—but when combined with self-deception, two are being deceived. In addition, by deceiving yourself, you are spoiling your own temple or structure. You are agreeing to base your own behavior on falsehoods, with negative downstream effects that may be very hard to guess yet intensify with time.

It is worth noting that we have also been selected to rape on occasion, to wage aggressive war when it suits us, and to abuse our own children if this brings us some compensating return benefit, yet I embrace none of these actions, regardless of whether they have been favored in the past. As one evolutionist told me, his genes could not care less about him, and he feels the same way toward them.

One variable that does enter my thinking is the concept of an evolutionarily stable strategy, defined as one that can't be driven out of a (well-defined) evolutionary game. As long as being honest, or trying to be, and as long as reducing one's self-deception, or trying to, are strategies that cannot be driven to extinction, then I am happy to leave the long-term evolutionary outcome to the future. If my strategy of attempted honesty leads by logic to its evolutionary disappearance for good, I need give special thought to the matter, but as long as it is merely evolutionarily stable—perhaps held at low frequencies but not driven extinct—I think I will go with anti-self-deception as my approach to life, my so-called internal strategy, not that I have much hope of achieving it.

A SERIES OF MINOR VICTORIES FOLLOWED BY A MAJOR DISASTER

In my life, self-deception is often experienced as a series of minor benefits followed by a major cost. I will be overly self-confident, project that image, and enjoy some of the illusions, only to suffer later on a sharp reversal, based in part on the blindness induced by this overconfidence. I may deny counterevidence to a happy relationship that is, in fact, deteriorating badly, each minor compromise with reality boosting mood temporarily while postponing the reckoning that may arrive with savage force. Denial, as we have seen, is often easy to get started but hard to stop. Put another way, self-deception often ends badly. This is as true of mega-events, such as misguided wars and economic policies, as it is for events in one's personal life. We may enjoy a temporary benefit of deceiving others and self, but we suffer a long-term cost.

I believe this is a general rule in life, that the cost of ignorance takes a while to kick in, while the benefit of self-deception may be immediate. Long ago, work on rats proved that these kinds of connections—that is, those with a time delay—are among the most difficult for an organism to learn. Immediate rewards and costs are obvious, long-term life effects much more difficult to discern. In addition, there is a strong tendency to discount future effects compared to current ones so that long-term negative effects may be especially difficult to register. In what follows, I will try to sketch out a few anti-self-deception devices that may prove useful in life. There must be many, many more.

SIGNALS OF UNDERLYING MENTAL SCREW-UPS

Imagine you are washing dishes and carelessly smash a wineglass against the sink bottom, splintering it. What were you thinking about while you did this? If you are like me, more often than not, you were thinking of something hostile and foolish to do to someone else. In this particular case, I was imagining telling a woman something she did not need to know or wish to hear. The splintered glass served as a warning to me. As I picked up the fragments, I meditated on the stupidity of what I had just been contemplating, vowing that whatever I did, it would not be what had been in my mind when I shattered the glass. Likewise, I once tore off half my lower lip while shaving and simultaneously calling (in my mind) someone a motherfucker. Motherfucker, indeed—he was capable of mutilating me at a distance of miles.

I think the power of this correlation first occurred to me when I was driving off the campus of the University of California, Santa Cruz, around sunset one evening. I was driving too fast and cursing out in my mind a colleague with whom I had had an argument. Just as I peaked by calling him a punk in my mind, I nearly ran over two students trying to cross the intersection. They cursed and shook their fists, and I shook mine back, but it soon occurred to me that I had nearly run over two completely innocent bystanders over my conflict with the colleague. It did not take long to realize that the behavior I was contemplating was almost as self-destructive in its domain as my actual behavior was dangerous to others. I vowed to temper my language. I have no idea what my near-victims vowed.

It is not just anger. The other day I managed to break my plastic door handle while trying to enter my car. It was enthusiasm that did it—overexcitement planning a premature and overly positive e-mail as interpreted after the fact. I stored the message, later rewrote it, and later still, sent it.

I have found this rule so often in my life that it is one of the few things I actually think I have learned, at least on a short-term basis: avoid actions that are being contemplated while you are screwing up during ongoing life. As I age, I find myself scrutinizing my errors more finely—not just broken wineglasses but an unexpected lurch or tripping over the curb or some minor social failure—for deeper correlated mental mistakes. Occasionally the problem is so well hidden that I may go through several mistakes, including a broken glass and a dropped computer, before I see it. For example, I may be slowing down my work because I am unconsciously afraid of negative reaction to it when it is completed. When conscious, the remedy is obvious: speed up the work, if necessary by frequently cursing out the individual causing the slowdown.

CORRECTING FOR OUR OWN BIASES

As we have just seen, it is possible consciously to correct for a bias in yourself that you have noticed—in that case, negating intended behavior associated with mishaps in ongoing behavior. Sometimes you can correct your biases quantitatively. For example, I long ago noticed that when asked for my straight-from-the-heart, no-thinking estimate of a variable, I tended to overshoot by 30 percent in the positive direction. So when I wanted to know the approximate truth, I just subtracted 30 percent from my first estimate.

Consider another example. In which order do you search for something? Do you start first with the most likely place to find it and then search each successive place in descending order of likelihood? Or do you do it the other way around—start with the least likely and move up? The only rational system is the first—you minimize costs by always searching where the expected returns are highest—but most of my life I have done it exactly the other way around. Why? I believe it may have been a response to relatively harsh paternal reaction when I failed to turn up with what I had been sent to find. If you are very fearful as you set out to search for something, you may be tempted to start with the last place you would expect to find it—having eliminated this, you then move to a more hopeful alternative, and so on. Your mood goes right up until the very last choice, whereas in the rational search, you try your best shot first. If it fails, you start to panic; as each succeeding one fails, your panic grows. In one situation, hope grows; in the second, panic. Whatever the cause of my aberrant behavior, I see the pattern and how foolish it is, so I act consciously to counteract the bias, forcing my brain to focus first on the most likely place to find what is missing and then move steadily down from there. Yet my very first move is often still in the wrong direction, and only then does the correction set in.

I also noticed a curious fact about my mind where arithmetic is concerned. I grew up before calculators and I learned numerous tricks to solve arithmetic problems quickly. But if you put a dollar sign in front of the numbers, my mind short-circuited. I added when I should have subtracted, multiplied when I should have divided. I had to remove the dollar signs and reinsert them only at the end. I also had to proofread my work more carefully. When you are copying a long number and want to make sure you have made no mistakes, you can read through the numbers again, comparing them directly, but the better way is to read through them backward. That way, unconscious mental biases that may prevent you from seeing the error twice in a row are very unlikely to do so. Professional proofreaders often use the same device.

Another example of noticing a pattern of behavior and acting consciously against it concerns displacement activities. It is a fact of human (and monkey) psychology that aggression is easily displaced onto others. Angry with your spouse, you may be harder on your children or kick the dog, often to their surprise. It is as if your anger is incited and, looking around for targets, is blocked from the logical one, so it looks for nearby victims, preferably those smaller and less able to retaliate. This is such a common occurrence that everyone sees it coming, including me, and yet often, as before, the initial impulse is to indulge the anger, even if shortly following it with contrition and apology.

WHY ARE WE SO COMPULSIVE?

Why do we repeat ourselves so often? Why do we have compulsions that reappear despite our every effort to suppress them? Why do we have lifelong arguments inside ourselves that hardly change and are never resolved? Why no learning? The details differ from case to case, but I believe that genetics is almost always involved.

As much as 60 percent of all our genes are active in the human brain, the most genetically diverse tissue in our body (see Chapter 6). Thus we expect enormous genetic variation affecting behavior, including deceit and self-deception. This means we may often differ one from one another psychologically on genetic grounds alone, with no environmental or social rhyme or reason accessible to us. Only by studying genealogies in our immediate environment—especially in our own extended family—could we glimpse the genes in action, and this is very difficult. Thus, for all we know, much of the variation in social complexity around us is beyond our ability to understand, at least in causal terms.

Our genes do not change, although their expression patterns may. If they continue to act in the same way, we may experience this as a compulsion we are unable to change. Likewise, genes may have laid down an early structure to our desires and impulses, a structure that is difficult to modify. This may well mean that we have repetitive features to our behavior that we wish we could do without but that are entrained in us by our particular genotype.

As for our internal conflicts, remember that the interests of our maternal and paternal genes are in conflict throughout our lives, so that internal conflict resulting from such genes may be hard to resolve (see Chapter 4). On the other hand, as we have noted, the older we get, the more symmetrically we are related to others on our maternal and paternal genes (more to children and grandchildren, less to siblings and parents), so we are expected to become more internally peaceful as we head into our sixties and experience (separately) the "positivity effect" of old age (see Chapter 6).

Regarding fighting our compulsion, few are as strong or regular (in a man, at least) as the compulsion to seek out sexual companionship late in the evening, with whomever and on whatever terms. One lesson I have learned in more recent years—a good forty years after it actually would have been useful—is that it is better to go to bed lonely than to wake up guilty. Formulating this as a simple rule has helped me to enforce it, not always but more often than not. And when not, I am more conscious that I am waking up guilty and that I'd better pray myself back into my own good graces and become more conscious. I also believe that there is strength in the new approach. No guilt morning after morning starts to build up a feeling of genuine confidence and relaxed strength. You can set yourself on a better path, and now you see the reinforcing benefits. How long this effect will last, of course, is another matter, but on the assumption that repetitive behavior leading to repetitive guilt is suboptimal, the goal seems worthy and obvious.

THE VALUE OF BEING CONSCIOUS

There are two great axes in human mental life: intelligence and consciousness. You can be very bright but unconscious, or slow but conscious, or any of the combinations in between. Of course, consciousness comes in many forms and degrees. We can deny reality and then deny the denial. We can be aware that someone in a group means us harm but not know who. We can know who, but not why, why but not when, and so on.

Regarding deceit and self-deception, lack of consciousness of such tendencies in others may victimize us. We may be too likely to believe them, especially when they are in positions of authority. We may believe what is printed in newspapers. We may believe con artists. And we may easily embrace false historical narratives. To be conscious is to be aware of possibilities, including those arising in a world saturated with deceit and self-deception.

Consciousness and ability to change are two different variables. I am prone to be moralistic, overconfident, and dismissive of alternative views, more or less as expected for an organism of my type, but I am also conscious that I am biased in this way. I can cite chapter and verse. Do I wish it were otherwise? Yes. Can I change it? No. This to me is the real paradox or tragedy of self-deception—we wish we could do better but we can't.

On the other hand, consciousness of deceit and self-deception allows us to enjoy it more, to understand it more deeply, to guard against it better (as it is directed against us), and, finally, to fight such tendencies in ourselves should we wish to. Mostly it gives us much greater insight into the social world surrounding us, everything from the lies of the government and the media to the deeper self-deceptions we tell ourselves and our loved ones.

THE DANGER OF FANTASY IN PROPAGATING DECEPTION

There is a kind of self-deception—indulging in fantasy—that makes deception less rational and less likely to succeed. Certainly for serious crimes, it is valuable to think the matter through consciously and carefully in some detail. Neither self-deception nor (especially) fantasy is apt to be of much use. Consider minor crimes. You are trying to sneak a small amount of illicit drugs through customs. The one thing you haven't thought through is what you are going to do when you get caught, perhaps because it is unpleasant to contemplate. You may also imagine that not thinking about the matter will be to your advantage, sailing through customs based on a pretense of innocence, bolstered by absence of fear. But exactly the opposite is likely to happen. Having failed to think about what you will do when caught, you become more and more nervous as you get closer to that moment. If you were calm in your knowledge of how you would handle this awkward situation, you would project much more nearly the innocence you would like to. The Times Square bomber was required to leave his engine running to set off his bomb properly, but he was not required to have his full set of house keys attached to the ring. Did he figure the ensuing fire would incinerate them or did he simply fail to carefully think through his crime?

Two absurd examples of the futility of letting fantasy guide deception were provided by the same individual, a distinguished mathematician who was an expert on chaos theory in the 1980s. In each case, he was trying to move small quantities of hashish across international borders, and in each case, he became persona non grata in the country he was visiting, unable to return for five years. In the UK he tried to send the hashish to his girlfriend in Germany, hollowing out a mathematics book, putting it in a university envelope, with her address and his, and marking it fourth class, as indeed it was (a book). But fourth class permits the postmaster to inspect the contents at will. The post office was in the basement of his building, and the package never left, but he did. The point is that his first job was to send off something that could not be traced to him—not to create the perfect pseudo-book for the Germans, complete with a real university stamp and fourth-class postage to show he had nothing to hide.

He then tried to bring hashish into Italy by train from France. He dressed up as a Catholic priest, on the theory perhaps that a priest could get away with murder in Italy, which may well be true, but first he had to convince the Italians that he was, in fact, a priest. Since he had a large, Karl Marx beard, appearance to match, and spoke no Italian, the customs officers naturally became suspicious. Neither he nor the drugs entered Italy. In each case, he appeared to be caught up in the fantasy of deception, producing elaborate hoaxes, which failed to either defend himself or fool the adversary.

THE BENEFITS OF PRAYER AND MEDITATION

Mindful meditation can produce long-term benefits in both mood and immune function. Prayer may have similar effects. Meditation and prayer may also be used against self-deception directly, but this may depend very much on the kind of prayer we use.

Although I had studied the gospels deeply by the time I was thirteen and had given myself over completely to this system of thought insofar as I understood it, I never knew that I never knew *how* to say the Lord's Prayer until I was on an airplane years later seated next to a "religious," that is, a person who had given himself over to the understanding and love of God. Beyond a priest or a monk, he was a lone soul, way out there in his religious knowledge. So we got to talking. Did I pray? he asked. Yes, I prayed. How did I pray? I mostly said the Lord's Prayer. And how did I say it? And here I burst forth with the old Presbyterian marching band version on which I had been raised. Out rolled the prayer as so much martial music and self-assertion:

Our father who *art* in heaven
Hallowed be thy *name.*
Thy *kingdom* come, thy *will* be done
On *earth* as it is in heaven.

It rolls right along as if you are telling God where and what she is. It even ends with an assertion that, spoken properly, inverts meaning—the way we act on earth, with your blessings, is the way you would have us act (as in heaven). No, no, no, said my new friend. Here is how you pray: the emphasis is on your own humility, on submitting to God's will—"*Thy* will be done, *Thy* kingdom come" with "thy" said very softly, and so on. I never prayed the old way again. Subject yourself to the will of the Lord and then be yourself.

If we really want to learn from experience in the sense of transforming the possibility that we will make the same mistake again, just looking at the phenomenon and saying "there goes good old self-deception again" does not do the trick. One has an anecdote for future amusement, but no change in the underlying dynamics. For this we need much deeper confrontations with ourselves and our inadequacies, ones often drenched in tears and humility. Even then it must usually be combined with a daily meditation contra the old behavior for it to have any chance of working. Seeing your self-deception in retrospect is one thing, decreasing its frequency in the future a much deeper matter.

VALUE OF FRIENDS AND COUNSELORS

As we saw in Chapter 6, disclosure of trauma, even if only to a private journal, produces both immune and related mood benefits, and this is probably at least equally true for sharing the trauma with a friend or a counselor. Sometimes the latter is necessary because we are unwilling to reveal deeply personal issues even to our close friends, but we will do so to a professional sworn to secrecy whom we typically encounter only at counseling sessions.

Friends are also useful as commentators on our ongoing life. I will talk with a friend about a recent unfortunate interpersonal interaction and tell him I am thinking of calling the person and giving him or her a good dose of personal abuse. He always argues against it and is much freer to do so than am I. He does not suffer the internal feelings I do; he simply asks what the consequence of my act will be. How will I feel afterward and what benefits will I thereby gain and what new pain will I receive in return for my spiteful behavior?

Friends have another advantage: they see the interaction from the outside, as if others were actors in a play. I am embedded in the play but they are not. They can see what I cannot. How often we look at a political leader and say, "But it is perfectly obvious what you should be doing," and yet often it is probably not obvious to the person caught up in the action. A tree among trees, they have a harder time seeing the forest. I have often thought the popularity of plays partly came from the fact that the audience could see all, while the actors were constrained by their position on the stage.

AN INVITATION TO SELF-DECEPTION AND PERSONAL DISASTER

Try to avoid overconfidence and unconsciousness. Each is dangerous; together they can be deadly, as we saw so vividly in several airplane crashes. Showing off is a special kind of behavior in which we tend to both be overconfident and deliberately exaggerate our behavior to impress others. This can create a very bad mismatch between behavior and reality. The closest I myself came to experiencing the potentially dreadful survival costs of showing off occurred on a lizard-collecting expedition high (above one thousand meters) in the Blue Mountains north of Kingston, Jamaica. My muscular young nephew-in-law was driving the car, itself a "muscle car" in having too small a steering wheel, requiring the quick application of real muscle to turn it properly. Among the team was a young woman, allegedly with me but she seemed to be admiring my nephew's muscular mastery of the road all too much for my comfort, so I took over the driving. We were soon rounding a corner too fast, one I was not sufficiently strong to handle, and the car drifted slowly toward the precipice, until it was caught by a small sand embankment, three wheels in the air, tilting downward, a tree six meters below that might have caught us, otherwise a clear one hundred-meter drop to a rocky death. The man behind me and I were the first ones out, and we had to reach down into the tilting car to pull out the other two, including my by then thoroughly terrified "girlfriend." Someone had white rum and we poured some on the ground, threw some ganja seeds on top, and thanked the Almighty for our survival. I do not remember seeing the young lady again.

For this and many other reasons, I regard showing off as one of the most dangerous things you can do. I believe your attention has shifted entirely to persuading others and away from current reality. While I am focused on the young woman next to me and on impressing her—wanting to divert her attention from my more muscular nephew—I am paying little attention to driving, careless and overconfident and completely unconscious that I am doing so, not at ground level but on some narrow road high in the mountains.

A NEVER-ENDING EXTRAVAGANZA

There is no doubt that deceit and self-deception—if it does nothing else—provides us with an unending extravaganza of nonsense, comedic and tragic, large and small. No human group has a monopoly on the disease, nor is anyone immune. How else can one explain that about 20 percent of US citizens in 2011 claim to believe that their president is a Muslim and 40 percent say he was not born in the United States. Or that the argument can be seriously advanced (and believed) that the same president has a "deep-seated hatred of white people," when his mother was white and he was raised entirely by her and her family. It has famously been said about the United States that no one ever lost a dollar underestimating the intelligence of its people. It could be said likewise that no one has lost a political position in the United States by underestimating the political intelligence of the US voter. In any case, the level of ignorance regarding fundamental facts is astonishing.

Consider other regular occurrences. Absent self-deception, how do we explain that Ponzi scheme after Ponzi scheme after Ponzi scheme marches across the pages of our newspapers despite a one hundred–year record of financial disaster (at least for all those rushing to join the scheme once it is under way). Or how—without self-deception—should we explain that every year in the United States another anti-homosexual politician or preacher is shown to have a hidden homosexual life?

Or to turn to real tragedy, without deceit and self-deception, how can we explain people across the world of various faiths murdering their daughters and sisters for often trivial infractions of local sexual mores in so-called honor killings? It is hard to believe that the feeble Y chromosome, with only a few dozen protein-coding genes, could do the job. But patriarchy—benefiting all or most of a male's genes even at the expense of his wife and female relatives—could, with proper mental adjustments, bring about this horror. Men (largely) who murder their daughters or sisters, or horribly disfigure them or drive them to suicide, do not appear to do so with a guilty conscience. Quite the contrary, they profess moral indignation and are outraged at the sins causing them to take such extreme measures. This appears to be a case of innocent women caught in the crosshairs of conflict between local patrilines, not quite out-groups or unrelated neighbors, but sets of related genes in paternally related people. As we have seen when discussing war and religion, these kinds of conflicts are especially likely to induce self-deception, along with heartless and cruel behavior.

The newspapers disgorge fresh material daily. It turns out that the safety culture at Japanese nuclear reactors was so bad, even NASA could admire it. All effort was put into a public relations campaign to convince the country that the reactors were safe, while no effort was spent on what to do in case of crisis. Although it's the world's leader in robotics—Japanese robots can run on two feet, sing, dance, and play the violin—none were designed to work in a crippled, radioactive plant, because "introducing them would inspire fear." Japan had to import them from a Massachusetts company better known for making vacuum cleaners. Nor did they have a way of introducing cooling water; they had to import a 203-foot-long water pump from China. But as we have seen so often in this book, when a company or government is in full-sale mode, mundane goals such as safety are cast aside.

Meanwhile, science continues to spit out examples. Metaphor is so strong that we wish to wash our hands when we have done something immoral (= dirty), but the form of our misbehavior can affect the disinfectant chosen: soap for a nasty e-mail sent by hand but mouthwash for a nasty message left on an answering machine. In principle, some of these subtle, unconscious associations are available for notice by others, especially those close by and motivated to do so.

And now conference calls by businesses about quarterly earnings have been subjected to linguistic analysis by economists for cues to deception, mostly using later performance restatements as the arbiter of truth. Sure enough, some of the usual villains reappear—people avoid first person references when lying, preferring "they" or impersonal pronouns, such as "people." People use fewer extreme positive and negative terms—as if moderating their position for the sake of plausibility—and fewer certainty and hesitation terms (as if having memorized their spiel). They also prefer references to general knowledge but avoid references to shareholder value and creation. Logic can go either way. Perhaps you hype shareholder value to fool others. But the evidence suggests otherwise. You shy away from the truth (shareholder value) because that is where you are weakest, but then you are stuck with weaker pleas to aspects of "general" knowledge. The above work is tentative but very appealing. At last we are moving out of the experimental psychology lab, a near-hopeless place in which to investigate deception and its consequences.

Finally, consider a clever experiment recently run on undergraduates. Although artificial in the extreme—telling an imaginary lie to a teacher (high status) versus an imaginary lie to a fellow student (equal status) — people forgot more simultaneously learned words in the high-status case than in the equal-status one, as if self-deception (including memory impairment) were more often practiced against high-status opponents.

One nice feature of the study of deceit and self-deception is that we will never run out of examples. Quite the contrary, they are being generated more rapidly than we can deconstruct them. At least we can enjoy the never-ending extravaganza while trying to deepen our consciousness. Everybody can join in, not just academics or scientists. The logic for understanding self-deception is simple and the phenomenon universal.

ACKNOWLEDGMENTS

I am grateful to the many research organizations that have supported my work over the years: the Harry Frank Guggenheim Foundation, the Ann and Gordon Getty Foundation, the John Simon Guggenheim Foundation, the Biosocial Research Foundation, and the Crafoord Foundation. I am especially grateful to the Royal Swedish Academy of Sciences for giving me the wonderful gift of the Crafoord Prize in 2007. I am also grateful to the University of the West Indies for making me in 2009 an Honorary Research Fellow for life.

The first draft of this book was written while I was a fellow at WIKO, the Institute for Advanced Studies in Berlin (2008–09). The institute provided a very warm and supportive environment in which to work, and I am grateful to all staff members at all levels, administrative, IT, library, and kitchen. Fellows who helped me that year were Roger Chickering, Holk Cruse, Thomas Metzinger, Srinivas Narayanan, and Ibrahima Thioub. I am especially grateful to Bill von Hippell for spending five months at WIKO teaching me social psychology and commenting on all aspects of the book.

For detailed, helpful comments on multiple chapters I am grateful to Nick Davies, Bernhard Fink, Norman Finkelstein, Steven Gangestad, Marc Hauser, Jody Hey, Srinivas Narayanan, Stephen Pinker, Richard Wrangham, Doron Zeilberger, and William Zimmerman. I am most grateful to David Haig for reading the entire book and engaging in numerous conversations on all aspects of the book. Finally, I am also grateful to Amy Jacobson and Darine Zaatari for their contributions to this book.

I thank my agent John Brockman for connecting me with three first-rate publishers. I thank my English-language editors for detailed comments and unfailing support—Will Goodland at Penguin and TJ Kelleher at Basic Books. I am also grateful for the critical support of Michele Luzzatto at Einaudi.

Finally, for early years filled with comic insight into the subject, I thank my brother Jonathan.

NOTES

NOTES TO CHAPTER 1

1 **Parent/offspring relations:** Trivers 1974; reciprocal altruism: Trivers 1971.

2 **Sex differences:** Trivers 1972; **sex ratio:** Trivers and Willard 1973, Trivers and Hare 1976.

7 **Within our genomes:** Burt and Trivers 2006.

8 **Stick insects:** Brock 1999; **at least fifty million years:** Wedman et al 2007.

10 **Behavioral cues to human deception:** DePaolo et al 2003, Vrij 2008; **accuracy of detection:** Bond and DePaolo 2006; **suppression versus faking**: Craig et al 1991, Larochette et al 2006; **cognitive load a critical variable:** Vrij 2004, Vrij et al 2006.

11 **By no means always a delay prior to lying:** Morgan et al 2009; **pitch of voice:** DePaolo and Kashy 2003, Vrij 2008; **displacement activities:** Troisi 2002; **nervousness as a weak factor:** Vrij 2004, 2008.

12 **Cognitive load and blurting out:** Wegner 2009; **common verbal features:** Newman et al 2003.

13 **Lies are detected 20 percent of the time:** DePaolo et al 1996; see also DePaolo et al 1998.

14 **Ignorance and confidence:** Ehrlinger et al 2008, Kruger and Dunning 1999.

15 **"Beneffectance," active voice, man and telephone pole:** Greenwald 1980; **BMW owners:** Johansson-Stenman and Martinsson 2006.

16 **Self-inflation effect, high school students, academics:** Greenwald 1980; general review: Alicke and Sedikides 2009, Guenther and Alicke 2010; **morphing faces:** Epley and Whitchurch 2008.

17 **Japan and China:** Alicke and Sedikides 2009, Kobayashi and Greenwald 2003; **area of the brain for self-inflation:** Kwan et al 2007; **narcissists high in dominance and power:** Campbell et al 2007; **overconfident, over-bet, and persistent in their delusions:** Campbell 2004.

18 **Contrast two sets of college students:** Fein and Spencer 1997.

19 **"Us" paired with nonsense syllables:** Perdue et al 1990; **generalize bad to out-group, good to in-group:** Maas et al 1989; **smiles imputed to in-group members:** Beaupre and Hess 2003; **smile intensity predicts longevity**: Abel and Kruger 2010.

20 **Monkeys:** Mahajan et al 2011.

21 **Power prime:** Galinksy et al 2006; **Churchill:** Mukerjee 2010.

22 **Moral hypocrisy:** Batson et al 1999; **under cognitive load, moral bias toward self vanishes:** Valdesolo and DeStano 2008; **predictable and unpredictable shocks in rats:** Weiss 1970; **effects of perceived control in humans:** Lykken et al 1972.

23 **Illusion of control:** Langer and Roth 1975; **in stockbrokers**: Fenton-O'Creavy et al 2003; **illusory pattern recognition:** Whitson and Galinksy 2008.

25 **Victim/perpetrator:** Baumeister et al 1990.

28 **Self-deception and inefficient mental systems include misapprehension of reality:** Peterson et al 2002; **failure to respond to error:** Peterson et al 2003.

NOTES TO CHAPTER 2

30 **Frequency-dependent selection in butterflies:** Sheppard 1959; **for matched pictures of the butterfly models and mimics**: Owen 1971.

31 **Brood parasites in birds:** Davies 2000; almost everything in this section can be found there.

32 **Individual species specialized to lay eggs:** Gibbs et al 2000; **advantageous to count eggs:** Lyon 2003; **single cuckoo's begging call mimicking entire brood:** Davies et al 1998; **hawk cuckoo in Japan:** Tanaka and Ubeda 2005, Tanaka et al 2005.

33 **Importance of recognition errors in reed warblers:** Davies et al 1996, Brooke et al 1998.

34 **"Mafia-like" behavior:** Soler et al 1995, Hoover and Robinson 2007; **cultural transmission:** Davies and Welbergen 2009.

35 **Ants:** Barbero et al 2009; **ant mimics:** Maderspracher and Stensmyr 2011.

37 **Monkey and ape brains:** Byrne and Corp 2004.

38 **Mimicry in general:** Wickler 1968; **fireflies:** El-Hani et al 2010, Lloyd 1986; **orchids:** Jersakova et al 2006; **deceptive orchids more outbred:** Cozzolino and Widmer 2005.

39 **Bluegill sunfish:** Dominey 1980, Gross 1982; **blister beetle:** Saul-Gershenz and Millar 2006.

40 **In mixed-species flocks:** Greig-Smith 1978; see also use by drongos to steal food from meerkats, Flower 2010; **to separate warring siblings:**

Spellerberg 1971, and more generally, Wiebe and Bartolotti 2000; **as a paternity guard:** Moller 1990; **antelopes:** Bro-Jorgensen and Pangle 2010; **each skin-color cell of an octopus:** Hanlon et al 2007.

41 **Adjusting its color to each new surface:** Barbosa et al 2007; **mimic flounders:** Hanlon and Conroy 2008; **randomly displaying variant phenotypes:** Hanlon et al 1999; **squid female mimic:** Hanlon et al 2005; **bird distraction displays:** Sordahl 1986.

42 **Crakes:** source unknown; **fake butterfly eggs:** Gilbert 1982; **pronghorn mother:** Byers and Byers 1983.

44 **Wasp status badges:** Tibbetts and Dale 2004; **key perceptual factor:** Tibbetts and Izzo 2010; **sparrow status badges:** Rohwer 1977, Rohwer and Rohwer 1978, Rohwer and Ewald 1981, Moller and Swaddle 1987.

46 **Ravens adjust behavior to context and competitors:** Bugnyar and Heinrich 2006; **cachers and raiders:** Bugnyar and Kotrschal 2002; **see around an obstacle:** Bugnyar et al 2004; **jays:** Dally et al 2004, 2006; **gray squirrels:** Leaver et al 2007; **chimps hiding erections:** de Waal 1982.

47 **Mantis shrimps:** Steger and Caldwell 1983, Caldwell 1986, Adams and Caldwell 1990; **fiddler crabs:** Lailvaux et al 2009.

48 **Hide objects behind back:** de Waal 1982; **throw object:** de Waal 1986; **cooperation modeled as prisoner's dilemma:** Axelrod and Hamilton 1981; more recent work reviewed: Trivers 2005.

49 I am most grateful to Karl Sigmund for suggesting this entire page.

NOTES TO CHAPTER 3

54 **Neurophysiology of action:** Libet 2004; we also have an illusion of conscious will, one we actively work to maintain: Wegner 2002.

55 **A novel experiment:** Soon et al 2008.

56 **Flip side:** Libet 2004, Wegner 2002; **thought suppression in the lab:** Anderson et al 2004.

57 **Ironic effects of thought suppression:** Wegner 1989, 2009, Wegner et al 2004.

58 **Neural inhibition of deception area:** Karim et al 2009.

59 **Brains of pathological liars:** Yang et al 2007; **jugglers:** Scholz et al 2009; **unconscious voice recognition experiments:** Gur and Sackeim 1979.

61 **Unconscious facial recognition:** Bobes et al 2004; **voice recognition in birds:** Margoliash and Konishi 1985.

62 **Anosognosia:** Ramachandran 2009.

63 **Response time to threatening words:** Nardone et al 2008; **dominance reversal in birds:** reviewed in Trivers 1985.

64 **IAT:** Greenwald et al 1998; **improvements in methodology:** Greenwald et al 2003; general IAT effects: Greenwald et al 2009.

65 **IAT for racial preferences:** Nosek et al 2002; **effects of racial prime on academic performance:** Steele and Aronson 1995; **effects of racial bias on executive control of the biased:** Richeson and Shelton 2003.
66 **False confessions:** Kassin 2005, Kassin and Gudjonsson 2005; **disassociation under torture:** Ray et al 2006; **high disassociators and interference on Stroop**: Freyd et al 1998.
67 **False memories of child abuse:** McNally 2003, Clancy 2009.
71 **Placebo effects in general:** Benedetti 2009, Price et al 2008; **rubbing is good**: Saradeth et al 1994; **so are sham devices:** Kaptchuk et al 2006; **homeopathic effects are placebo effects:** Shang et al 2005.
72 **Color of pills:** de Craen et al 1996; **angina surgery**: Cobb et al 1959; **arthroscopic surgery**: Moseley et al 1996.
73 **Placebo and pain**: Wager et al 2004, Benedetti 2009; I thank Anders Moller for the quote from unpublished work; **meta-analysis of placebo and depression**: Fournier et al 2010.
74 **Auto-stimulatory effects on female sexuality:** Palace 1995; **caffeine and cyclists**: Beedie et al 2006; **placebo effect out of a placebo effect**: Kaptchuk et al 2010.
75 **Placebo and suggestibility:** Benedetti 2009; **hypnosis and Stroop test**: Raz et al 2002; **Stroop test:** Stroop 1935; **immune benefits of hypnosis:** Gruzeller 2002.

NOTES TO CHAPTER 4

77 **Hamilton's rule:** Hamilton 1964.
78 **Self-deception regarding parental investment:** Eibach and Mock 2011
79 **Paternal grandmothers:** Fox et al 2010.
80 **Parent/offspring conflict:** Trivers 1974, Trivers 1985.
82 **High disassociators and interference on Stroop:** Freyd et al 1998; **early discovery of imprinted genes:** Haig and Westoby 1989.
83 **Conflict between *Igf2* and *Igf2r*:** Haig and Graham 1991; **evidence for Haig's rule:** Haig 2004, Burt and Trivers 2006; **chimeric mice:** Keverne et al 1996.
84 **Selves-deception:** Burt and Trivers 2006; **imprinting and genes in the brain:** Gregg et al 2010; **paternal genes for maternal behavior:** Li et al 1999, Curley et al 2004; **incest:** Haig 1999.
86 I am indebted to David Haig for the notion that imprinting becomes less important with increasing adult age.
87 **Children's reaction to a new half-sibling:** Schlomer et al 2010.
88 **Variety of children's deception:** Reddy 2007.
89 **How often children lie:** Wilson et al 2003; **white lies:** Talwar et al 2007; **temper tantrums in chimps and pelicans:** reviewed in Trivers 1985; **fetal deception during pregnancy:** Haig 1993.

90 **Intelligence and deception in children:** Lewis 1993.

91 **Smiling at victim and deception in children:** Talwar et al 2007; **dominance and deception:** Keating and Heitman 1994.

NOTES TO CHAPTER 5

96 **Investment and genes:** Trivers 1972; **asexual species small, frequent extinction:** Bell 1982.

97 **Bluegill sunfish single-siders:** Gross et al 2007.

99 **Human female and male choice:** Thornhill and Gangestad 2008.

100 **Attribution of relatedness:** Daly and Wilson 1982.

101 **Creating artificial parental resemblance:** Platek et al 2004; **male sexual jealousy:** Daly et al 1982.

103 **Duck re-raped by mate:** Barash 1977; **women and men respond to infidelity:** Daly et al 1982; **women are more attractive at the time of ovulation:** Thornhill and Gangestad 2008; **derogate the looks of other women more:** Fisher 2004; **in several clubs in Vienna:** Grammer et al 2004.

104 **Preferences shift at ovulation to signs of genetic quality:** Thornhill and Gangestad 2008; **lap dancers:** Miller et al 2007; **genetic matching lowers female sexual interest:** Garver-Apgar et al 2006; **women's sense of smell more acute:** Yousem et al 1999; **especially at ovulation:** Thornhill et al 2003.

105 **Women are better at reading facial expressions:** Williams and Mattingley 2006; **women's brains tend to act more symmetrically:** Kovalev et al 2003; **men deceive themselves about women's sexual interest:** Haselton 2003.

106 **Two sexes introduced together for ten minutes:** Grammer et al 2000; **male denial of homosexual tendencies:** Adams et al 1996; for a possible alternative view of the latter, Meier et al 2006.

108 **People have a bias toward seeing improvement:** Karney and Coombs 2000; **both spouses reported steady improvement:** Frye and Karney 2004; **self-justification as assassin of marriage:** Tavris and Aronson 2007.

111 **Elin Woods:** *National Enquirer* April 2010, *National Enquirer* December 2009, Vecsey 2010.

NOTES TO CHAPTER 6

115 **Parasites arrayed against immune systems:** for an excellent general review of the two-sided interaction, Schmidt-Hempel 2011.

116 **The immune system sends many cellular types:** Murphy et al 2008; **immune system as sixth sense:** Blalock and Smith 2007.

117 **Immune system is expensive:** Murphy et al 2008.

118 **Metabolic cost of fever and immune response:** Lochmiller and Deerenberg 2000, Baracos et al 1987.
119 **Sickness behavior:** Dantzer and Kelley 2007.
120 **Sleep beneficial for immune function:** Cohen et al 2009, Bryant et al 2004; **different species of mammals:** Preston et al 2009.
121 **The lowest testosterone levels:** Gray and Campbell 2009, Burnham et al 2003, Muller et al 2009; **males with higher testosterone are more likely to become infected:** Muhlenbein et al 2006, Muhlenbein 2006, Muhlenbein 2008.
122 **Degree of fat-free muscle mass:** Lassek and Gaulin 2009; **stress:** Segerstrom and Miller 2004; **arithmetic:** Sokoloff et al 1955; **brain's resting energy cost remains virtually constant:** Raichle and Gusnard 2002, Clarke and Sokoloff 1999.
123 **Brain is the most genetically active tissue:** Hsiao et al 2001.
124 **Brightly colored males chosen for their parasite-resistant genes:** Hamilton and Zuk 1982; **honeybee associative learning:** Mallon et al 2003; **bird brain size and immunity:** Moller et al 2005.
125 **River otters and nematode worms, Scherr and Bowman 2009; series of experiments writing about trauma:** Pennebaker 1997, Petrie et al 1998; for effects on HIV: Petrie et al 2004.
126 **A recent review of about 150 studies:** Frattaroli 2006; **in New World Amerindian religions; as one psychologist drily notes:** Pennebaker 1997.
127 **Emotion words and pronouns:** Ramirez-Esparza and Pennebaker 2006; **undisclosed trauma and sexual trauma:** Pennebaker 2011; **suicide support groups:** Pennebaker and O'Heeron 1984; **chance of reemployment:** Spera et al 1994.
128 **Expressive group therapy:** Belanoff et al 2004; **deny HIV-positive statu.s:** Strachan et al 2007; **higher survival of HIV-positive men who are out of the closet:** Cole et al 1996a (study corrected for unsafe sex).
129 **HIV-positive women:** Eisenberger et al 2003; **hide your heterosexual identity:** Sullivan, 2010.
130 **Better health for HIV-negative men out of closet:** Cole et al 1996b; **rejection-sensitive men:** Cole et al 1997; **direct experimental tests:** Rosenkrantz et al 2003; **response to vaccines:** Marsland et al 2006, Cohen et al 2006; in general: Marsland et al 2007.
131 I am grateful to Srinivas Narayanan for help producing the first and third paragraphs; **effects on older people:** Pennebaker 1997.
132 **Monkey music:** Snowdon and Tele 2009; **Musak, jazz, and noise:** Charnetski and Brennan 1998; **injecting cancer cells into mice:** Nunez et al

2002; **Bach's music:** le Roux et al 2007; **playing music appears to work even better than listening to it**: Kuhn 2002.

133 **The original experiment on positivity:** Mather and Carstenson 2003; **results are true among Asians:** Kwon et al 2009; **positive remembered better:** Charles et al 2003; **amygdala:** Mather et al 2004.

134 **Older people preferentially look at:** Isaacowitz et al 2008; **until at exactly sixty:** Nosek et al 2002.

135 **Old people cranky:** Henry et al 2009.

136 **Immune response and survival in birds:** Moller and Saino 2004, Hanssen et al 2004; **optimism:** Segerstrom et al 1998, Segerstrom and Miller 2004; **a recent study:** Segerstrom 2010.

NOTES TO CHAPTER 7

139 **The psychology of self-deception:** covered in von Hippell and Trivers 2011, Hallinan 2009, and Tavris and Aronson 2007.

141 **Testing a strip:** Ditto and Lopez 2002; **listening to a tape describing the dangers of smoking; avoid taking HIV tests:** Dawson et al 2006; **chosen for a prospective date:** Wilson et al 2004.

142 **Capital B or the number 13:** Balcetis and Dunning 2006; **children draw coins larger:** Bruner and Goodman 1947; **thirst primed, gardening made fun:** Veltkamp et al 2008; **capital punishment:** Lord et al 1979.

143 **More easily remember positive information:** D'Argembeau et al 2008, Green et al 2008; **telling others:** Coman et al 2009, Cuc et al 2007; **differential rehearsal and biased memory:** Gonsalves et al 2004, Gonsalves and Paller 2000; **biased memory of skills:** Conway and Ross 1984; **men and women both remember:** Tavris and Aronson 2007.

144 **Memory continually re-created:** Loftus 1996; **health information distorted:** Croyle et al 2006; **inventive memory:** Mark Twain, numerous times.

145 **The illusion of improvement:** Ross and Wilson 2002, Wilson and Ross 2001; **biased reporting and argumentation:** Mercier and Sperber 2011; **deterministic view and cheating:** Vohs and Schooler 2008; **"unintentional" cheating:** von Hippell et al 2005.

146 **Sit next to the handicapped:** Snyder et al 1979; **predicting future feelings:** Gilbert 2006.

148 **Sounds that are coming toward us:** Neuhoff 1998, 2001; **general rules that work well in most situations:** Kahneman and Tversky 1971; Haselton and Nettle 2006; **in the words of one psychologist:** Wegner 2009.

150 **A coauthor decides that an article is not fraudulent:** Chapter 12 in Trivers et al 2009.

151 **Beetle facing backwards, *Oreodera glauca*; fish:** Wickler 1968; for an excellent review of cognitive dissonance and self-justification, covering much of the material here, see Tavris and Aronson 2007.
154 **Two experts:** Tavris and Aronson 2007.
155 **Cognitive dissonance in monkeys and young children:** Egan et al 2007, Egan et al 2010.

NOTES TO CHAPTER 8

158 **Stock trading by amateurs:** Barber and Odean 2001; **overconfidence and trading volume:** Glaser and Weber, 2007.
159 **Overconfidence in currency markets:** Oberlechner and Osler 2008; **arithmetic contests:** Vandegrift and Yavas 2009; **thrill seeking and over-trading in Finnish men:** Grinblatt and Keloharju 2008.
160 **Agent versus object metaphors:** Morris et al 2007.
161 **Metaphor is a key part of language:** Pinker 2007, Thibodeau and Boroditsky 2011; **invert meaning:** Hochschild, 2004; **euphemism tread-mill:** Pinker 1994.
162 **Concepts are in charge, not words; you relax when euphemism tread-mill stops:** Pinker 1994.
163 **Switch from "sex" to "gender":** Haig 2004b; **the "awful" German language:** http://www.crossmyt.com/hc/linghebr/awfgrmlg.html; see also Boroditsky et al 2003.
164 **Name-letter effect:** Nuttin 1985, 1987.
165 **Young Japanese women:** Kitayama and Karasawa 1997; letters that are **rarely encountered:** Nuttin 1987; **alleged widespread important effects:** Pelham et al 2002; **would create a spurious correlation:** Simonsohn 2011.
166 **Name-letter effect on academic performance:** Nelson and Simmons 2007; **aspects of early parental style:** DeHart et al 2006.
167 **Daily events can affect one's name-letter bias:** DeHart and Pelham 2007; **deceiving down:** Hartung 1988.
168 **Face-ism is lower in African Americans:** Zuckerman and Kieffer 1994.
169 **Face-ism and the two sexes:** Archer et al 1983; **men shown in relatively intellectual professions:** Matthews 2007; **politicians' self-presentations:** Konrath and Schwartz 2007.
170 **George W. Bush's head:** Calogero and Mullen 2008.
171 **Anti-spam:** Stone 2006.
173 **Self-deception and humor:** Lynch 2010, Lynch and Trivers 2011; **tick-ling a rat will produce laughter-like sounds:** Panksepp and Burgdorf 2003; **chimpanzees will pant-laugh:** Matsusaka 2004.
174 **Split personalities:** Hilgard 1977.

176 **Bernie Madoff:** Henriques 2010.
177 **A book on gullibility:** Greenspan 2009.
178 **You're experiencing the ride:** *New York Times.*
181 **Classic lie detector test** (accuracy overstated): Reid and Inbau 1977; **neurophysiological lie detectors:** Harris 2010, Abe et al 2007, Nunez et al 2005, Kozel et al 2005.

NOTES TO CHAPTER 9

184 **For the French counterpart of the NTSB with less exacting standards:** Traufetter 2010; **Air Florida Flight 90:** Trivers and Newton 1982.
187 **More accidents occur when the pilot is flying:** NTSB 1994.
188 **Korea Airlines:** Gladwell 2008; **nurses and surgeons:** Pronovost and Vohr 2010.
189 **Gol Flight 1907:** Langewiesche 2009.
191 **Eldar takes command:** http://en.wikipedia.org/wiki/Aeroflot_Flight_593.
192 **Pilot error said to cause 80 percent of accidents:** http://www.aopa.org/asf/publications/05nal.pdf; **John F. Kennedy Jr. crash:** Vulliamy 1999.
193 **Kirksville, Missouri, crash:** Wald 2006.
194 **The current FAA rules:** Goo 2006; **Delta and forty-eight hours rest:** Negroni 2009.
195 **Long history of alarming behavior under icing conditions:** Engelberg and Bryant 1995.
198 **Buffalo crash:** Freeman and Wilber 2009.
199 **Air Transport Association opposing safety improvements:** Hall 2005.
200 **Bush administration dropped the ball:** Lichtblau 2005, Ridgeway 2004.
201 **Feynman concluded:** Feynman 1988; see also Vaughan 1996.
202 **Example of how not to do statistics:** Kitchens 1998.
203 **Evidence presented to invite rebuttal:** Tufte 1997.
205 ***Columbia* disaster:** Langewiesche 2003, Sanger 2003.
206 **Low-level engineers alarmed:** Glanz and Schwartz 2003.
208 **Safety office a fraud:** Sanger 2003; **round table in the conference room:** Schwartz 2005; **self-criticism in organizations:** Wildavsky 1972.
209 **EgyptAir:** Langewiesche 2001; http://www.ntsb.gov.publictn/2000/AAB0201.htm
212 **Saved by *lack* of self-deception:** Wilson 2009.

NOTES TO CHAPTER 10

218 **Slaughter and dispossession of an entire people:** Stannard 1992.

219 **Columbus in Hispaniola:** Stannard 1992.

221 **Deliberate infection via smallpox blankets:** http://www.fordham.edu/halsall/mod/smallpox1.html; **quotations from great leaders:** Diamond 2006.

222 **Frequent US wars against nearby countries:** Loewen 2007.

223 **United States alone now spends almost as much on warfare as rest of world:** Norton and Taylor 2009.

224 **US history textbooks:** Loewen 2007.

226 **Civil War:** Faust 2008; **later history of African Americans was in some ways more dreadful:** Blackmon 2008.

227 **Rewriting of Japanese history:** Rose 2006, Masalski 2001.

228 **Cromwell:** Siochru 2008.

229 **Deletion of all references to the role of the Japanese Imperial Army in Okinawa:** Oshini 2007.

230 **Genocide against Armenians:** Fisk 2005; **elementary-school students forced to watch film:** Rainsford 2009.

231 **Genocide is presumably never pretty:** Fisk 2005.

233 **Even by 1920:** McCarthy 1990; **Zionist project set from the beginning:** Flapan 1987.

234 **"Could also affect the history of the future":** Peretz 1984; ***From Time Immemorial*:** Peters 1984; **for a detailed unraveling of the hoax:** Finkelstein 2003; **denial of Palestinian history built into curricula:** for 1948, Chomsky and Pappe 2010.

235 **Founding of the state of Israel:** Flapan 1987.

236 **All the land, part of which their distant ancestors may once have occupied:** El-Haj 2001; **Arabs reject 1928 assembly:** Porath 1974; Eban 1973.

237 **War of 1947–1948:** Flapan 1987; **secret agreement with Abdullah of Transjordan:** Shlaim 1988.

238 **Ethnic cleansing:** Morris 1987, Pappe 2006; **mal-genetics:** Finkelstein 2005.

239 **Compensation for property stolen by Nazis:** Finkelstein 2003.

241 **US is the elephant in the room:** Chomsky 1983; **Christian Zionism:** Collins 2007.

242 **In 1891, four hundred people:** Mead 2008, Oren 2007; **Harry Truman and Israel:** Radosh and Radosh 2009.

243 **Rumsfeld and Bush:** Draper 2009; for a raving US Christian Zionist, see Hagee 2006.

244 **With full bells and whistles:** see organisms described by Dwyer 2011 and Finkelstein 2005; **the missing "the":** Fisk 2008.

245 **The truth about Israel's theft of Arab land and water:** Zertal and Eldar 2007; **what can be said about Israel in Israel:** Carey and Shainin 2002.

NOTES TO CHAPTER 11

247 **Four main causes:** Wrangham 1999.
248 **Napoleon's invasion of Russia:** Lieven 2010.
249 **Chimpanzee raiding:** Wrangham and Peterson, 1996, Wrangham 2006; **recently, remarkable evidence:** Mitani et al 2010.
250 **14 percent of human mortality:** Bowles 2009; **a percentage that thankfully has declined steadily:** Pinker 2011.
252 **Self-deception encourages warfare:** Tuchman 1984.
253 **Swimming:** Starek and Keating 1991.
254 **Men less likely to read facial expressions correctly:** Williams and Mattingley 2006.
255 **Men less likely to remember emotional information:** Bloise and Johnson 2007, Canli et al 2002; **in experiments less compassionate:** Singer et al 2006, but see Vol et al 2008.
256 **Home-field advantage in sports:** Moskowitz and Wertheim 2011; **boost in testosterone on home field:** Neave 2003; **human holocaust such as World War I:** Johnson 2004; for self-deception and coping by the soldiers on the front, see Watson 2006.
258 **Avocadoes and tomatoes, Noam Chomsky numerous times; a blunder into a catastrophe:** two books give most of the details cited, Ricks 2006, Packer 2005.
259 **An ambitious five-year plan:** Wesley Clark, democracynow.org. http://www.democracynow.org/2007/3/2/gen_wesley_clark_weighs_presidential_bid
260 **Deliberative versus instrumental phase:** Fujita et al 2007.
261 **The decision had to be sold to the larger country:** Rich 2006; **"could there be 100 percent certainty?":** Lelyveld 2011.
263 **Contrast with World War II is instructive:** Fallow 2004; **working groups walled off from decision-makers:** Fallow 2004.
264 **Identical forces were at work in the UK:** Price 2010.
269 As shown conclusively later, http://www.amnesty.org/en/library/asset/MDE15/042/1996/en/dbadaf6a-eaf6-11dd-aad1-ed57e7e5470b/mde150421996en.pdf; **second Qana massacre**, http://www.hrw.org/en/reports/2007/09/05/why-they-died?print.
270 **Israeli commentators:** Milne 2008; **talking heads on US television:** Governor Corzine, George Will, Governor Sanford, ex-Governor Romney, Senator McConnell.

271 **Savagery continued for seventeen days:** Finkelstein 2010; **someone forgot to tell the soldiers:** Macintyre 2009, Breaking the Silence 2009; for historical background and logic of the Gaza front see Shlaim 2009.

272 **Sold under patently false pretenses:** Siegman 2009.

274 **Congo rapes:** Peterman et al 2011.

NOTES TO CHAPTER 12

278 **Religion is a viral meme:** Dawkins 2006, Dennett 2006; **one such proponent of the meme-centered view:** Dennett 2007.

280 **Religion expected to increase group cooperation:** Norenzayan and Shariff 2008; **pair of eyes on computer screen:** Burnham and Hare 2007; **God prime:** Shariff and Norenzayan 2007; **fear of God may reduce cost of selfish behavior:** Johnson 2009.

281 **Religiosity and higher false opinions of self:** Trimble 1997; **religious sects outlast secular:** Sosis and Alcorta 2003; **more costly requirements better:** Sosis and Bressler 2003.

283 **Male God and anti-female behavior:** see Ruether 2009.

285 **Religion and health:** Lee and Newberg 2005.

286 **Church attendance predicts later healthy behavior:** Gillings et al 1996.

288 **Less ACC firing and religious belief:** Inzlicht et al 2009; **parasites and religious diversity:** Fincher and Thornhill 2008a.

289 **Linguistic diversity:** Fincher and Thornhill 2008b.

290 **High-parasite-load societies appear to be more xenophobic:** Fincher et al 2008, Thornhill et al 2009, Schaller and Murray 2008, Faulkener et al 2004; **the mere sight of someone else's disease symptoms increases immune response of perceiver**: Schaller et al 2010.

292 **Christianity's change after Constantin:** Wright 2009.

293 **Islam provides a nice example:** Wright 2009.

295 **Biases in between-group transfers and Jewish in-group bias:** see Shahak 1994.

299 **Effects of intercessory prayer:** Benson et al 2006.

300 **Support for suicide attacks:** Ginges et al 2009.

NOTES TO CHAPTER 13

306 **Gravity bends light:** shown by Arthur Eddington in 1919; **ant ratios of investment:** Trivers and Hare 1976; **recent work reviewed**: West 2009.

308 **Langur male infanticide:** reviewed in Trivers 1985; **DNA evidence**: Borries et al 1999.

309 **Aggression good for species:** Lorenz 1966.

311 **Market forces will naturally constrain the cost of deception:** for an account of recent resurrections of this fallacy, see Krugman and Wells 2011 and Madrick 2011; see also Krugman 2009.

312 **Trait evolved to fit the laboratory:** see relevant chapters in Hammerstein 2003; for a review of the fallacy, Trivers 2005.

313 **Infatuation with beautiful mathematics:** Krugman 2009.

318 **"Wolf Man":** Schatzman 1973; **for the malignant cultural phenomenon that Freudianism was in the US**: Torrey 1992.

319 **Freud's quote:** Rosenzweig, 1997; **Feynman's 1964 lecture given to Cornell students:** http://www.youtube.com/watch?v=b240PGCMwV0

NOTES TO CHAPTER 14

323 David Haig's genes do not care about him.

331 **Mindful meditation:** Davidson et al 2003.

335 **Japanese nuclear safety:** Onishi 2011.

336 **Soap or mouthwash:** Lee and Schwarz 2010.

336 **Business conference calls:** Larcker and Zakolyukina 2010; **experiment run on undergraduates:** Lu and Chang 2011.

BIBLIOGRAPHY

Abe, N., Suzuki, M., Mori, E., Itoh, M., & Fujii, T. (2007). Deceiving others: distinct neural responses of the prefrontal cortex and amygdala in simple fabrication and deception with social interactions. *Journal of Cognitive Neuroscience* 19:287–295.

Abel, E. L., & Kruger, M. L. (2010). Smile intensity in photographs predicts longevity. *Psychological Science*:1–3.

Adams, E. S., & Caldwell, R. L. (1990). Deceptive communication in asymmetric fights of the stomatopod crustacean *Gonodactylus bredini. Animal Behavior* 39:706–716.

Adams, H. E., Wright, L. W., & Lohr, B. A. (1996). Is homophobia associated with homosexual arousal? *Journal of Abnormal Psychology* 105:440–445.

Alicke, M. D., & Sedikides, C. (2009). Self-enhancement and self-protection: what they are and what they do. *European Review of Social Psychology* 20:1–48.

Anderson, M. C. et al. (2004). Neural systems underlying the suppression of unwanted memories. *Science* 303:232–235.

Archer, D., Iritani, B., Kimes, D. D., & Barrios, M. (1983). Face-ism: five studies of sex differences in facial prominence. *Journal of Personality and Social Psychology* 45:725–735.

Balcetis, E., & Dunning, D. (2006). See what you want to see: motivational influences on visual perception. *Journal of Personality and Social Psychology* 91:612–625.

Baracos, V. E., Whitmore, W. T., & Gale, R. (1987). The metabolic cost of fever. *Journal of Physiology and Pharmacology* 65:1248–1254.

Barash, D. P. (1977). Sociobiology of rape in mallards (*Anas platyrhynchos*): responses of mated male. *Science* 197(4305):788–789.

Barber, B. M., & Odean, T. (2001). Boys will be boys: gender overconfidence and common stock investment. *Quarterly Journal of Economics* 112(2):261–292.

Barbero, F., Thomas, J. A., Bonelli, S., Balleto, E., & Schonrogge, K. (2009). Queen ants make distinctive sounds that are mimicked by a butterfly social parasite. *Science* 323:782–785.

Barbosa, A. et al. (2007). Disruptive coloration in cuttlefish: a visual perception mechanism that regulates ontogenetic adjustment of skin patterning. *Journal of Experimental Biology* 210:1139–1147.

Batson, C. D., Thompson, E. R., Seuferling, G., Whitney, H., & Strongman, J. A. (1999). Moral hypocrisy: appearing moral to oneself without being so. *Journal of Personality and Social Psychology* 77(3):525–537.

Baumeister, R. F., Stillwell, A., & Wotman, S. R. (1990). Victim and perpetrator accounts of interpersonal conflict: autobiographical narratives about anger. *Journal of Personality and Social Psychology* 59:994–1005.

Beaupre, M. G., & Hess, U. (2003). In my mind, *we* all smile: A case of in-group favoritism. *Journal of Experimental Social Psychology* 39:371–377.

Bechara, A., Damasio, H., Tranel, D., & Damasio, A. R. (1997). Deciding advantageously before knowing the advantageous strategy. *Science* 275:1293–1295.

Beedie, C. J., Stuart, E. M., Coleman, D. A., & Foad, A. J. (2006). Placebo effects of caffeine on cycling performance. *Medicine & Science in Sports & Exercise* 38:2159–2164.

Belanoff, J. K. et al. (2005). A randomized trial of the efficacy of group therapy in changing viral load and CD4 counts in individuals living with HIV infection. *International Journal of Psychiatry in Medicine* 35:349–362.

Bell, G. (1982). *The Masterpiece of Nature.* Berkeley: University of California Press.

Benedetti, F. (2009) *Placebo Effects: Understanding the Mechanisms in Health and Disease.* New York: Oxford University Press.

Benson, H. et al. (2006). Study of the therapeutic effects of intercessory prayer (STEP) in cardiac bypass patients: a multicenter randomized trial of uncertainty and certainty of receiving intercessory prayer. *American Heart Journal* 151:934–942.

Blackmon, D. A. (2008). *Slavery by Another Name: The Re-enslavement of Black Americans from the Civil War to World War II.* New York: Anchor Books.

Blalock, J. E., & Smith, E. M. (2007). Conceptual development of the immune system as a sixth sense. *Brain, Behavior, and Immunity* 21:23–33.

Bloise, S. M., & Johnson, M. K. (2007). Memory for emotional and neutral information: gender and individual differences in emotional sensitivity. *Memory* 15(2):192–204.

Bobes, M. A. et al. (2004). Brain potentials reflect residual face processing in a case of prosopagnosia. *Cognitive Neuropsychology* 21:691–718.

Bond, C. F., & DePaulo, B. M. (2006). Accuracy of deception judgments. *Personality Social Psychology Review* 10:214–234.

Boroditsky, L., Schmidt, L. A., & Phillips, W. (2003). Sex, syntax, and semantics. *Language in Mind: Advances in the Study of Language and Thought*, eds. Gentner, D., & Goldin-Meadow, S., 61–80. Cambridge, MA: MIT Press.

Borries, C., Launhardt, K., Epplen, C., Epplen, J., & Winkler, P. (1999). DNA analyses support the hypothesis that infanticide is adaptive in langur monkeys. *Proceedings of the Royal Society B* 266:901–904.

Bowles, S. (2009). Did warfare among ancestral hunter-gatherers affect the evolution of human social behaviors? *Science* 324:1293–1298.

BreakingtheSilence. (2009). *Breaking the Silence: Soldiers' Testimonies from Operation Cast Lead, Gaza.* Jerusalem: Breaking the Silence.

Bro-Jørgensen, J., & Pangle, W. M. (2010). Male topi antelopes alarm snort deceptively to retain females for mating. *American Naturalist* 176:E33–E39.

Brock, P. D. (1999). *The Amazing World of Stick and Leaf Insects.* London: Amateur Entomologists Society.

Brock, T. C., & Balloun, J. L. (1967) Behavioral receptivity to dissonant information. *Journal of Personality and Social Psychology* 6:413–428.

Brooke, M. L., Davies, N. B., & Noble, D. G. (1998). Rapid decline of host defences in response to reduced cuckoo parasitism: behavioural flexibility of reed warblers in a changing world. *Proceedings of the Royal Society B* 265:1277–1282.

Bruner, J. S., & Goodman, C. C. (1947). Value and need as organizing factors in perception. *Journal of Abnormal Social Psychology* 42:33–41.

Bryant, P. A., Trinder, J., & Curtis, N. (2004). Sick and tired: does sleep have a vital role in the immune system? *Nature Reviews Immunology* 4:457–467.

Bugnyar, T., & Heinrich, B. (2006). Pilfering ravens, *Corvus corax*, adjust their behavior to social context and identity of competitors. *Animal Cognition* 9:369–376.

Bugnyar, T., & Kotrschal, K. (2002). Observational learning and the raiding of food caches in ravens, *Corvus corax*: is it "tactical" deception? *Animal Behaviour* 64:185–195.

Bugnyar, T., Stowe, M., & Heinrich, B. (2004). Ravens, *Corvus corax*, follow gaze direction of humans around obstacles. *Proceedings of the Royal Society B* 271:1331–1336.

Burnham, T. C. et al. (2003). Men in committed, romantic relationships have lower testosterone. *Hormones and Behavior* 44:119–122.

Burnham, T. C., & Hare, B. (2007). Engineering human cooperation: does involuntary neural activation increase public goods contribution? *Human Nature* 18:88–108.

Burt, A., & Trivers, R. (2006). *Genes in Conflict: The Biology of Selfish Genetic Elements.* Cambridge, MA: Harvard University Press.

Byers, J. A., & Byers, K. Z. (1983). Do pronghorn mothers reveal the locations of their hidden fawns? *Behavioral Ecology and Sociobiology* 13:147–156.

Byrne, R. W., & Corp, N. (2004). Neocortex size predicts deception rate in primates. *Proceedings of the Royal Society B* 271:1693–1699.

Caldwell, R. L. (1986). The deceptive use of reputation by stomatopods. *Deception: Perspectives on Human and Non-Human Deceit*, eds. Mitchell, R. W., & Thompson, N. S., 129–145. Albany: State University of New York Press.

Calogero, R. M., & Mullen, B. (2008). About face: facial prominence of George Bush in political cartoons as a function of war. *Leadership Quarterly* 19:107–116.

Campbell, W. K., Bosson, J. K., Goheen, T. W., Lakey, C. E., & Kernis, M. H. (2007). Do narcissists dislike themselves "deep down inside"? *Psychological Science* 18:227–229.

Campbell, W. K., Goodie, Q. W., & Foster, J. D. (2004). Narcissism, confidence and risk attitude. *Journal of Behavioral Decision Making* 17:297–311.

Canli, T., Desmond, J. E., Zhao, Z., & Gabrieli, J. D. E. (2002). Sex differences in the neural basis of emotional memories. *PNAS* 99(16):10789–10794.

Carey, R., & Shainin, J. (2002). *The Other Israel: Voices of Refusal and Dissent.* New York: New Press.

Carrico, A. W., & Antoni, M. H. (2008). Effects of psychological interventions on neuroendocrine hormone regulation and immune status in HIV-positive persons: a review of randomized controlled trials. *Psychosomatic Medicine* 70:575–584.

Charles, S. T., Mather, M., & Carstensen, L. L. (2003). Aging and emotional memory: the forgettable image of negative images for older adults. *Journal of Experimental Psychology: General* 132:310–324.

Charnetski, C. J., & Brennan Jr., F. X. (1998). Effect of music and auditory stimulation on secretory immunoglobin A (IgA). *Perceptual and Motor Skills* 87:1163–1170.

Chiao, C. C., Kelman, E. J., & Hanlon, R. T. (2005). Disruptive body coloration of cuttlefish (*Sepia officianalis*) requires visual information regarding edges and contrast of objects in natural substrate backgrounds. *Biological Bulletin* 208:7–11.

Chomsky, N. (1983). *Fateful Triangle: The United States, Israel and the Palestinians.* Cambridge: South End Press.

Chomsky, N., & Pappe, I. (2010). *Gaza in Crisis.* Chicago: Haymarket Books.

Clarke, D. D., & Sokoloff, L. (1999). Circulation and energy metabolism of the brain. *Basic Neurochemistry: Molecular, Cellular and Medical Aspects*, 6th ed., eds. Siegel, G. J., Agranoff, B. W., Albers, R. W., Fisher, S. K., & Uhler, M. D. Philadelphia: Lippincott-Raven.

Cobb, L. A., Thomas, G. I., Dillard, D. H., Merendino, K. A., & Bruce, R. A. (1959). An evaluation of internal-mammary-artery ligation by a double-blind technique. *New England Journal of Medicine* 260:1115–1118.

Cohen, S., Alper, C. M., Doyle, W. J., Treanor, J. J., & Turner, R. B. (2006). Positive emotional style predicts resistance to illness after experimental exposure to rhinovirus or Influenza A virus. *Psychosomatic Medicine* 68:809–815.

Cohen, S., Doyle, W. J., Alper, C. M., Janicki-Deverts, D., & Turner, R. B. (2009). Sleep habits and susceptibility to the common cold. *Archives of Internal Medicine* 169(1):62–67.

Cole, S. W., Kemeny, M., & Taylor, S. E. (1997). Social identity and physical health: accelerated HIV progression in rejection-sensitive gay men. *Journal of Personality and Social Psychology* 72:320–335.

Cole, S. W., Kemeny, M. E., Taylor, S. E., & Visscher, B. R. (1996a). Elevated physical health risk among gay men who conceal their homosexual identity. *Health Psychology* 15:243–251.

Cole, S. W., Kemeny, M. E., Taylor, S. E., Visscher, B. R., & Fahey, J. L. (1996b). Accelerated course of human immunodeficiency virus infection in gay men who conceal their homosexual identity. *Psychosomatic Medicine* 58:219–231.

Collins, C. (2007). *Homeland Mythology: Biblical Narratives in American Culture.* University Park: Pennsylvania State University Press.

Coman, A., Manier, D., & Hirst, W. (2009). Forgetting the unforgettable through conversation: socially shared retrieval-induced forgetting of September 11 memories. *Psychological Science* 20:627–633.

Conway, M., & Ross, M. (1984). Getting what you want by revising what you had. *Journal of Personality and Social Psychology* 47:738–748.

Cozzolino, S., & Widmer, A. (2005). Orchid diversity: an evolutionary consequence of deception? *Trends in Ecology and Evolution* 20:487–494.

Craig, K. D., Hyde, S. A., & Patrick, J. C. (1991). Genuine, suppressed, and faked facial behavior during exacerbation of chronic low back pain. *PAIN* 46:161–171.

Creswell, J., & Baja, V. (2006, November 26). The High Cost of Too Good to be True. *New York Times.*

Croyle, R. T. et al. (2006). How well do people recall risk factor test results? Accuracy and bias among cholesterol screening participants. *Health Psychology* 25:425–432.

Cuc, A., Koppel, J., & Hirst, W. (2007). Silence is not golden: a case for socially shared retrieval-induced forgetting. *Psychological Science* 18:727–733.

Curley, M. J., Barton, S., Surani, A., & Everne, E. B. (2004). Coadaptation in mother and infant regulated by a paternally expressed imprinted gene. *Proceedings of the Royal Society of Biological Sciences* 271:1301–1309.

D'Argembeau, A., & Van der Linden, M. (2008). Remembering pride and shame: self-enhancement and the phenomenology of autobiographical memory. *Memory* 16:538–547.

Dally, J. M., Emery, N. J., & Clayton, N. S. (2004). Cache protection strategies by western scrub-jays (*Aphelocoma californica*): hiding food in the shade. *Biology Letters* 271:S387–S390.

Dally, J. M., Emery, N. J., & Clayton, N. S. (2006). Food-caching western scrub-jays keep track of who was watching them. *Science* 312:1662–1665.

Daly, M., & Wilson, M. (1982). Whom are newborn babies said to resemble? *Ethology and Sociobiology* 3:69–68.

Daly, M., Wilson, M., & Weghorst, S. J. (1982). Male sexual jealousy. *Ethology and Sociobiology* 3:11–27.

Dantzer, R., & Kelley, K. W. (2007). Twenty years of research on cytokine-induced sickness behavior. *Brain, Behavior, and Immunity* 21:153–160.

Davidson, R. J. et al. (2003). Alterations in brain and immune function produced by mindfulness meditation. *Psychosomatic Medicine* 65:564–570.

Davies, N. B. (2000). *Cuckoos, Cowbirds and Other Cheats.* London: T. & A. D. Poyser.

Davies, N. B., Brooke, L., & Kacelnik, A. (1996). Recognition errors and probability of parasitism determine whether reed warblers should accept or reject mimetic eggs. *Proceedings of the Royal Society B* 263:925–931.

Davies, N. B., Kilner, R. M., & Noble, D. G. (1998). Nestling cuckoos, *Cuculus canorus*, exploit hosts with begging calls that mimic a brood. *Proceedings of the Royal Society B* 265:673–678.

Davies, N. B., & Welbergen, J. A. (2009). Social transmission of a host defense against cuckoo parasitism. *Science* 324:1318–1320.

Dawkins, R. (2006). *The God Delusion.* New York: Houghton Mifflin Harcourt.

Dawson, E., Savitsky, K., & Dunning, D. (2006). "Don't tell me, I don't want to know": understanding people's reluctance to obtain medical diagnostic information. *Journal of Applied Social Psychology* 36:751–768.

de Craen, A. J. M., Roos, P. J., de Vries, A. L., & Kleijnen, J. (1996). Effect of color of drugs: systematic review of perceived effect of drugs and of their effectiveness. *British Medical Journal* 313:1624–1626.

de Waal, F. (1982). *Chimpanzee Politics: Power and Sex Among Apes.* New York: Harper and Row.

de Waal, F. (1986). Deception in the natural communication of chimpanzees. *Deception: Perspectives on Human and Nonhuman Deceit*, eds. Mitchell, R. W., & Thompson, N. S. Albany, New York: SUNY Press.

DeHart, T., & Pelham, B. W. (2007). Fluctuations in state of implicit self-esteem in response to daily negative events. *Journal of Experimental Social Psychology* 43:157–165.

DeHart, T., Pelham, B. W., & Tennen, H. (2006). What lies beneath: parenting style and implicit self-esteem. *Journal of Experimental Social Psychology* 42:1–17.

Dennett, D. C. (2006). *Breaking the Spell: Religion as a Natural Phenomenon.* New York: Viking Adult.

Dennett, D. C. (2007). The Evaporation of the Powerful Mystique of Religion. http://www.edge.org/q2007/q07_1.html.

DePaolo, B. M., & Kashy, D. A. (1998). Everyday lies in close and casual relationships. *Journal of Personality and Social Psychology* 74:63–79.

DePaolo, B. M., Kashy, D. A., Kirkendol, S. E., Wyer, M. M., & Epstein, J. A. (1996). Lying in everyday life. *Journal of Personality and Social Psychology* 70:979–995.

DePaolo, B. M. et al. (2003). Cues to deception. *Psychological Bulletin* 129:74–118.

Diamond, J. (2006). *Guns, Germs, and Steel: A Short History of Everybody for the Last 13,000 Years.* New York: W. W. Norton.

Dickerson, S. S., Kemeny, M. E., Aziz, N., Kim, K. H., & Fahey, J. L. (2004). Immunological effects of induced shame and guilt. *Psychosomatic Medicine* 66:124–131.

Ditto, P. H., & Lopez, D. E. (1992). Motivated skepticism: use of differential decision criteria for preferred and nonpreferred conclusions. *Journal of Personality and Social Psychology* 63:568–584.

Ditto, P. H., & Lopez, D. E. (2003). Spontaneous scepticism: the interplay of motivation and expectation in response to favorable and unfavorable medical diagnoses. *Personality and Social Psychology Bulletin* 29:1120–1132.

Dominey, W. J. (1980). Female mimicry in male bluegill sunfish: a genetic polymorphism? *Nature* 284:546–548.

Draper, R. (June 2009). And he shall be judged. *GQ*.

Dunning, D., Johnson, K., Ehrlinger, J., & Kruger, J. (2003). When people fail to recognize their own incompetence. *Current Directions in Psychological Science* 12:83–87.

Dwyer, J. (May 6, 2011). A CUNY trustee expands on his views of what is offensive. *New York Times*.

Eban, A. (1973). http://wikiquote.org/wiki/Abba_Eban.

Egan, L. C., Bloom, P., & Santos, L. R. (2010). Choice-induced preferences in the absence of choice: evidence from a blind two choice paradigm with young children and capuchin monkeys. *Journal of Experimental Social Psychology* 46:204–207.

Egan, L. C., Santos, L. R., & Bloom, P. (2007). The origins of cognitive dissonance: evidence from children and monkeys. *Psychological Science* 18:978–983.

Ehrlinger, J., Johnson, K., Banner, M., Dunning, D., & Kruger, J. (2008). Why the unskilled are unaware: further explorations of (absent) self-insight among the incompetent. *Organizational Behavior and Human Decision Processes* 105(1):98–121.

Eibach, R. P., & Mock, S. E. 2011. Idealizing parenthood to rationalize parental investments. *Psychological Sciences 22:* 203-208.

Eisenberger, N. I., Kemeny, M. E., & Wyatt, G. E. (2003). Psychological inhibition and CD4 T-cell levels in HIV-seropositive women. *Journal of Psychosomatic Research* 54:213–224.

El-Haj, N. A. (2001). *Facts on the Ground.* Chicago: University of Chicago Press.

El-Hani, C. N., Queiroz, J., & Stjernfelt, F. (2010). Firefly femmes fatales: a case study in the semiotics of deception. *Biosemiotics* 3(1):33–55.

Engelberg, S., & Bryant, A. (February 26, 1995). FAA's fatal fumbles on computer plane's safety. *New York Times.*

Epley, N., & Whitchurch, E. (2008). Mirror, mirror on the wall: enhancement in self-recognition. *Personality and Social Psychology Bulletin.*

Eppig, C., Fincher, C. L., & Thornhill, R. (2010). Parasite prevalence and the worldwide distribution of cognitive ability. *Proceedings of the Royal Society B* 277(1701): 3801–3808.

Fallow, J. (2004, January/February) Blind into Baghdad. *Atlantic Monthly.*

Faulkner, J., Schaller, M., Park, J. H., & Duncan, L. A. (2004). Evolved disease-avoidance mechanisms and contemporary xenophobic attitudes. *Group Processes and Intergroup Relations* 7:333–353.

Faust, D. G. (2008). *This Republic of Suffering: Death and the American Civil War.* New York: Vintage Books.

Fein, S., & Spencer, S. J. (1997). Prejudice as self-image maintenance: affirming the self through derogating others. *Journal of Personality and Social Psychology* 73(1):31–44.

Fenton-O'Creevy, M., Nicholson, N., Soane, E., & Willman, P. (2003). Trading on illusions: unrealistic perceptions of control and trading performance. *Journal of Occupational and Organizational Psychology* 76:53–68.

Feynman, R. (1988). *What Do You Care What Other People Think? Further Adventures of a Curious Character.* New York: W. W. Norton.

Fincher, C. L., & Thornhill, R. (2008a). Assortative sociality, limited dispersal, infectious disease and the genesis of global pattern of religion diversity. *Proceedings of the Royal Society London B.*

Fincher, C. L., & Thornhill, R. (2008b). A parasite-driven wedge: infectious diseases may explain language and other biodiversity. *Oikos* 117:1289–1297.

Fincher, C. L., Thornhill, R., Murray, D. R., & Schaller, M. (2008). Pathogen prevalence predicts human cross-cultural variability in individualism/collectivism. *Proceedings of the Royal Society B* 275:1279–1285.

Finkelstein, N. G. (2000). *The Holocaust Industry: Reflections on the Exploitation of Jewish Suffering.* London: Verso.

Finkelstein, N. G. (2003). *Image and Reality of the Israel-Palestine Conflict*, 2nd ed. London: Verso.

Finkelstein, N. G. (2005). *Beyond Chutzpah: On the Misuse of Anti-Semitism and the Abuse of History.* Berkeley: University of California Press.

Finkelstein, N. G. (2010). *This Time We Went Too Far.* New York: Orbooks.

Fisher, M. (2004). Female intrasexual competition decreases female attractiveness. *Proceedings of the Royal Society B* 271(supp):283–285.

Fisk, R. (2005). *The Great War for Civilization: The Conquest of the Middle East.* New York: Knopf.

Fisk, R. (December 20, 2008). How the absence of one tiny word sowed the seeds of catastrophe. *The Independent.*

Flapan, S. (1987). *The Birth of Israel: Myths and Realities.* New York: Pantheon.

Flower, T. (2010). Fork-tailed drongos use deceptive mimicked alarm calls to steal food. *Proceedings of the Royal Society B* doi: 10.1098/rspb.2010.1932.

Fournier, J. C. et al. (2010). Antidepressant drug effects and depression severity. *Journal of the American Medical Association* 303:47–53.

Forster, S. (1999). Dreams and nightmares: German military leadership and the images of future warfare, 1871–1914. *Anticipating Total War: The German and American Experiences, 1971–1914.* Boemeke, M.F., Chickering, R, & Forster, S., eds. Cambridge: Cambridge University Press, 343–376.

Fox, M. et al. (2010). Grandma plays favourites: X-chromosome relatedness and sex-specific childhood mortality. *Proceedings of the Royal Society B* 277:567–573.

Frattaroli, J. (2006). Experimental disclosure and its moderators: a meta-analysis. *Psychological Bulletin* 132(6):823–865.

Freeman, S., & Wilber, D. Q. (February 14, 2009). Pilots spoke of ice on wings before deadly crash in New York. *Washington Post.*

Freyd, J. J., Martorello, S. R., Alvarado, J. S., Hayes, A. E., & Christman, J. C. (1998). Cognitive environments and dissociative tendencies: performance on the standard Stroop task for high versus low dissociators. *Applied Cognitive Psychology* 12:S91–S103.

Frye, N. E., & Karney, B. R. (2004). Revision in memories of relationship development: do biases persist over time? *Personal Relationships* 11(1):79–97.

Fujita, K., Gollwitzer, P. M., & Oettingen, G. (2007). Mind-sets and pre-conscious open-mindedness to incidental information. *Journal of Experimental Social Psychology* 43:48–61.

Galinsky, A. D., Magee, J. C., Inesi, M. E., & Greenfeld, H. (2006). Power and perspectives not taken. *Psychological Science* 17:1068–1074.

Garver-Apgar, C. E., Gangestad, S. W., Thornhill R., & Miller, R. D. (2006). MHC alleles, sexual responsivity, and unfaithfulness in romantic couples. *Psychological Science* 17:830–835.

Gibbs, H. L. et al. (2000). Genetic evidence for female host-specific races of the common cuckoo. *Nature* 407:183–185.

Gilbert, D. (2006). *Stumbling on Happiness.* New York: Alfred A. Knopf.

Gilbert, L. E. (1982). The co-evolution of a butterfly and a vine. *Scientific American* 247:110–121.

Gillings, V., & Joseph, S. (1986). Religiosity and social desirability: impression management and self-deceptive positivity. *Personality Processes and Individual Differences* 21(6):1047–1050.

Ginges, J., Hansen, I., & Norenzayan, A. (2009). Religion and support for suicide attacks. *Psychological Science* 20:224–230.

Gladwell, M. (2008). *Outliers: The Story of Success.* New York: Little, Brown and Company.

Glanz, J., & Schwartz, J. (September 26, 2003). Dogged engineer's effort to assess shuttle damage. *New York Times.*

Glaser, M., & Weber, M. (2007). Overconfidence and trading volume. *Geneva Risk and Insurance Review* 32:1–36.

Gonsalves, B., & Paller, K. A. (2000). Neural events that underlie remembering something that never happened. *Nature Neuroscience* 3:1316–1321.

Gonsalves, B., Reber, P. J., Gitelman, D. R., Parrish, T. B., & Mesulam, M. M. (2004). Neural evidence that vivid imagining can lead to false remembering. *Psychological Science* 15:655–660.

Goo, S. K. (January 25, 2006). Poor behavior, fatigue led to '04 plane crash. *Washington Post.*

Graber, M. L., Franklin, N., & Gordon, R. (2005). Diagnostic error in internal medicine. *Archives of Internal Medicine* 165:1493–1499.

Grammer, K., Kruck, K., Juette, A., & Fink, B. (2000). Non-verbal behavior as courtship signals: the role of control and choice in selecting partners. *Evolution and Human Behavior* 21:371–390.

Grammer, K., Renninger, L. A., & Fischer, B. (2004). Disco clothing, female sexual motivation, and relationship status: is she dressed to impress? *Journal of Sex Research* 41:66–74.

Gray, P. B., & Campbell, B. (2009). Human male testosterone, pair-bonding, and fatherhood. *Endocrinology of Social Relationships*, eds. Ellison, P. T., & Gray, P. B. Cambridge, MA: Harvard University Press.

Gray, P. B., Ellison, P. T., & Campbell, B. C. (2007). Testosterone and marriage among Ariaal men of Northern Kenya. *Current Anthropology* 48:750–755.

Gray, P. B. et al. (2002). Human male pair bonding and testosterone. *Human Nature* 15:119–131.

Green, J. D., Sedikides, C., & Gregg, A. P. (2008). Forgotten but not gone: the recall and recognition of self-threatening memories. *Journal of Experimental Social Psychology* 44:547–561.

Greenspan, S. (2009). *Annals of Gullibility: Why We Get Duped and How to Avoid It.* Westport, CT: Praeger.

Greenwald, A. G. (1980). The totalitarian ego: fabrication and revision of personal history. *American Psychologist* 35:603–618.

Greenwald, A. G., McGhee, D. E., & Schwartz, J. L. K. (1998). Measuring individual differences in implicit cognition: the Implicit Association Test. *Journal of Personality and Social Psychology* 74:1464–1480.

Greenwald, A. G., Nosek, B. A., & Banaji, M. R. (2003). Understanding and using the Implicit Association Test: I. An improved scoring algorithm. *Journal of Personality and Social Psychology* 85(2):197–216.

Greenwald, A. G., Poehlman, T. A., Uhlmann, E., & Banaji, M. R. (2009). Understanding and using the Implicit Association Test: III. Meta-analysis of predictive validity. *Journal of Personality and Social Psychology* 97(1):17–41.

Gregg, C. et al. (2010). High-resolution analysis of parent-of-origin allelic expression in the mouse brain. *Science* 329:643–648.

Greig-Smith, P. W. (1978). Imitative foraging in mixed-species flocks of Seychelles birds. *Ibis* 120:233–235.

Grinblatt, M., Keloharju, M., & Ikaheimo, S. (2008). Social influence and consumption: evidence from the automobile purchases of neighbors. *Review of Economics and Statistics* 90:735–753.

Gross, M. (1982). Sneakers, satellites, and parentals: polymorphic mating strategies in North American sunfishes. *Z. Tierpsychol.* 60:1–26.

Gross, M. R., Suk, H. Y., & Robertson, C. T. (2007). Courtship and genetic quality: asymmetric males show their best side. *Proceedings of the Royal Society B* 274:2115–2122.

Gruzeller, J. H. (2002) A review of the impact of hypnosis, relaxation, guided imagery and individual differences on aspects of immunity and health. *Stress* 5:147–163.

Guenther, C. L., & Alicke, M. D. (2010). Deconstructing the better-than-average effect. *Journal of Personality and Social Psychology* 99:755–770.

Gur, R., & Sackeim, H. A. (1979). Self-deception: a concept in search of a phenomenon. *Journal of Personality and Social Psychology* 37:147–169.

Hagee, J. (2006). *Jerusalem Countdown.* Lake Mary, FL: FrontLine.

Haig, D. (1993). Genetic conflicts in human pregnancy. *The Quarterly Review of Biology* 68:495–532.

Haig, D. (1999). Asymmetric relations: internal conflicts and the horror of incest. *Evolution and Human Behavior* 20:83–98.

Haig, D. (2004a). Genomic imprinting and kinship: how good is the evidence? *Annual Review of Genetics* 38:553–585.

Haig, D. (2004b). The inexorable rise of gender and the decline of sex: social change in academic titles, 1945–2001. *Archives of Sexual Behavior* 33:87–96.

Haig, D. (2006). Intrapersonal conflict. *Conflict*, eds. Jones, M., & Fabian, A. Cambridge: Cambridge University Press.

Haig, D., & Graham, C. (1991). Genomic imprinting and the strange case of the insulin-like growth factor II receptor. *Cell* 64:1045–1046.

Haig, D., & Westoby, M. (1989). Parent-specific gene expression and the triploid endosperm. *American Naturalist* 134:147–154.

Hall, J. (February 23, 2005). Paying the price for safety. *New York Times*.

Hallinan, J. T. (2009). *Why We Make Mistakes: How We Look Without Seeing, Forget Things in Seconds, and Are All Pretty Sure We Are Above Average.* New York: Broadway Books.

Hamilton, W. D. (1964). The genetical evolution of social behaviour. *Journal of Theoretical Biology* 7:1–52.

Hamilton, W. D., & Zuk, M. (1982). Heritable true fitness in birds: a role for parasites. *Science* 218:384–387.

Hammerstein, P., ed. (2003). *Genetic and Cultural Evolution of Cooperation.* Cambridge, MA: MIT Press.

Hanlon, R. T., & Conroy, L. A. (2008). Mimicry and foraging behavior of two tropical sand-flat octopus species off North Sulawesi, Indonesia. *Biological Journal of the Linnean Society* 93:23–38.

Hanlon, R. T., Forsythe, J. W., & Joneschild, D. E. (1999). Crypsis, conspicuousness, mimicry, and polephenism as antipredator defenses of foraging octopuses on Indo-Pacific coral reefs, with a method of quantifying crypsis from video tapes. *Biological Journal of the Linnaean Society* 66:1–22.

Hanlon, R. T. et al. (2007). Adaptable night camouflage by cuttlefish. *American Naturalist* 169:543–551.

Hanlon, R. T., Naud, M. J., Shaw, P. W., & Havenhand, J. N. (2005). Transient sexual mimicry leads to fertilization. *Nature* 430:212.

Hanlon, R. T., Watson, A. C., & Barbosa, A. (2010). A "mimic octopus" in the Atlantic: flatfish mimicry and camouflage by *Macrotritopus defilippi*. *Biological Bulletin* 218:15–24.

Hanssen, S. A., Hasselquist, D., Folstad, I., & Erikstad, K. E. (2004). Costs of immunity: immune responsiveness reduces survival in a vertebrate. *Proceedings of the Royal Society B* 271:925–930.

Harris, M. (2010). MRI lie detectors. *IEEE Spectrum.* http://spectrum.ieee.org/biomedical/imaging/mri-lie-detectors

Hartung, J. (1988). Deceiving down: conjectures on the management of subordinate status. *Self-deception: An Adaptive Mechanism?* eds. Lackhard, J. S., & Paulhus, D. L. Englewood Cliffs, NJ: Prentice-Hall.

Haselton, M. G. (2003). The sexual overperception bias: evidence of a systematic bias in men from a survey of naturally occurring events. *Journal of Research in Personality* 37:34–47.

Haselton, M. G., & Nettle, D. (2006). The paranoid optimist: an integrative evolutionary model of cognitive biases. *Personality and Social Psychology Review* 10:47–66.

Heinrich, B., & Pepper, J. W. (1998). Influence of competitors on caching behavior in common ravens, *Corvus corax. Animal Behaviour* 56:1083–1090.

Henriquez, D. B. (2010). *The Wizard of Lies: Bernie Madoff and the Death of Trust.* New York: Holt.

Henry, J. D., von Hippel, W., & Baynes, K. (2009). Social inappropriateness, executive control, and aging. *Psychology and Aging* 24:239–244.

Hilgard, E. R. (1977). *Divided Consciousness: Multiple Controls in Human Thought and Action.* New York: John Wiley.

Hochschild, A. (May 23, 2004). What's in a word? Torture. *New York Times.*

Hodson, G., & Olson, J. M. (2005). Testing the generality of the name letter effect: name initials and everyday attitudes. *Personality and Social Psychology Bulletin* 31:1099–1111.

Hoover, J. P., & Robinson, S. K. (2007). Retaliatory Mafia behavior by a parasitic cowbird favors host acceptance of parasitic eggs. *PNAS* 104:4479–4483.

Hsiao, L. et al. (2001). A compendium of gene expression in normal human tissues. *Physiological Genomics* 7:97–104.

Inzlicht, M., McGregor, I., Hirsh, J. B., & Nash, K. (2009). Neural markers of religious conviction. *Psychological Science* 20(3):385–392.

Irwin, M. R. (2007). Human psychoneuroimmunology: 20 years of discovery. *Brain, Behavior, and Immunity* 22:129–139.

Isaacowitz, D. M, Toner, K., Goren, D., & Wilson, H. R. (2008). Looking while unhappy: mood-congruent gaze in young adults, positive gaze in older adults. *Psychological Science* 19(9):848–853.

Isaacowitz, D.M., Wadlinger, H. A., Goren, D., & Wilson, H. R. (2006) Is there an age-related positivity effect in visual attention? A comparison of two methodologies. *Emotion* 6:511–516.

Jersakova, J., Johnson, S. D., & Kindlmann, P (2006). Mechanisms and evolution of deceptive pollination in orchids. *Biological Reviews* 81:219–235.

Johansson-Stenman, O., & Martinsson, P. (2006). Honestly, why are you driving a BMW? *Journal of Economic Behavior and Organization* 60:129–146.

Johnson, D. D. P. (2004). *Overconfidence and War: the Havoc and Glory of Positive Illusions.* Cambridge, MA: Harvard University Press.

Johnson, D. D. P. (2009). The error of God: error management theory, religion, and the evolution of cooperation. *Games, Groups and the Global Good*, ed. Levin, S. A. Berlin: Springer.

Kahneman, D., & Twersky, A. (1996). On the reality of cognitive illusions. *Psychological Review* 103:582–591.

Kaptchuk, T. J. et al. (2010). Placebos without deception: a randomized controlled trial in irritable bowel syndrome. *PLoS ONE* 5(12):1–7.

Kaptchuk, T. J. et al. (2006). Sham device *v* inert pill: randomized controlled trial of two placebo treatments. *British Medical Journal* 332:391–397.

Karim, A. A. et al. (2009). The truth about lying: inhibition of the anterior prefrontal cortex improves deceptive behavior. *Cerebral Cortex.*

Karney, B. R., & Coombs, R. H. (2000). Memory bias in long-term close relationships: consistency or improvement? *Personality and Social Psychology Bulletin* 26:959–970.

Kassin, S. M. (2005). On the psychology of confessions: does innocence put innocents at risk? *American Psychologist* 60:215–228.

Kassin, S. M., & Gudjonsson, G. H. (2005). True crimes, false confessions: why do innocent people confess to crimes they did not commit? *Scientific American Mind*: 24–31.

Keating, C. F., & Heltman, K. R. (1994). Dominance and deception in children and adults: are leaders the best misleaders? *Personality and Social Psychology Bulletin* 20:312–321.

Keverne, E. B., Fundele, R., Narasimha, M. E. B., Barton, S. C., & Surani, M. A. (1996). Genomic imprinting and the differential roles of parental genomes in brain development. *Developmental Brain Research* 92:91–100.

Kitayama, S., & Karasawa, M. (1997). Implicit self-esteem in Japan: name letters and birthday numbers. *Journal of Stress Physiology and Biochemistry* 23:736–742.

Kitchens, L. J. (1998). *Exploring Statistics: A Modern Introduction to Data Analysis and Inference,* 2nd ed. Pacific Grove, CA: Duxbury Press.

Klein, S. et al. (2003). The influence of gender and emotional valence of visual cues on fMRI activation in humans. *Pharmacopsychiatry* 36 Suppl 3:S191–S194.

Kobayashi, C., & Greenwald, A. G. (2003). Implicit-explicit differences in self-enhancement for Americans and Japanese. *Journal of Cross-Cultural Psychology* 34(5):522–541.

Konrath, S. H., & Schwartz, N. (2007). Do male politicians have big heads? Face-ism in online self-representations of politicians. *Media Psychology* 10:436–448.

Kovalev, V. A., Kruggel, F., & von Cramon, Y. (2003). Gender and age effects in structural brain asymmetry as measured by fMRI texture analysis. *Neuroimage* 19:895–905.

Kozel, F. A. et al. (2005). Detecting deception using functional magnetic resonance imaging. *Biological Psychiatry* 58:605–613.

Kruger, J., & Dunning, D. (1999). Unskilled and unaware of it: how difficulties in recognizing one's own incompetence lead to inflated self-assessments. *Journal of Abnormal Social Psychology* 77:1121–1134.

Krugman, P. (September 6, 2009). How did economists get it so wrong? *New York Times Magazine,* pp. 36–44.

Krugman, P. (December 19, 2010). When zombies win. *New York Times.*

Krugman, P., & Wells, R. (2011). The busts keep getting bigger: why? *New York Review of Books* 58:28–29.

Kuhn, D. (2002). The effects of active and passive participation in musical activity on the immune system as measured by salivary immunoglobin A (SigA). *Journal of Music Therapy* 39:30–39.

Kwan, V. S. Y. et al. (2007). Assessing the neural correlates of self-enhancement bias: a transcranial magnetic stimulation study. *Experimental Brain Research* DOI 10.1007/s00221-007-0992-2.

Kwon, Y., Scheibe, S., Samanez-Larkin, G. R., Tsai, J. L., & Carstensen, L. L. (2009). Replicating the positivity effect in picture memory in Koreans: evidence for cross-cultural generalizability. *Psychology and Aging* 24:748–754.

Lailvaux, S. P., Reaney, L. T., & Blackwell, P. R. Y. (2009). Dishonest signaling of fighting ability and multiple performance traits in the fiddler crab *Uca mjoeberg. Functional Ecology* 23:359–366.

Langer, E. J., & Roth, J. (1975). Heads I win, tails it's chance: the illusion of control as a function of the sequence of outcomes in a purely random task. *Journal of Personality and Social Psychology* 32:951–955.

Langewiesche, W. (November 2001). The crash of EgyptAir 990. *The Atlantic Monthly.*

Langewiesche, W. (November 2003). Columbia's last flight. *The Atlantic Monthly.*

Langewiesche, W. (January 2009). The devil at 37,000 feet. *Vanity Fair.*

Larcker, D. F., & Zakolyukina, A. A. (2010). Detecting deceptive discussions in conference calls. Stanford, CA: Stanford Graduate School of Business Research Paper No. 2060, pp 1–33.

Larochette, A. C., Chambers, C. T., & Craig, K. D. (2006). Genuine, suppressed, and faked facial expressions of pain in children. *PAIN* 126:64–71.

Lassek, W. D., & Gauilin, S. J. C. (2009). Costs and benefits of fat-free muscle mass in men: relationship to mating success, dietary requirements, and native immunity. *Evolution and Human Behavior* 30:322–328.

Le Roux, F. H., Bouic, P. J. D., & Bester, M. M. (2007). The effect of Bach's Magnificat on emotions, immune, and endocrine parameters during physiotherapy treatment of patients with infectious lung conditions. *Journal of Music Therapy* 44(2):156–168.

Leaver, L. A., Hopewell, L., Caldwell, C., & Mallarky, L. (2007). Audience effects on food caching in grey squirrels (*Sciurus carolonensis*): evidence for pilferage avoidance strategies. *Animal Cognition* 10:23–27.

Lee, B. Y., & Newberg, A. B. (2005). Religion and health: a review and critical analysis. *Zygon* 40(2):443–468.

Lee, S. W. S., & Schwarz, N. (2010). Dirty hands and dirty mouths: embodiment of the moral-purity metaphor is specific to the motor modality involved in moral transgression. *Psychological Science* 20:1–3.

Lelyveld, J. (2011). Curveball. *New York Review of Books* 58:4.

Lewis, M. (1992). *Shame: The Exposed Self.* New York: Free Press.

Lewis, M. (1993). The development of deception. *Lying and Deception in Everyday Life*, eds. Lewis, M., & Saarni, C., 90–105. New York: Guilford Press.

Li, L. L. et al. (1999). Regulation of maternal behavior and offspring growth by paternally expressed *Peg3*. *Science* 284:330–333.

Libet, B. (2004). *Mind Time: The Temporal Factor in Consciousness.* Cambridge, MA: Harvard University Press.

Lichtblau, E. (February 10, 2005). 9/11 report cites many warnings about hijackings. *New York Times.*

Lieven, D. (2010). *Russia Against Napoleon: The True Story of the Campaigns of War and Peace.* New York: Viking.

Lloyd, J. E. (1986). Firefly communication and deception: "Oh, what a tangled web." *Deception: Perspectives on Human and Nonhuman Deceit*, eds. Mitchell, R. W., & Thompson, N. S., 113–128. Albany: State University of New York.

Lochmiller, R. L., & Deerenberg, C. (2000). Trade-offs in evolutionary immunology: just what is the cost of immunity? *Oikos* 88:87–98.

Loewen, J. W. (2007). *Lies My Teacher Told Me: Everything Your American History Textbook Got Wrong.* New York: New Press.

Loftus, E. (1996). *Eyewitness Testimony.* Cambridge: Harvard University Press.

Lord, C. G., Ross, L., & Lepper, M. R. (1979). Biased assimilation and attitude polarization: the effects of prior theories on subsequently considered evidence. *Journal of Personality and Social Psychology* 37:2098–2108.

Lorenz, K. (1966). *On Aggression.* New York: Harcourt, Brace & World.

Lu, H. J., & Chang, L. *Deceiving Yourself to Achieve High But Not Necessarily Equal Status.* Hong Kong: The Chinese University of Hong Kong.

Lykken, D. T., Macindoe, I., & Tellegen, A. (1972). Perception: autonomic response to shock as a function of predictability in time and locus. *Psychophysiology* 9(3):318–333.

Lynch, R. (2010). It's funny because we think it's true: laughter is augmented by implicit preferences. *Evolution and Human Behavior* 31:141–148.

Lynch, R., & Trivers, R. (2011). Self-deception inhibits laughter and humor appreciation. Unpublished manuscript.

Lyon, B. E. (2003). Egg recognition and counting reduce costs of avian conspecific brood parasitism. *Nature* 422:495–499.

Maas, A., Salvi, D., Arcuri, L., & Semin, G. (1989). Language use in intergroup contexts: the linguistic intergroup bias. *Journal of Personality and Social Psychology* 57(6):981–993.

Macintyre, D. (July 15, 2009). Israeli soldiers reveal the brutal truth of Gaza attack. *The Independent.*

Maderspacher, F., & Stensmyr, M. (2011). Myrmecomorphomania. *Current Biology* 21:R291–R293.

Madrick, J. (2011). *Age of Greed: The Triumph of Finance and the Decline of America, 1970 to the Present.* New York: Knopf.

Mahajan, N. et al. (2011). The evolution of intergroup bias: perceptions and attitudes in Rhesus Macaques. *Journal of Personality and Social Psychology* 100:387–405.

Mallon, E. B., Brockmann, A., & Schmid-Hempel, P. (2003). Immune response inhibits associative learning in insects. *The Royal Society* 270:2471–2473.

Mann, S., & Vrij, A. (2006). Police officers' judgments of veracity, cognitive load, and attempted behavioral control in real-life police interviews. *Psychology Crime & Law* 12:307–319.

Mann, S., Vrij A., & Bull, R. (2004). Detecting true lies: police officers ability to detect deceit. *Journal of Applied Psychology* 89:137–149.

Margoliash, D., & Konishi, M. (1985). Auditory representation of autogenous song in the song system of white-crowned sparrows. *Proceedings of the National Academy of Sciences* 82:5997–6000.

Marsland, A. L., Cohen, S., Rabin, B. S., & Manuck, S. B. (2006). Trait positive effect and antibody response to hepatitis B vaccination. *Brain, Behavior, and Immunity* 20:261–269.

Marsland, A. L., Pressman, S., & Cohen, S. (2007). Positive affect and immune function. *Psychoneuroimmunology* 4E(II):761–779.

Masalski, K. W. (November 2001). Examining the Japanese history textbook controversies. *Japan Digest.*

Mather, M. et al. (2004). Amygdala responses to emotionally valenced stimuli in older and younger adults. *Psychological Science* 15:259–263.

Mather, M., & Carstensen, L. L. (2005). Aging and motivated cognition: the positivity effect in attention and memory. *Trends in Cognitive Sciences* 9:496–502.

Matsusaka, T. (2004). When does play panting occur during social play in wild chimpanzees? *Primates* 45(4):221–229.

Matthews, A. U. (2007). Hidden sexism: facial prominence and its connections to gender and occupational status in popular print media. *Sex Roles* 57(7–8):515–525.

McCarthy, J. (1990). *The Population of Palestine: Population History and Statistics of the Late Ottoman Period and the Mandate.* New York: Columbia University Press.

McNally, R. J. (2003). *Remembering Trauma.* Cambridge, MA: The Belknap Press of Harvard University Press.

Mead, W. R. (2008). The new Israel and the old: why gentile Americans back the Jewish state. Washington, DC: Council on Foreign Relations.

Meier, B. P., Robinson, M. D., Gaither, G. A., & Heinert, N. J. (2006). A secret attraction or defensive loathing? Homophobia, defense, and implicit cognition. *Journal of Research in Personality* 40:377–394.

Mercier, H., & Sperber, D. (2010). Why do humans reason? Arguments for an argumentative theory. *Behavioral and Brain Sciences,* in press.

Miller, G., Tybur, J. M., & Jordan, B. D. (2007). Ovulatory cycle effects on tip earnings by lap dancers: economic evidence for human estrus? *Evolution and Human Behavior* 28:375–381.

Milne, S. (December 30, 2008). Israel's onslaught on Gaza is a crime that cannot succeed. *Guardian.*

Mitani, J. C., Watts, D. P., & Amsler, S. J. (2010). Lethal intergroup aggression leads to territorial expansion in wild chimpanzees. *Current Biology* 20(12):R507–508.

Moller, A. P. (1990). Deceptive use of alarm calls by male swallows, *Hirundo rustica*: a new paternity guard. *Behavioral Ecology* 1:1–6.

Moller, A. P., Erritzoe, J., & Garamszegi, L. Z. (2005). Covariation between brain size and immunity in birds: implications for brain size evolution. *Journal of Evolutionary Biology* 18:223–237.

Moller, A. P., & Saino, N. (2004). Immune response and survival. *Oikos* 104:299–304.

Moller, A. P., & Swaddle, J. P. (1987). Social control of deception among status signaling house sparrows *Passer domesticus. Behavioral Ecology and Sociobiology* 20(307–311).

Morgan, C. J., LeSage, J. B., & Kosslyn, S. M. (2009). Types of deception revealed by individual differences in cognitive ability. *Social Neuroscience* 4:554–569.

Morris, B. (1987). *The Birth of the Palestinian Refugee Question.* Cambridge: Cambridge University Press.

Morris, M. W., Sheldon, O. J., Ames, D. R., & Young, M. J. (2007). Metaphors and the market: consequences and preconditions of agent and object metaphors in stock market commentary. *Organizational Behavior and Human Decision Processes* 102:174–192.

Moseley, J. B., Wray, N. P., Kuykendall, D., Willis, K., & Landon, G. (1996). Arthroscopic treatment of osteoarthritis of the knee: a prospective, randomized, placebo-control trial. *American Journal of Sports Medicine* 24:28–34.

Moskowitz, T. J., & Wertheim, L. J. (2011). *Scorecasting: The Hidden Influences Behind How Sports are Played and Games are Won.* New York: Crown Archetype.

Muehlenbein, M. P. (2006). Intestinal parasite infections and fecal steroid levels in wild chimpanzees. *American Journal of Physical Anthropology* 130:546–550.

Muehlenbein, M. P. (2008). Adaptive variation in testosterone levels in response to immune activation: empirical and theoretical perspectives. *Social Biology* 53:13–23.

Muehlenbein, M. P., Cogswell, F. B., James, M. A., Koterski, J., & Ludwig, G. V. (2006). Testosterone correlates with Venezuelan equine encephalitis virus infection in macaques. *Virology Journal* 3:19.

Mukerjee, M. (2010). *Churchill's Secret War: The British Empire and the Ravaging of India During World War II.* New York: Basic Books.

Muller, M. N., Marlowe, F. W., Bugumba, R., & Ellison, P. (2009). Testosterone and paternal care in East African foragers and pastoralists. *Proceedings of the Royal Society B* 276:347–354.

Murphy, K., Travers, P., & Walport, M. (2008). *Immunobiology,* 7th ed. New York: Garland.

Nardone, I. B., Ward, R., Fotopoulou, A., & Turnbull, O. H. (2008). Attention and emotion in anosognosia: evidence of implicit awareness and repression. *Neurocase* 13:438–445.

Neave, N. (2003). Testosterone, territoriality, and the "home advantage." *Psychology and Behavior* 78:269–275.

Negroni, C. (March 2, 2009). How long should air crews rest? *International Herald Tribune.*

Nelson, L. D., & Simmons, J. P. (2007). Moniker maladies: when names sabotage success. *Psychological Science* 18(12):1106–1112.

Nesse, R. M., Silverman, A., & Bortz, A. (1990). Sex differences in ability to recognize family resemblance. *Ethology and Sociobiology* 11:11–21.

Neuhoff, J. G. (1998). Perceptual bias for rising tones. *Nature* 395:123–124.

Neuhoff, J. G. (2001). An adaptive bias in the perception of looming auditory motion. *Ecological Psychology* 13:87–110.

Newman, M. L., Pennebaker, J. W., & Richards, J. M. (2003). Lying words: predicting deception from linguistic styles. *Personality and Social Psychology Bulletin* 29:665–675.

Niederle, M., & Veterlund, L. (2007). Do women shy away from competition? Do men compete too much? *Quarterly Journal of Economics* 122:1067–1101.

Norenzayen, A., & Shariff, A. F. (2008). The origin and evolution of religious prosociality. *Science* 322:58–62.

Nosek, B. A., Banaji, M. R., & Greenwald, A. W. (2002). Harvesting implicit group attitudes and beliefs from a demonstration website. *Group Dynamics: Theory, Research and Practice* 6:101–115.

NTSB. (1994). *A Review of Flightcrew-Involved, Major Accidents of U.S. Air Carriers 1978 Through 1990.* Washington, DC: National Transportation Safety Board.

Nunez, J. M., Casey, B. J., Egner, T., Hare, T., & Hirsch, J. (2005). Intentional false responding shares neural substrates with response conflict and cognitive control. *NeuroImage* 25:267–277.

Nunez, M. J. et al. (2002). Music, immunity, and cancer. *Life Sciences* 71:1047–1057.

Nuttin, J. M. J. (1985). Narcissism beyond Gestalt and awareness: the name letter effect. *European Journal of Social Psychology* 15:353–361.

Nuttin, J. M. J. (1987). Affective consequences of mere ownership: the name letter effect in twelve European languages. *European Journal of Social Psychology* 17:381–402.

Oberlechner, T., & Osler, C. L. (2008). Overconfidence in currency markets. *Working Paper.* Brandeis University.

Okado, Y., & Stark, C. E. L. (2005). Neural activity during encoding predicts false memories created by misinformation. *Learning and Memory* 12:3–11.

Onishi, N. (October 7, 2007). Okinawans protest Japan's plan to revise bitter chapter of WWII. *New York Times.*

Onishi, N. (June 24, 2011). "Safety Myth" left Japan ripe for nuclear crisis. *New York Times.*

Oren, M. B. (2007). *Power, Faith and Fantasy: The United States in the Middle East, 1776 to the Present.* New York: W. W. Norton.

Owen, D. F. (1971). *Tropical Butterflies.* Oxford: Clarendon.

Packer, G. (2005). *The Assassins' Gate: America in Iraq.* New York: Farrar, Straus and Giroux.

Palace, E. M. (1995). Modification of dysfunctional patterns of sexual response through autonomic arousal and false physiological feedback. *Journal of Consulting Clinical Psychology* 63:604–615.

Panksepp, J., & Burgdorf, J. (2003). "Laughing" rats and the evolutionary antecedents of human joy? *Physiology and Behavior* 79:533–547.

Pappe, I. (2006). *The Ethnic Cleansing of Palestine.* Oxford: Oneworld.

Pelham, B. W., Mirenberg, M. C., & Jones, J. T. (2002). Why Susie sells seashells by the seashore: implicit egotism and major life decisions. *Journal of Personality and Social Psychology* 82:469–487.

Pennebaker, J. W. (1997). *Opening Up: The Healing Power of Expressing Emotions.* New York: Guilford Press.

Pennebaker, J. W., & O'Heeron, R. C. (1984). Confiding in others and illness rates among spouses of suicide and accidental death. *Journal of Abnormal Psychology* 93:473–476.

Penrod, S. D., & Cutler, B. L. (1995). Witness confidence and witness accuracy: assessing their forensic relation. *Psychology, Public Policy, and Law* 1:817–845.

Perdue, C. W., Dovidio, J. F., Gurtman, M. B., & Tyler, R. B. (1990). Us and them: social categorization and the process of intergroup bias. *Journal of Personality and Social Psychology* 59:475–486.

Peretz, M. (1984). *From Time Immemorial* (book review). *New Republic* 191:3/4.

Pérez-Benítez, C. I., O'Brien, W. H., Carels, R. A., Gordon, A. K., & Chiros, C. E. (2007). Cardiovascular correlates of disclosing homosexual orientation. *Stress and Health* 23:141–152.

Peterman, A., Palermo, T., & Bredenkamp, C. (2011). Estimates and determinants of sexual violence against women in the Democratic Republic of Congo. *American Journal of Public Health* 101:1060–1067.

Peters, J. (1984). *From Time Immemorial: the origins of the Arab-Jewish conflict over Palestine.* Chicago: JKAP.

Peterson, J. B. et al. (2003). Self-deception and failure to modulate responses despite accruing evidence of error. *Journal of Research in Personality* 37:205–223.

Peterson, J. B., Driver-Linn, E., & DeYoung, C. G. (2002). Self-deception and impaired categorization of anomaly. *Personality Processes and Individual Differences* 33:327–340.

Petrie, K. J., Booth, R. J., & Pennebaker, J. W. (1998). The immunological effects of thought suppression. *Journal of Personality and Social Psychology* 75:1264–1272.

Petrie, K. J., Booth, R. J., Pennebaker, J. W., & Davison K. P. (2004). Disclosure of trauma and immune response to a hepatitis B vaccination program. *Journal of Consulting Clinical Psychology* 63:787–792.

Petrie, K. J., Fontanilla, I., Thomas, M. G., Booth, R. J., & Pennebaker, J. W. (2004). Effect of written emotional expression on immune function in patients with human immunodeficiency virus infection: a randomized trial. *Psychosomatic Medicine* 66:272–275.

Pinker, S. (April 5, 1994). The game of the name. *New York Times.*

Pinker, S. (2007). *The Stuff of Thought: Language as a Window into Human Nature.* New York: Viking.

Pinker, S. (2011). *The Better Angels of Our Nature: Why Violence has Declined.* New York: Viking.

Platek, S. M. et al. (2004). Reactions to children's faces: males are more affected by resemblance than females are, and so are their brains. *Evolution and Human Behavior* 25:394–405.

Porath, Y. (1974). *The Emergence of the Palestinian National Movement, 1918–1929.* London: Frank Cass Publisher.

Preston, B. T., Capellini, I., McNamara, P., Barton, R. A., & Nunn, C. L. (2009). Parasite resistance and the adaptive significance of sleep. *BMC Evolutionary Biology* 9:7.

Price, D. D., Finniss, D. G., & Benedetti, F. (2008). A comprehensive review of the placebo effect: recent advances and current thought. *Annual Review of Psychology* 59:565–590.

Price, L. (February 2, 2010). Short shrift for Blair at Chilcot. *Guardian.*

Pronovost, P. J., & Vohr, E. (2010). *Safe Patients, Smart Hospitals: How One Doctor's Checklist Can Help Us Change Health Care from the Inside Out.* New York: Hudson Street Press.

Radosh, R., & Radosh, A. (2009). *A Safe Haven: Harry S. Truman and the Founding of Israel.* New York: Harper Perennial.

Raichle, M. E., & Gusnard, D. A. (2002). Appraising the brain's energy budget. *PNAS* 99:10237–10239.

Rainsford, S. (March 21, 2009). Turkish children drawn into Armenia row. *BBC News.*

Ramachandran, V. S. (1996). What neurological syndromes can tell us about human nature: some lessons from phantom limbs, Capgras Syndrome, and Anosognosia. *Cold Spring Harbor Symposia on Quantitative Biology* 61:115–134.

Ramirez-Esparza, N., & Pennebaker, J. W. (2006). Do good stories produce good health? Exploring words, language, and culture. *Narrative Inquiry* 16(1):211–219.

Ray, W. J. et al. (2006). Decoupling neural networks from reality: dissociative experiences in torture victims are reflected in abnormal brain waves in left frontal cortex. *Psychological Science* 17:825–829.

Raz, A., Shapiro, T., Fan, J., & Posner, M. I. (2002). Hypnotic suggestion and the modulation of Stroop interference. *Archives of General Psychiatry* 59:1155–1161.

Reddy, V. (2007). Getting back to the rough ground. *Philosophical Transactions of the Royal Society B* 362:621–637.

Reid. J. E., & Inbau, F. E. (1977) *Truth and Deception: The Polygraph ("Lie Detector") Technique.* Baltimore: Williams & Wilkins.

Rich, F. (2006). *The Greatest Story Ever Sold: The Decline and Fall of Truth in Bush's America.* New York: Penguin.

Richeson, J. A., & Shelton, J. N. (2003). When prejudice does not pay: effects of interracial contact on executive function. *Psychological Science* 14:287–290.

Ricks, T. E. (2006). *Fiasco: The American Military Adventure in Iraq.* New York: Penguin Press.

Ridgeway, J. (July 13, 2004). Flying in the face of facts: lots of people dialed 911 to the U.S. before 9/11. Who put them on hold? *Village Voice.*

Rohwer, S. (1977). Status signaling in Harris sparrows: some experiments in deception. *Behaviour* 61:107–129.

Rohwer, S., & Ewald, P. W. (1981). The cost of dominance and advantage of subordination in a badge signaling system. *Evolution* 35:441–454.

Rohwer, S., & Rohwer, F. A. (1978). Status signaling in Harris sparrows: experimental deception achieved. *Animal Behavior* 62:1012–1022.

Rose, C. (2006). The battle for hearts and minds: patriotic education in Japan in the 1990s and beyond. *Nationalism in Japan*, ed. Shimazu, N. New York: Routledge.

Rosenkranz, M. A. et al. (2003). Affective style and *in vivo* immune response: neurobehavioral mechanisms. *PNAS* 100(19):11148–11152.

Rosenzweig, S. (1997). Letters by Freud on experimental psychodynamics. *American Psychologist* 52:571.

Ross, M., & Wilson, A. E. (2002). It feels like yesterday: self-esteem, valence of personal past experiences, and judgments of subjective distance. *Journal of Personality and Social Psychology* 80:572–584.

Ruether, R. R. (2009). The politics of God in the Christian tradition. *Feminist Theology* 17(3):329–338.

Sanger, D. E. (August 27, 2003). Report on loss of shuttle focuses on NASA blunders and issues somber warning. *New York Times.*

Saradeth, T., Resch, K. L., & Ernst, E. (1994). Placebo treatment for varicosity: don't eat it, rub it! *Phlebology* 9:63–66.

Saul-Gershenz, L. S., & Millar, J. G. (2006). Phoretic nest parasites use sexual deception to obtain transport to their host's nest. *PNAS* 103:14039–14044.

Schaller, M., Miller, G. E., Gervais, W. M., Yager, S., & Chen, E. (2010). Mere visual perception of other people's disease symptoms facilitates a more aggressive immune response. *Psychological Science* 21:649–652.

Schaller, M., & Murray, D. R. (2008). Pathogens, personality, and culture: disease prevalence predicts worldwide variability in sociosexuality, extroversion, and openness to experience. *Journal of Personality and Social Psychology* 95:212–221.

Schatzman, M. (1973). *Soul Murder: Persecution in the Family.* New York: Random House.

Scherr, H., & Bowman, J. (2009). A sex-biased effect of parasitism on skull morphology in river otters. *Ecoscience* 16:119–114.

Schlomer, G. L., Ellis, B. J., & Garber, J. (2010). Mother-child conflict and sibling relatedness: a test of the hypotheses from parent-offspring conflict theory. *Journal of Research on Adolescence* 20(2):287–306.

Schmidt-Hempel, P. (2011). *Evolutionary Parasitology.* London: Oxford University Press.

Scholz, J., Klein, M. C., Behrens, T. E. J., & Johansen-Berg, H. (2009). Training induces changes in white-matter architecture. *Nature Neuroscience* 12:1370–1371.

Schwartz, J. (April 4, 2005). Some at NASA say its culture is changing, but others say problems still run deep. *New York Times.*

Segerstrom, S. C., & Miller, G. E. (2004). Psychological stress and the human immune system: a meta-analytic study of 30 years of inquiry. *Psychological Bulletin* 130(4):601–630.

Segerstrom, S. C., & Sephton, S. E. (2010). Optimistic expectancies and cell-mediated immunity: the role of positive affect. *Psychological Science* 21:448–455.

Segerstrom, S. C., Taylor, S. E., Kemeny, M. E., & Fahey, J. L. (1998). Optimism is associated with mood, coping, and immune change in response to stress. *Journal of Personality and Social Psychology* 74:1646–1655.

Shahak, I. (1994). *Jewish History, Jewish Religion: The Weight of Three Thousand Years.* London: Pluto Press.

Shang, A. et al. (2005). Are the clinical effects of homeopathy placebo effects? Comparative study of placebo-controlled trials of homeopathy and allopathy. *Lancet* 366:726–732.

Shariff, A. F., & Norenzayan, A. (2007). God is watching you: priming God concepts increases prosocial behavior in an anonymous economic game. *Psychological Science* 19(9):803–809.

Sheppard, P. M. (1959). The evolution of mimicry: a problem in ecology and genetics. *Cold Spring Harbor Symposia on Quantitative Biology* 24:131–140.

Shiv, B., Carmon, Z., & Ariely, D. (2005). Placebo effects of marketing actions: consumers may get what they pay for. *Journal of Marketing Research* 42:383–393.

Shlaim, A. (1988). *Collusion Across the Jordan.* New York: Columbia University Press.

Shlaim, A. (January 7, 2009). How Israel brought Gaza to the brink of humanitarian catastrophe. *Guardian.*

Siegman, H. (January 29, 2009). Israel's lies. *London Review of Books.*

Simonsohn, U. (2011). Spurious? Name similarity effects (implicit egotism) in marriage, job, and moving decisions. *Journal of Personality and Social Psychology* 101(1):1–24.

Singer, T. et al. (2006). Empathic neural responses are modulated by perceived fairness of others. *Nature* 439:466–469.

Siochru, M. O. (2008). *God's Executioner: Oliver Cromwell and the Conquest of Ireland.* London: Faber and Faber.

Snowdon, C. T., & Tele, D. (2009). Affective responses in tamarins elicited by species-specific music. *Biology Letters* 6:30–32.

Snyder, M. L., Kleck, R. E., Strent, A., & Mentzer, S. J. (1979). Avoidance of the handicapped: an attributional ambiguity analysis. *Journal of Personality and Social Psychology* 37:2297–2306.

Sokoloff, L., Mangold, R., Wechsler, R. L., Kennedy, C., & Kety, S. S. (1955). The effect of mental arithmetic on cerebral circulation and metabolism. *Journal of Clinical Investigation* 34:1101–1108.

Soler, M., Soler, J. J., Martinez, J. G., & Moeller, A. P. (1995). Magpie host manipulation by Great Spotted Cuckoos: evidence for an avian Mafia? *Evolution* 49:770–775.

Soon, C. S., Brass, M., Heinze, H. J., & Haynes, J. D. (2008). Unconscious determinants of free decisions in the human brain. *Nature Neuroscience* 11:543–545.

Sordahl, T. A. (1986). Evolutionary aspects of avian distraction display: variation in American Avocet and Black-necked Stilt antipredator behavior. *Deception: Perspectives on Human and Nonhuman Deceit*, eds. Mitchell, R. W., & Thompson, N. S., 87–112. Albany: State University of New York.

Sosis, R., & Alcorta, C. (2003). Signaling, solidarity, and the sacred: the evolution of religious behavior. *Evolutionary Anthropology* 12:264–274.

Sosis, R., & Bressler, E. R. (2003). Cooperation and commune longevity: a test of the costly signaling theory of religion. *Cross-Cultural Research* 37:211–239.

Spellerberg, I. F. (1971). Breeding behaviour of the McCormick skua *Catharacta maccormicki* in Antarctica. *Ardea* 59:189–230.

Spera, S. P., Buhrfeind, E. D., & Pennebaker, J. W. (1994). Expressive writing and coping with job loss. *Academy of Management Journal* 37:722-733.

Stannard, D. (1992). *American Holocaust: The Conquest of the New World.* New York: Oxford University Press.

Starek, J. E., & Keating, C. F. (1991). Self-deception and its relationship to success in competition. *Basic and Applied Social Psychology* 12:145–155.

Steele, C. M., & Aronson, J. (1995). Stereotype threat and the intellectual performance of African Americans. *Journal of Personality and Social Psychology* 69:797–811.

Steger, R., & Caldwell, R. L. (1983). Intraspecific deception by bluffing: a defense strategy of newly molted stomatopods (Arthropoda: Crustacea). *Science* 221:558–560.

Stone, B. (December 6, 2006). Spam doubles, finding new ways to deliver itself. *New York Times.*

Strachan, E. D., Bennett, W. R. M., Russo, J., & Roy-Byrne, P. P. (2007). Disclosure of HIV status and sexual orientation independently predicts increased absolute

CD4 cell counts over time for psychiatric patients. *Psychosomatic Medicine* 69:74–80.

Stroop, J. R. (1935). Studies of interference in serial reactions. *Journal of Experimental Psychology* 14:25–39.

Sullivan, A. (February 8, 2010). In the bunker. *The Atlantic.*

Talwar, V., Murphy, S. M., & Lee, K. (2007). White lie-telling in children for politeness purposes. *International Journal of Behavioral Development* 31:1–11.

Tanaka, K. D., Morimoto, G., & Ueda, K. (2005). Yellow wing patch of a nestling Horsfield's hawk-cuckoo *Cuculus fugax* induces misrecognition by hosts: mimicking a gape? *Journal of Avian Biology* 36:461–464.

Tanaka, K. D., & Ueda, K. (2005). Horsfield's hawk-cuckoo nestlings simulate multiple gapes for begging. *Science* 308:653.

Tavris, C., & Aronson, E. (2007). *Mistakes Were Made (But Not by Me).* New York: Harcourt.

Taylor, S. E., Lerner, J. S., Sherman, D. K., Sage, R. M., & McDowell, N. K. (2003). Portrait of the self-enhancer: well adjusted and well liked or maladjusted and friendless? *Personality Processes and Individual Differences* 84:165–176.

Tenney, E. R., & MacCoun, R. J. (2006). Calibration trumps confidence as a basis for witness credibility. *JSP/Center for the Study of Law and Society Faculty Working Papers* (University of California, Berkeley) 40:1–16.

Thibodeau, P. H., & Boroditsky, L. (2011). Metaphors we think with: the role of metaphors in reasoning. *PLoS ONE* 6:1–11.

Thornhill, R., Fincher, C. L., & Aran, J. (2009). Parasites, democratization, and the liberalization of values across contemporary countries. *Biological Reviews* 84:113–131.

Thornhill, R., & Gangestad, S. (2008). *The Evolutionary Biology of Human Female Sexuality.* New York: Oxford University Press.

Thornhill, R. et al. (2003). Major histocompatibility complex genes, symmetry, and body scent attractiveness in males and females. *Behavioral Ecology* 15(5):668–678.

Tibbetts, E. A., & Dale, J. (2004). A socially enforced signal of quality in a paper wasp. *Nature* 432:218–222.

Tibbetts, E. A. & Izzo, A. (2010). Social punishment of dishonest signalers caused by mismatch between signal and behavior. *Current Biology* 20:1637–1640.

Tiger Woods confession. (December 2, 2009). *National Enquirer.*

Tiger Woods sex romp with neighbor's young daughter. (April 7, 2010). *National Enquirer.*

Torrey, E. F. (1992). *Freudian Fraud: The Malignant Effect of Freud's Theory on American Thought and Culture.* New York: HarperCollins.

Traufetter, G. (February 25, 2010). The last four minutes of Air France flight 447. *Spiegel Online.*

Trimble, D. E. (1997). The religious orientation scale: review and meta-analysis of social desirability effects. *Educational and Psychological Measurement* 57:970–986.

Trivers, R. (1971). The evolution of reciprocal altruism. *Quarterly Review of Biology* 46:35–57.

Trivers, R. (1972). Parental investment and sexual selection. *Sexual Selection and the Descent of Man, 1871–1971*, ed. Campbell, B. Chicago: Aldine-Atherton.

Trivers, R. (1974). Parent-offspring conflict. *American Zoologist* 14:249–264.

Trivers, R. (1985). *Social Evolution.* Menlo Park, CA: Benjamin Cummings.

Trivers, R. (2005). Reciprocal altruism: 30 years later. *Cooperation in Primates and Humans: Mechanisms and Evolution*, eds. van Schaik, C. P., & Kappeler, P. M., 67–83. Berlin: Springer-Verlag.

Trivers, R., & Hare, H. (1976). Haplodiploidy and the evolution of the social insects. *Science* 191:249–263.

Trivers, R., & Newton, H. P. (1982). The crash of Flight 90: doomed by self-deception. *Science Digest* November:66–67, 111.

Trivers, R., Palestis, B. G., & Zaatari, D. (2009). *The Anatomy a Fraud: Symmetry and Dance.* Antioch, CA: TPZ Publishers.

Trivers, R., & Willard, D. (1973). Natural selection of parental ability to vary the sex ratio of offspring. *Science* 179:90–92.

Troisi, A. (2002). Displacement activities as a behavioral measure of stress in nonhuman primates and human subjects. *Stress* 5:47–54.

Tuchman, B. (1984). *The March of Folly: From Troy to Viet Nam.* New York: Ballantine Books.

Tufte, E. R. (1997). *Visual Explanations: Images and Quantities, Evidence and Narrative.* Cheshire, CT: Graphics Press.

Tversky, A., & Kahneman, D. (1973). A heuristic for judgment frequency and probability. *Cognitive Psychology* 5:207–232.

Valdesolo, P., & DeSteno, D. (2008). The duality of virtue: deconstructing the moral hypocrite. *Journal of Experimental Social Psychology* 44:1334–1338.

Van Evera, S. (2003). Why states believe foolish ideas: non-self-evaluation by states and societies. *Perspectives on Structural Realism,* ed. Hanami, A. K., 163–198. New York: Palgrave.

Vandegrift, D., & Yavas, A. (2009). Men, women, and competition: an experimental test of behavior. *Journal of Economic Behavior and Organization* 72:554–570.

Vaughan, D. (1996). *The Challenger Launch Decision: Risky Technology, Culture and Deviance at NASA.* Chicago: University of Chicago Press.

Vecsey, G. (August 11, 2010). Woods's downfall is as gripping as his reign. *New York Times.*

Veltkamp, M., Aarts, H., & Custers, R. (2008). Perception in the service of goal pursuit: motivation to attain goals enhances the perceived size of goal-instrumental objects. *Social Cognition* 26.

Vohs, K. D., & Schooler, J. W. (2008). The value of believing in free will: encouraging a belief in determinism increases cheating. *Psychological Science* 19:49–54.

von Hippel, W., Lakin, J. L., & Shakarchi, R. J. (2005). Individual differences in motivated social cognition: the case of self-serving information processing. *Personality and Social Psychology Bulletin* 31:1347–1357.

von Hippel, W., & Trivers, R. (2011). The evolution and psychology of self-deception. *Behavioral and Brain Sciences* 34:1–56.

Vrij, A. (2004). Why professionals fail to catch liars and how they can improve. *Legal Criminological Psychology* 9:159–181.

Vrij, A. (2008). *Detecting Lies and Deceit: Pitfalls and Opportunities,* 2nd ed. Chichester, UK: Wiley.

Vrij, A., & Heaven, S. (1999). Vocal and verbal indicators of deception as a function of lie complexity. *Psychology, Crime and Law* 5:203–315.

Vrij, A., Mann, S., Robbins, E., & Robinson, M. (2006). Police officers' ability to detect deception in high stakes situations and in repeated lie detection tests. *Applied Cognitive Psychology* 20:741–755.

Vulliamy, E. (July 25, 1999). Why Kennedy crashed. *The Observer.*

Wager, T. D. et al. (2004). Placebo-induced changes in fMRI in the anticipation and experience of pain. *Science* 303(1162–1167).

Wald, M. L. (January 25, 2006). Voice recorder shows pilots in '04 crash shirked duties. *New York Times.*

Watson, A. (2006). Self-deception and survival: mental coping strategies on the western front, 1914–18. *Journal of Contemporary History* 41:247–268.

Wedmann, S., Bradler, S., & Rust, J. (2007). The first fossil leaf insect: 47 million years of specialized cryptic morphology and behavior. *PNAS* 104:565–569.

Wegner, D. M. (1989). *White Bears and Other Unwanted Thoughts.* New York: Viking.

Wegner, D. M. (2002). *The Illusion of Conscious Will.* Cambridge, MA: MIT Press.

Wegner, D. M. (2009). How to think, say, or do precisely the worst thing for any occasion. *Science* 325:48–50.

Wegner, D. M., Wenzlaff, R. M., & Kozak, M. (2004). Dream rebound: the return of suppressed thoughts in dreams. *Psychological Science* 15:232–236.

Weiss, J. M. (1970). Somatic effects of predictable and unpredictable shock. *Psychosomatic Medicine* 32:397–408.

Wessel, E., Drevland, G. C. B., Eilertsen, D. E., & Magnussen, S. (2006). Credibility of the emotional witness: a study of ratings by court judges. *Law and Human Behavior* 30:221–230.

West, S. (2009). *Sex Allocation.* Princeton: Princeton University Press.

Whitson, J. A., & Galinsky, A. D. (2008). Lacking control increases illusory pattern perception. *Science* 322:115–117.

Wickler, W. (1968). *Mimicry in Plants and Animals.* New York: McGraw-Hill.

Wildavsky, A. (1972) The self-evaluating organization. *Public Administration Review* 32:509–520.

Wiebe, K. L., & Bartolotti, G. R. (2000). Parental interference in sibling aggression in birds: what should we look for? *Ecoscience* 7:1–9.

Williams, M. A., & Mattingley, J. B. (2006). Do angry men get noticed? *Current Biology* 16(11):R402–R404.

Wilson, A. E., & Ross, M. (2001). From chump to champ: people's appraisals of their earlier and present selves. *Journal of Personality and Social Psychology* 80:572–584.

Wilson, A. E., Smith, M. D., & Ross, H. S. (2003). The nature and effects of young children's lies. *Social Development* 12:21–45.

Wilson, M. (February 9, 2009). Flight 1549 pilot tells of terror and intense focus. *New York Times.*

Wilson, T. D., Wheatley, T. P., Kurtz, J. L., Dunn, E. W., & Gilbert, D. T. (2004). When to fire: anticipatory versus post-event reconstrual of uncontrollable events. *Personality and Social Psychology Bulletin* 30:340–351.

Wrangham, R. (1999). Is military incompetence adaptive? *Evolution and Human Behavior* 20:3–12.

Wrangham, R. (2006). Why apes and humans kill. *Conflict*, eds. Jones, M., & Fabian, A. Cambridge: Cambridge University Press.

Wrangham, R., & Peterson, D. (1996). *Demonic Males: Apes and the Origins of Human Violence.* Boston: Houghton Mifflin.

Wright, R. (2009). *The Evolution of God.* New York: Little, Brown.

Yang, Y. L. et al. (2007). Localisation of increased prefrontal white matter in pathological liars. *British Journal of Psychiatry* 190:174–175.

Yousem, D. M. et al. (1999). Gender effects on odor-stimulating functional magnetic resonance imaging. *Brain Research* 818(2):480–487.

Zertal, I., & Eldar, A. (2007). *Lords of the Land: The War Over Israel's Settlements in the Occupied Territories, 1967–2007.* New York: Nation Books.

Ziv, Y., & Schwartz, M. B. (2008). Immune-based regulation of adult neurogenesis: implications for learning and memory. *Brain, Behavior, and Immunity* 22:167–176.

Zuckerman, M., & Kieffer, S. C. (1994). Race differences in face-ism: does facial prominence imply dominance? *Journal of Personality and Social Psychology* 66:86–92.

INDEX

Abdullah, King, 237, 240
Abuse, 66–67, 81–82
Academia, academics, 16, 304, 314
ACC. *See* Anterior cingulate cortex
Aeroflot Flight 593, 191–192
Afghanistan, 256, 259, 260, 267
African Americans, 64–65, 146, 167, 168–169, 169–170, 226
Aggression, 43–45, 151
AIDS, 116, 128, 131
Air Florida Flight 90, 184–189, 191, 193
Air France, 188
Air Transport Association, 194
Airline Pilots Association, 194
Al-Qaeda, 267
Altruism, 77, 81, 281, 309, 313
Alzheimer's disease, 73
American Eagle Flight 4184, 195
American exceptionalism, 220, 246, 273
Amerindians, 217, 218–222, 226, 287
Anger, 43–45, 151
Anterior cingulated cortex (ACC), 288
Anterior prefrontal cortex (aPFC), 58
Anthropology, cultural, 305
Anti-semitism, 243–245, 271
Ants, 31, 35
Apes, 37, 90
aPFC. *See* Anterior prefrontal cortex
Arabs, 233, 235, 236, 237–239, 239–241
Argentina, 222, 223
Armenian genocide, 216, 230–233
Athletics, 74, 253
Australia, 163, 169
Automobile industry, 199
Aviation disasters, 6
  Aeroflot Flight 593, 191–192
  Air Florida Flight 90, 184–189, 191, 193
  American Eagle Flight 4184, 195
  causes of, 184
  EgyptAir Flight 990, 209–212
  FAA response to, 194–198
  Gol Flight 1907, 189–191
  ice and, 194–198
  overconfidence and, 191–192
  pilot error and, 191–194
  September 11 and, 198–200
  tombstone technology and, 195–196

Belgium, 165
Beneffectance, 15
Berlin Wall, 259
Betrayal, 95, 101–103, 110–113
Bible, 242, 243, 278, 290, 294, 301
Bin Laden, Osama, 149, 200, 258, 265

Biology, 3–4, 4–5, 6, 77, 308–310
camouflage and, 40–41
coevolutionary struggle and, 29–36
death and near-death acts and, 41–42
deception, consciousness of and, 45–48
deception and anger and, 43–45
false alarm calls and, 39–40
female mimicry and, 38–39
frequency dependence and, 30–31
game theory and, 48–50
intelligence and deception and, 36–38
parasite-host relationships and, 31–36
predator-prey relationships and, 41–42
randomness and, 42–43
religion and, 280–281
Birds, 31–36, 39–40, 45–47, 63–64, 133, 135–136, 155
Black Panthers, 162, 168
Black Plague, 290
Blair, Tony, 264
Blick, Hans, 261
Bolivia, 262
Bombing, 265–268, 268–270
Brain
happiness and, 135–136
immune system and, 122–125
of monkeys and apes, 37
right vs. left, 61–63
sex differences and, 125
sleep, importance of and, 122–123
Brazil, 189–191, 222, 223, 289
Breast cancer, 131
Brood parasitism, 32–36
Bush, George W., 147, 170, 200, 223, 243, 260, 261, 298
Butterflies, 30–31, 42

Caching, 46–47
California, 67
Cambodia, 149, 217, 232, 251, 267–268
Cambodians, 216
Canada, 102, 169, 289
Cancer cells, 116, 119
Capital punishment, 143
Catholic Church, Catholicism, 287, 290, 293–294
Central America, 149, 217
Central Intelligence Agency (CIA), 200, 223, 262, 263
Challenger disaster, 201–205, 263
Children
cognitive dissonance in, 155–156
deception in, 88–92
disclosing trauma and, 126–127
marital conflict and, 86–87
name-letter effect and, 164
parental investment and, 2–3
parent-offspring relationship and, 2–3
sexual abuse of, 66–67
Chile, 222, 223
Chimpanzees, 47–48, 249–252, 274
China, 17, 59
Chinese, 226
Chomsky, Noam, 269
Christian Zionism, 241–243, 273
See also Zionism, Zionists
Christianity, 273, 277, 283, 292–293
Churchill, Winston, 21
CIA. *See* Central Intelligence Agency

Civil War, US, 226, 272
Clarke, Richard, 200, 259
Class, 18
Clinton, Bill, 25, 248, 267
Coevolutionary struggle, 29
  deceiver and deceived and, 29–30
  frequency dependence and, 30
  intelligence and, 29
  parasite-host relationships and, 31–36
  predator-prey relationships and, 30–31
Cognitive dissonance, 6, 47, 72, 108, 140
  in children, 155–156
  in monkeys, 155–156
  rationalization and, 153
  religion and, 281–282
  self-justification and, 151–154
  social effects of reduction of, 154–155
Cognitive load, 4, 9–13, 22, 47
Colombia, 222
Colonialism, 290
Columbia disaster, 191, 204, 205–209, 263
Columbus, Christopher, 218–219
Compulsion, 327–329
Con artists, 176–180
Congo, 232, 248
Consciousness, 4, 12, 53, 329–330
  deception and, 3
  knowledge and, 54–56
  language and, 157
  memory and, 9
  neurophysiology and, 5
Contras, 222–223
Control, 10, 11, 22–24, 69, 90, 253
Counselors, counseling, 332–333
Courtship, 6, 13, 15, 99–100, 147, 254
Crime, 7, 12
Cromwell, Oliver, 228
Cuba, 222
Cultural anthropology, 305, 313–315

Daily life, 6
  con artists and, 176–180
  deceiving up and dummying down and, 167–168
  drugs and, 173–175
  face-ism and, 168–170
  humor and laughter and, 172–173
  lie-detector tests and, 180–181
  metaphors, use of in, 161–164
  name-letter effect and, 164–167
  sex differences in overconfidence and, 157–160
  spam vs. anti-spam and, 170–172
  stock market trading and, 160–161
  vulnerability to manipulation and, 175–176
Darwin, Charles, 166, 278, 308, 311
Death, 41–42
Deceit. *See* Deception; Self-deception
Deceived, 5, 29–30, 36–37
Deceiver, 5, 29–30, 37
Deception
  anger and, 43–45
  behavioral, 37
  children and, 88–92
  cognitive load and, 9–13
  consciousness of, 45–48
  control and, 10, 11
  costs of, 7–8

Deception (*continued*)
denial and, 27–28
detection of, 5, 7, 9–13, 29–30, 36, 42, 43–45
economics and, 7–8
evolutionary approach to, 29
fantasy and, 330–331
game theory and, 48–50
general attributes of, 10–12
intelligence and, 36–38, 46
natural selection and, 8
in nature, 29–51
nervousness and, 10, 11–12
nonverbal forms of, 37
pervasiveness of, 6–8
propagation of, 5, 7, 330–331
selves, 5, 84
truth and, 7
unconscious, 3
See also Self-deception
Defense Department, US, 206, 262
Defense mechanisms, 2, 5
Delta Airlines, 188
Denial, 6, 12, 62, 140, 324
aggression and, 151
of deception, 27–28
false historical narratives and, 246, 255
homosexuality and, 128–130
projection and, 148–150
as self-reinforcing, 150
truth and, 2
war and, 255, 261–263
Depression, 69, 73–74, 117, 120
Derogation of others, 15, 18, 25, 103, 255–257
Determinism, 146
Disassociation, 81–82
Discrimination, 32, 34, 36, 172
Disease, 115, 116, 128, 131, 279
Dishonesty. *See* Deception; Self-deception
Displacement activities, 11, 327
Divorce, 154
DLPFC. *See* Dorsolateral prefrontal cortex
Dominance, 17, 98
Dominican Republic, 222
Dopamine, 131–132
Dorsolateral prefrontal cortex (DLPFC), 56–57
Drugs, 173–175
Dummying up, 167–168

Eban, Abba, 236
Economics, 6, 7–8, 310–312, 312–313
Egypt, 183, 240
EgyptAir Flight 990, 209–212
Einstein, Albert, 306
Eisenhower, Dwight, 223
El Salvador, 222
England, 233
Ethnic cleansing, 237–239, 241
Ethnicity, 18
Euphemism treadmill, 161–163
Evolution
deception and, 29
happiness, regulation of and, 69
of intelligence, 5
natural selection and, 1
of religion, 278
self-deception and, 1–28
of sex differences, 2, 313
of sexes, 95
of social relationships, 1–3
of war, 248, 249–252, 274–275

FAA. *See* Federal Aviation Administration

Face-ism, 168–170
Faith, 279, 284
False alarm calls, 39–40
False confessions, 66–67
False historical narratives, 6
  Amerindians, eradication of, 218–222
  counterattack and, 255
  denial and, 246, 255
  expansion at other's expense and, 255
  history textbooks and, 224–225
  Israel and, 233–245
  Japanese history and, 227–230
  purpose of, 245–246
  rationalization and, 218
  religion and, 246, 279
  Turkish genocide and, 230–233
  United States and, 218–226
  war and, 255, 273
  Zionism and, 233–235
False internal narratives, 15, 25–26
False personal narratives, 25–26
Family, 5, 110, 111–112
  children, deception in and, 88–92
  extreme abuse and, 81–82
  genetic conflict and, 86–87
  genetic relatedness and, 77–79
  genomic imprinting and, 82–83, 84–85, 87–88
  inbreeding and, 84–85
  internal conflict in, 84–85
  marital conflict and, 86–87
  natural selection and, 81
  parent-offspring conflict and, 80–81
  parent-offspring relationship and, 85–86
  self-deception and, 77–94
Fantasy, 95, 104, 107, 109–110, 330–331
Federal Aviation Administration (FAA), 193, 194–198, 200
Females. *See* Women
Feynman, Richard, 201–202, 204
Finland, 159–160
Flattery, 66–67
Founding Fathers, 25, 221
Frequency dependence, 30–31, 97
Freud, Sigmund, 126, 317–319, 320
Friendship, 332–333
From Time Immemorial (Peters), 234
Future of Iraq project (State Department), 263, 265

Galvanic skin response (GSR), 60, 180
Game theory, 48–50
Garner, Jay, 264–265
Gates, Bill, 171
Gaza, 234, 240, 270–274
Genocide, 149, 216–217, 230–233
Genomic imprinting, 82–83, 84–85, 87–88
God, 280, 283, 284, 293, 332
Golan Heights, 240–241
Good Samaritan, 297
Gorillas, 47–48
GSR. *See* Galvanic skin response
Guatemala, 222
Guilt, 10, 12, 117

Haiti, 222, 223
Hamas, 240, 271–272
Happiness, 68–69, 135–137, 147
Harrison, Benjamin, 242
Harrison, William Henry, 221
Hezbollah, 240–241, 266, 272

History. *See* False historical narratives
Hitler, Adolf, 232, 251, 275
HIV, 7, 116, 128–129, 141, 172
Holocaust, 230–233
Homicide, 102
Homosexuality, 102, 106–107, 111, 128–130, 291
Humor, 25, 172–173
Hussein, Saddam, 258, 259, 261, 265
Hypnosis, 67, 70–71, 75

Immunology, immune system
  brain and, 122–125
  cost of, 117–120
  happiness and, 135–137
  homosexuality and denial and, 128–130
  humor and laughter and, 172
  marriage and, 117
  music and, 132–133
  positive affect and, 130–132
  positivity in old age and, 133–135
  psychology and, 6, 68–70, 137
  religion and, 279
  sleep, importance of and, 120–121
  trade-offs and, 121–125
  writing about trauma and, 125–128
Inbreeding, 84–85, 295, 318
India, 169, 257
In-group associations, 14, 15, 19–20, 291
Intelligence, 5, 29, 36–38, 46
Investment, 2–3
Iran, 259, 266–267
Iraq war (2003), 223, 243, 247, 251, 257–263
Iraq Working Group, 264
Islam, 283, 293, 298, 301
Israel, 257, 266
  anti-Semitism and, 243–245
  Arab deceit and self-deception and, 239–241
  Arabs, flight of and, 237–239
  Arabs, voluntary flight of and, 237–239
  Armenian genocide and, 232
  Christian Zionism and, 241–243
  false historical narratives and, 233–243
  founding of state of, 235–237
  Gaza, assault on of, 248, 270–274
  Lebanon, assault on by, 268–270

Jackson, Andrew, 221
Jackson, Michael, 72
Jamaica, 36
Japan, 17, 335
Japanese, 165, 216, 217, 227–230
Japanese-Americans, 153
Jefferson, Thomas, 221
Jesus, 242, 282, 283–284, 293, 296, 300
Jews, 231, 236–237
Jews-4-Peace, 273
Jihad, 298, 301
Johnson, Lyndon B., 251
Josiah, King, 293
Judaism, 273, 283
Justice, 304–305
Justice in Palestine, 273

Kennedy, John F., 251
Kennedy, John F., Jr., 191–192
Kenya, 169
Kerry, John, 147
Khmer Rouge, 251, 268

Kinship theory, 79, 313
Kissinger, Henry, 251, 268
Koran, 278
Korea Airlines, 188
Korean War, 216
Koreans, 216

Land of Promise (textbook), 225
Language
  deceit and, 14
  euphemism treadmill and, 161–162
  metaphors and, 161–164
  name-letter effect and, 164–167
  parental investment and, 79
  religion and, 275, 289
  self-deception and, 13–27
  unconscious use of, 157
  war and, 274–275
Laos, 251
Laotians, 216
Laughter, 172–173
Lebanese, 149
Lebanon, 234, 244, 259, 266–267, 268–270
Liberation theology, 293–294
Lie-detector tests, 180–181
Lord's Prayer, 297–298, 331–332

Madoff, Bernie, 176–177
Males. *See* Men
Manifest destiny, 220, 301
Marriage, 107–109, 117, 154, 295, 304
Marshall, John, 221
Medial prefrontal cortex (MPFC), 17
Meditation, 280, 331–332
Melville, Herman, 241
Memory
  abuse and, 66–67
  biased, 143–145
  construction of false, 145
  distortions of, 145
  false, 66–67
  readjustment of, 108
  suppression of, 2, 5, 9, 56–57, 57–58, 82
Men
  betrayal, response to of, 101–103
  celibacy and, 291–292
  deception, detection of and, 98
  female mimicry and, 39
  female sexual interest and, 95, 105–106
  homosexual tendencies in, 106–107
  overconfidence and, 98, 157–160
  war and, 253–255
Metaphors, 160–161, 161–164
Mexico, 169, 217, 222
Mimicry, 30–31, 34, 38–39, 90, 97
Monkeys, 20, 37, 90, 155–156
Monotheism, 277, 280, 290, 293
Monroe Doctrine, 222
Moral superiority, 22, 253
Morality, 17, 22, 253
MPFC. *See* Medial prefrontal cortex
Ms. magazine, 169
Mubarak, Hosni, 240
Muhammad, 282, 293
Murder, 95, 102
Music, 132–133
Muslim Brotherhood, 240

Napoleon, 248
Narcissism, narcissists, 17–18
National Aeronautics and Space Administration (NASA), 183, 184, 200, 257
  Challenger disaster and, 201–205
  Columbia disaster and, 205–209

National Transportation Safety Board (NTSB), 184, 194, 196, 197, 198, 209, 211
Nationalism, 256
Natural selection
  deception and, 8
  evolution and, 1
  family and, 81
  randomness and, 42
  self-deception and, 1–3, 251
  social relationships and, 1–2
  war and, 251
Nature. *See* Biology
Neurophysiology, 53–75
  conscious knowledge and, 54–56
  consciousness and, 5
  imposed self-deception and, 53–54, 63–67
  improving deception through neural inhibition and, 58–59
  placebo effect and, 70–75
  psychological immune system and, 68–70
  religion and health and, 287–288
  right vs. left brain and, 61–63
  self-esteem, implicit vs. explicit and, 64–66
  thought suppression and, 56–57, 57–58
  unconscious self-recognition and, 59–61
Newton, Huey, 174, 176
Nicaragua, 222–223
Nixon, Richard, 251, 268
Norway, 169
Novelty, 29, 31
NTSB. *See* National Transportation Safety Board
Nuclear war, 261

Obama, Barack, 231–232, 334
Old Testament, 242, 243
Optimism, 69, 135–137
Out-group associations, 19–20
Overconfidence, 14, 98, 324
  aviation disasters and, 191–192
  knowledge and, 14–15
  narcissism and, 17
  sex differences and, 14, 157–160
  war and, 247–248, 253, 254, 255–257, 263

Pain, 73, 75, 112–113
Palestine, Palestinians, 149, 216, 233–245, 272
Panama, 168, 222
Parasite load, 288–291
Parasite-host relationships, 7, 30–31, 31–36, 69, 96–97, 116
Parental investment, 2–3, 78, 96, 97, 313
Parent-offspring relationships, 95–96, 99, 313
  children, deception in and, 88–92
  conflict in, 2–3, 80–81
  family and, 84–85
  internal bifurcation and, 78, 79–80
  language and self-deception and, 13
  manipulation and, 78, 79, 84–85
  misrepresentation and, 78
Paternity, uncertain, 95, 100–101
Paul XXIII, Pope, 293
Pentagon, 261
Pillar, Paul, 262
Pitch of voice, 12
Placebo effects
  neurophysiology and, 70–75

prayer and, 299–300
religion and, 281
self-deception, benefits of and, 54
Polytheism, 277, 279
Positivity effect, 133–135
Power, 15, 17, 20–21, 254, 292–294
Prayer, 280, 297–298, 299–300, 331–332
Predator-prey relationships, 30, 41–42
Predictability, 22
Primates, 47–48
Prisoner's dilemma, 48–49
Projection, 2, 6, 106, 140, 148–151
Protestantism, 290
Psychology, 6, 315–316
avoiding and seeking information and, 140–141
biased memory and, 143–145
biasing and, 140–148
cognitive dissonance and, 140, 151–156
denial and, 140, 148–151
encoding and interpretation of information and, 142–143
immunology and, 6, 68–70, 137
prediction of future feelings and, 146–147
projection and, 140, 148–151
psychoanalysis and, 317–319
psychological immune system and, 68–70
rationalization and biased reporting and, 145–146
self-deception, advantages of and, 3–4
self-deception, defensive function of, 68–69
social, 6, 18, 22, 53, 316

Qana massacre (1996), 269

Racism, 222, 243, 256, 293
Raiding, 249–252
Randomness, 42–43
Rationalization, 2, 3, 4, 62, 72, 145–146, 152, 154, 218, 261
Reason, 279, 284
Relatedness. *See* Family
Relationships
conflict in, 308
courtship and, 13
deception and, 7
evolution of, 1–3
low-love, 107
parasite-host, 7, 30–31, 69, 96–97, 116
parent-offspring, 1–3, 7, 13, 78–85, 88–92, 313
prayer and, 299–300
predator-prey, 7, 30, 41–42
prophet, deification of and, 283–284
religion and, 294–295
sexual, 5–6, 95–113
social, 1–3
Religion, 6, 74
believer, status of and, 279, 282
cognitive dissonance and, 281–282
disease and, 279
evolutionary theory of, 278
faith and reason and, 279, 284
false historical narratives and, 246, 279
health and, 285–288
immunology and, 279
in-group cooperation and, 280–282
language and, 275, 289

Religion (*continued*)
liberation theology and, 293–294
mating systems and, 294–295
meditation and, 280, 331–332
meme-centered view of, 278
miracles and, 282–283
parasites and diversity and, 288–291
power and, 292–294
prayer and, 280, 297–298, 331–332
splits in, 290
suicide attacks and, 300–301
truth and, 278–279
war and, 275, 301
women, bias against and, 291–292
Reproduction, 3–4, 5–6, 8, 97, 115
Republican Party, Republicans, 24–25
Reputation, 45, 280
Risk taking, 253, 254
Rock, Chris, 25, 56, 112
Roman Catholicism. *See* Catholic Church, Catholicism
Roosevelt, Franklin D., 153, 222
Roosevelt, Theodore, 221
Rumsfeld, Donald, 243, 259
Russia, 248
Rwanda, 217, 232

Schwarzenegger, Arnold, 111
Screw-ball, 260
Selective recall, 81
Self-deception
aviation disasters and, 6, 183–200, 209–212
benefits of, 3, 54, 324
biological approach to, 3–4
biology and, 4–5, 6
costs of, 3, 4, 28, 324
daily life and, 6, 157–181
deceit and, 3, 27–28
deception of others and, 4
defensive function of, 54, 66, 68
definition of, 8–9, 13
economics and, 6
evolution of, 3–6
evolutionary logic of, 1–28
false historical narratives and, 6, 215–246
family and, 77–94
fighting, 321–336
group, 184, 215, 279
hallmarks of, 27–28
immunology and, 6, 54, 115–137
imposed, 53–54, 63–67
language and, 13–27
natural selection and, 1–3, 251
neurophysiology and, 53–75
as offensive strategy, 54
primary function of, 3
psychological approach to, 3–4
psychology and, 6
psychology of, 139–156
rationalization and, 3
religion and, 6, 277–301
science of, 4–5
sex and, 95–113
sexual relationships and, 5–6
social sciences and, 303–320
space disasters and, 6, 183–184, 201–209
war and, 6, 247–275
See also Deception
Self-esteem, 64–66, 167
Self-inflation, 15–18, 25, 98, 167
Self-justification, 151–154
Self-opinion, 2
September 11, 149–150, 198–200, 236, 257, 258, 259, 263

Serotonin, 131
Sex
  extra-pair, 95
  imposed self-deception and, 64
  meaning of, 95
  shame and, 111
  See also Sex differences; Sexual relationships
Sex differences, 95
  brain and, 125
  evolution of, 2, 313
  overconfidence and, 14, 157–160
  power and, 21
  war and, 253–255
  See also sex; Sexual relationships
Sexual dysfunction, 64, 74
Sexual orientation, 116
Sexual relationships, 5–6, 95–113
  betrayal and, 95, 101–103, 110–113
  conflict in, 6
  courtship and, 6, 13, 99–100, 254
  extra-pair sex and, 95
  fantasy and, 95, 104, 107, 109–110
  female monthly cycle and, 95
  female sexual interest and, 95, 105–106
  frequency dependence and, 97
  marriage and, 107–109
  men, homosexual tendencies in and, 106–107
  murder and, 95, 102
  purpose of, 96–97
  reproduction and, 5–6
  sexual relationships and, 5–6
  uncertain paternity and, 95, 100–101
  women, monthly cycle of and, 103–105
  See also Sex; Sex differences
Shame, 45, 117
Silberglied, Robert, 184
Slavery, 226, 272
Sleep, 120–121, 126–127
Social psychology, 6, 18, 22, 53, 316
Social sciences
  biology and, 308–310
  cultural anthropology and, 305, 313–315
  development of, 307–308
  economics and, 310–313
  justice and truth and, 304–305
  psychology and, 315–316, 317–319
Social theories
  biased, 24–25
  false, 15
Somalia, 259
Soviet Union, 224
Space disasters, 6, 183–184
  Challenger disaster, 191, 201–205, 263
  Columbia disaster, 191, 204, 205–209, 263
Spam vs. anti-spam, 170–172
Spencer, Herbert, 251
Sports, 74, 253
Squid, 40–41
Squirrels, 45, 46
State Department, US, 242, 263
Stick insects (Phasmatodea), 8
Stock market, 155, 160–161
Sudan, 232, 259
Suicide attacks, 300–301
Suppression, 2, 5, 9, 11, 12, 17, 56–57, 57–58, 82
Syria, Syrians, 149, 240–241, 259

Terrorism, 217, 261
Thrill seeking, 253, 254

Time magazine, 169
Times Square bomber, 331
Torah, 278
Torture, 66–67, 121, 265
Trading, 155
Triumph of the American Nation (textbook), 225
Truman, Harry, 218, 242
Truth
- deception and, 7
- denial of, 2
- justice and, 304–305
- religion and, 278–279
- science and, 322
- war and, 247

Turkish genocide, 216–217, 230–233

UG. *See* Ultimatum game
UK. *See* United Kingdom
Ultimatum game (UG), 49, 312
UN. *See* United Nations
Unconscious modules, 26–27
United Kingdom (UK), 163, 233
United Nations (UN), 235, 237, 244, 261, 269
United States, 102, 163, 169
- false historical narratives and, 218–226
- foreign policy in, 149–150
- safety approach of, 198–200
- war and, 222–224

UNSCOM, 261
USAID, 263, 264

Vatican II, 293
Vietnam War, 216, 217, 251, 255, 267–268
Virginia, University of, 263
Viruses, 7, 69, 119

War
- bombing and, 265–268, 268–270
- control and, 253
- creating and walling off information and, 263–265
- denial and, 255, 261–263
- derogation of others and, 18, 255–257
- evolution of, 248, 249–252, 274–275
- false historical narratives and, 255, 273
- faulty decisions in, 247
- "home field" advantage and, 253, 255–256
- Iraq war (2003) and, 247, 257–263
- language and, 274–275
- modern, 253
- moral superiority and, 253
- natural selection and, 251
- overconfidence and, 247–248, 253, 254, 255–257, 263
- power and, 254
- primitive, 249–252
- rationalization and, 261
- religion and, 275, 301
- risk taking and, 253, 254
- sex differences and, 253–255
- small, 222–224
- truth and, 247

Washington, George, 221
Weapons of mass destruction (WMDs), 258, 261–262, 265
West Africa, 30–31
West Bank, 234
Wetherbee, James, 208
White lies, 89
WMDs. *See* Weapons of mass destruction

Wolf Man, 318
Wolfowitz, Paul, 261, 262
Women
 deception, detection of and, 98
 infidelity and, 101–103
 monthly cycle of, 95, 103–105
 overconfidence and, 157–160
 religion and, 291–292
 sexual abuse of, 66–67
 sexual interest of, 95, 105–106
 sexuality of, 74
 war and, 253–255
Woods, Elin, 111
Woods, Tiger, 111
World Bank, 313
World War I, 222, 230–231, 248, 250, 256, 265–266, 290
World War II, 21, 153, 216, 217, 227, 229, 242, 263

Zionism, Zionists, 216, 220, 233–243, 273
 See also Christian Zionism